Sustainable Production and Environmentally Responsible Consumption

Sustainable Production and Environmentally Responsible Consumption

Editor

Dimitrios A. Georgakellos

MDPI • Basel • Beijing • Wuhan • Barcelona • Belgrade • Manchester • Tokyo • Cluj • Tianjin

Editor
Dimitrios A. Georgakellos
University of Piraeus
Greece

Editorial Office
MDPI
St. Alban-Anlage 66
4052 Basel, Switzerland

This is a reprint of articles from the Special Issue published online in the open access journal *Energies* (ISSN 1996-1073) (available at: https://www.mdpi.com/journal/energies/special_issues/Sustainable_Production_and_Environmentally_Responsible_Consumption).

For citation purposes, cite each article independently as indicated on the article page online and as indicated below:

LastName, A.A.; LastName, B.B.; LastName, C.C. Article Title. *Journal Name* **Year**, *Volume Number*, Page Range.

ISBN 978-3-0365-7544-5 (Hbk)
ISBN 978-3-0365-7545-2 (PDF)

Cover image courtesy of Dimitris Georgakellos

Contents

About the Editor

Dimitrios A. Georgakellos

Dimitrios A. Georgakellos is a Professor of Technology, Energy and Environmental Management at the department of Business Administration of the University of Piraeus. He is Chairman of the Department of Business Administration, Director of the Executive MBA Postgraduate Program, Deputy President of the University of Piraeus Research Centre (UPRC) and Director of the Health Technology Assessment and Health Analytics Laboratory (HTA & HA Lab). He is, currently, President of the Scientific Committee of the National Centre of Public Administration and Local Government (EKDDA). He has more than fifteen years' of project experience in the private sector. His current academic and research interests focus on Environmental Economics and Management, Technology and Innovation Management, Business Plans Preparation & Entrepreneurship and Energy Management.

Preface to "Sustainable Production and Environmentally Responsible Consumption"

Achieving economic growth and sustainable development requires that we urgently reduce our ecological footprint by changing the way we produce and consume goods and resources. We consume more resources than the planet can generate, and growing rates of pollution and waste only worsen the problem. In this context, globally, a significant number of frameworks and programs on sustainable consumption and production have been introduced. Most of them concern the environmentally sound management of energy and raw materials and all wastes throughout their life cycle; the substantial reduction of waste generation through prevention, reduction, recycling, and reuse; the adoption of sustainable practices and the integration of sustainability information into their reporting cycle; the promotion of procurement practices that are sustainable; the enhancement of scientific and technological capacity intending to achieve more sustainable patterns of consumption and production; etc. This special issue on "Sustainable Production and Environmentally Responsible Consumption" takes a broad view of all these issues and aims to inform about ways contributing to sustainable consumption and production by reducing our environmental impact, promoting the use of renewable sources of energy, and encouraging responsible purchasing decisions.

Dimitrios A. Georgakellos
Editor

energies

MDPI

Article

Forward Modeling of Natural Fractures within Carbonate Rock Formations with Continuum Damage Mechanics and Its Application in Fuman Oilfield

Ming Yi [1,2], Xinpu Shen [3,*], Lixin Jin [4], Jianliang Wang [5], Zhiqiang Huang [1] and Guoyang Shen [3]

1 School of Petroleum Engineering, Changjiang University, Wuhan 430100, China
2 Research Institute of Engineering Technology, CNPC West Drilling, Karamay 834000, China
3 Institute of Reservoir Engineering, College of Petroleum Engineering, China University of Petroleum (Huadong), Qingdao 266580, China
4 Exploration Division, PetroChina Xinjiang Oilfield Company, Karamay 834000, China
5 Changqing Drilling Subcompany, CNPC Chuanqing Drilling, Xi'an 710016, China
* Correspondence: 20180011@upc.edu.cn

Abstract: Accurate information about the distribution of natural fractures is a key factor for the success of the exploration and development of oil and gas in carbonate rock formations. Forward modeling of natural fractures generated by tectonic movement within carbonate rock formations was investigated by jointly using the continuum damage model and finite element numerical technology. Geological analysis of natural fractures was used as the basis of the geomechanical finite element calculation. A workflow of numerical calculations for natural fractures was proposed. These achievements were applied to investigate natural fractures' distribution within Ordovician carbonate rock formations of the Fuman Oilfield, Xinjiang, in the west of China. Finite element sub-modeling technology was used to further investigate natural fractures within key target reservoir formations with a finer mesh. The contour of natural fractures represented by the localization band of continuum damage variables was obtained. A comparison of the numerical results of the natural fractures' distribution represented by continuum damage variables with those of natural fractures interpreted from seismic data shows that: (1) the numerical solution of natural fractures matches the measured data, and their orientations are in good accordance; (2) their distribution and locations are basically the same, with some small differences in local details; (3) the numerical results indicate that the maximum value of the damage variable SDEG within the zones of natural fractures is 0.2686, and the widths of the bands of natural fractures/faults are in the range of 500 m to 1000 m. Validation of the results of the distribution of natural fractures was performed indirectly via the distribution of the minimum horizontal stress gradient ShG.

Keywords: natural fracture; carbonate rock formation; Caledonian period; tectonic movement; continuum damage mechanics; Ordovician period; Fuman Oilfield

Citation: Yi, M.; Shen, X.; Jin, L.; Wang, J.; Huang, Z.; Shen, G. Forward Modeling of Natural Fractures within Carbonate Rock Formations with Continuum Damage Mechanics and Its Application in Fuman Oilfield. *Energies* **2022**, *15*, 6318. https://doi.org/10.3390/en15176318

Academic Editor: Dimitrios A. Georgakellos

Received: 11 August 2022
Accepted: 26 August 2022
Published: 30 August 2022

Publisher's Note: MDPI stays neutral with regard to jurisdictional claims in published maps and institutional affiliations.

1. Introduction

Information about the distribution of natural fractures has special significance in the development of oil and gas resources within carbonate rock formations in the Tarim Basin [1–3]. Oil existing in the carbonate reservoirs in the Tarim Basin was filled into the current locations from underlying formations. These carbonate rock reservoirs are not source rock [3]. Because oil and gas resources from carbonate rock reservoirs in the Fuman Oilfield only exist in natural fractures, distribution information about natural fractures determines the selection of the location of a wellbore trajectory. In addition, in the exploration and development of limited oil and gas resources such as shale oil and gas, as well as that of the conglomerate reservoir in the Junggar Basin [4], information about the distribution of natural fractures within a reservoir is one of the factors that determines the success of hydraulic fracturing. Therefore, it is one of the research hotspots in the field of petroleum engineering to accurately obtain the information on the distribution of natural fractures within carbonate rock formations and other tight oil and gas formations.

A conventional method for natural fracture interpretation and identification in engineering is to use seismic data to identify the distribution of natural fractures/faults in a target reservoir. It has been used widely in the current petroleum industry. However, its weakness is the need for obvious amounts of displacement of a fault (hereby, the term 'displacement of a fault' represents the amount of displacement discrepancy across the fault). When the displacement of a fault is not obvious, it is difficult to accurately identify natural fractures by using seismic data.

In addition to the aforementioned technology of identifying natural fractures and faults by the interpretation of seismic data, there are a variety of technologies for the identification and calculation of natural fractures. An effective method adopted by the international engineering community is to calculate the process of orogenic tectonic movement experienced by a carbonate rock formation by using geomechanical forward modeling combined with the formation sedimentary history, and to obtain the formation and distribution of faults/fracture zones [5–7].

In this paper, the forward modeling method of geomechanics and continuum damage mechanics is used to predict the distribution of faults and natural fractures in carbonate rock formations in the Fuman area according to the sedimentary history/generation sequence and the history of crustal movement/main activities of orogeny.

Natural fractures in Ordovician carbonate rock formations exist in the form of sets of fractures in rock as quasi-brittle material, not in the form of a single fracture. For sets of natural fractures in rock as quasi-brittle material, the model of fracture mechanics is unable to describe the mechanical behavior of a rock formation with multiple sets of natural fractures; instead, continuum damage mechanics is a suitable mechanical tool to describe the behavior of sets of this type of natural fracture in rock as quasi-brittle material [8,9].

Damage mechanics theory uses damage variables to mathematically formulate the crushing degree of materials [7]. A damage variable is particularly suitable for quantifying the degree of crushing/fragmentation of natural fractures of rock: 0 indicates that the rock is intact and 1 indicates that it is extremely broken. The examples presented in this paper show that the degree of crushing/fragmentation of rock with natural fractures is in the range of 0.3 to 0.

In the following sections, the following content is presented:

- Firstly, we introduce the calculation principle and process of the numerical solution of natural fracture distribution by using the method of geomechanics and damage mechanics.
- Then, we introduce the 3D geological model based on the interpretation of seismic data of the Fuman Oilfield. A finite element model for this block is established on the basis of this geological model: it includes the model grid, load and boundary conditions, and material model of damage mechanics, which includes damage evolution, etc., material parameters, and other model details.
- At the end, the numerical results of the contour of damage variables and the analysis of natural fractures are presented.

It should be noted that the distribution of natural fractures/faults obtained by the method of continuum damage mechanics with the calculation of forward modeling does not include natural fractures in the form of karst caves. In addition, natural fractures obtained by forward modeling technology include not only natural fractures for a fault with a width of several kilometers, but also small fractures with a width at the meter scale, as well as micro-fractures.

2. Materials and Methods

2.1. Principle of Calculation and Workflow

Figure 1 shows the flow chart for performing the numerical solution of the natural fractures' distribution by using the integrated method of geomechanics and damage mechanics. This process includes the following steps.

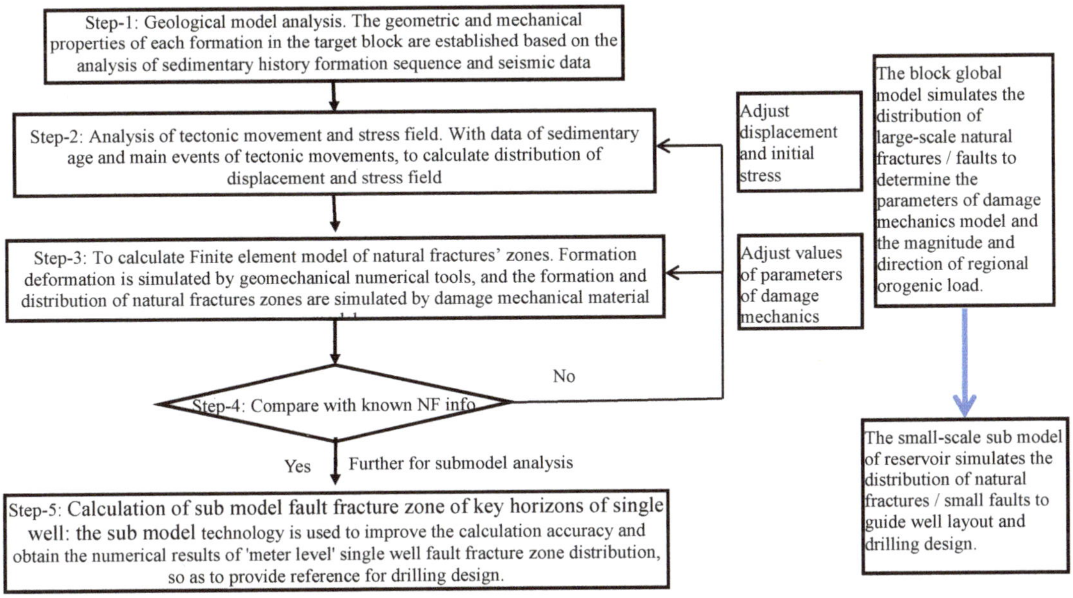

Figure 1. Flow chart of forward modeling of plastic damage calculation for natural fractures (NF) generated by tectonic movement.

(1) Analysis of geological model. Analyze the sedimentary history and formation's sequence, establish the geometry of each formation in the target block, and calculate the values of parameters of mechanical properties.

(2) Analysis of stratigraphic tectonic movement and characteristics of stress field. For a given tectonic movement (here, it is the middle Caledonian period for an Ordovician formation), determine the values of displacement vector or stress field distribution according to the history of tectonic movement/main events in orogeny. This is an iterative process, which requires the adjustment of the parameter values of related parameters according to the damage variable field distribution of numerical results and the known natural fracture difference information observed.

(3) Finite element model for calculation of natural fractures by forward modeling. The formation and deformation generated by tectonic movements are calculated by geomechanical numerical calculation tools, and the distributions of natural fractures/faults are simulated by a material model of mechanical damage.

(4) Submodels are built and further numerical calculations are performed for key target formations. In key target formations, submodel technology is used to improve the accuracy of the numerical results of the natural fracture distribution with a refined mesh.

(5) Parameter identification for the finite element model of the target block. Values of parameters of the finite element model for the target block include parameters of loads and boundary conditions, as well as parameters of the plastic damage model of the carbonate rock. By comparison, between the numerical results of the distribution of the main large-scale fractures/faults and the results of the distribution of the observed faults/fractures, modification to the initial input data of the model used can

be performed for better fitting between the numerical solution and measured solution for the distribution of natural fractures. Iteratively modify the values of input data with reference to the comparison results until the calculation results are consistent.

The right-hand side of Figure 1 is an explanation of the functions of steps 1 to 5 described in the left-hand side. The right-hand side is parallel to the left-hand side and thus they do not connect to each other. The first four steps in the workflow shown in Figure 1 are actually the process of establishing the geomechanical model, as well as a model of the damage mechanics, identifying the model parameters, and determining the values of loads and material parameters. A global model is used in this process. The last step is to apply the calibrated model to predict the three-dimensional distribution of natural fractures. A submodel with a finer mesh is used in this process. Target reservoir formation is the object of analysis in this step.

2.2. Geological Model and Natural Fracture Analysis of Target Formation

The reservoir in the Fuman Oilfield is a carbonate rock reservoir that consists of natural fractures, faults, and karst caves. The reservoir is buried at a true vertical depth (TVD) of value 7500–8800 m and is an ultra-deep formation. Preliminary geological analysis indicates that the horizon of the reservoir formation is relatively smooth, the fault drop is small, and the identification accuracy of natural fractures/fault zones obtained by seismic data interpretation is relatively low. The characteristics of the fault structure and natural fractures in the Fuman area have been the focus of regional geological research in recent years. Several researchers have dedicated efforts to these topics [10,11], and the research results obtained by using different technical methods vary and have their own characteristics.

Despite the works reported in the aforementioned references, there are still problems to be solved in the research on the classification of natural fractures in the Fuman Oilfield, as well as its mechanical characteristics: (1) the classification of natural fractures in Ordovician carbonate reservoirs is not detailed enough—there are multiple tectonic movements related to these fractures, but their impacts on these fractures have not been investigated; (2) the mechanical characteristics of these tectonic movements are not very clear—most of the existing conclusions on these tectonic movements are aimed at the model on the basin scale and are qualitative estimations—and there is a lack of quantitative analysis.

The difficulties in determining the mechanical characteristics of natural fractures in the Fuman Oilfield are as follows: (1) the characteristics of the distribution of natural fractures within carbonate rock formations are complicated, and there are few data from experimental measurements; (2) this research content involves many aspects, such as sedimentary history and orogeny, with many uncertain factors and great difficulty; (3) related research needs to include a three-dimensional numerical simulation with a large amount of calculations and more complex material modeling technology, which is difficult to carry out.

With the rapid development of technology for geological measurement and computation, the technology for modeling and analyzing a three-dimensional geological model and mechanical model of a target block of an oilfield has seen great progress in recent years, which provides a good technical basis for solving the aforementioned problems. In Section 2.2.1, the three-dimensional geological model of block F8B in the Fuman Oilfield is presented first. Then, information on the natural fracture analysis of each layer in this block is introduced. Next, we focus on the Ordovician carbonate target layer and provide the characteristics of the natural fracture system in the target layer.

2.2.1. 3D Geological Model of Block F8B in Fuman Oilfield

Variation in the depth of the formation top of the target reservoir is relatively smooth. Figure 2 shows the three-dimensional geological model of block F8B in the Fuman Oilfield based on seismic data and formation horizon information obtained through well drilling (pipe-stuck was encountered when drilling across the interface between two types of formations at some wells; mud-log was performed at some wells to confirm the locations of these interfaces). The length, width, and height of the geological model shown in

Figure 2 are 55 km, 26 km, and approximately 9980 m, respectively. The surface elevation is approximately 985 m, which is nearly flat. The lowest layer is a layer added to introduce displacement constraints to the model, and the lower bottom surface is set as a plane.

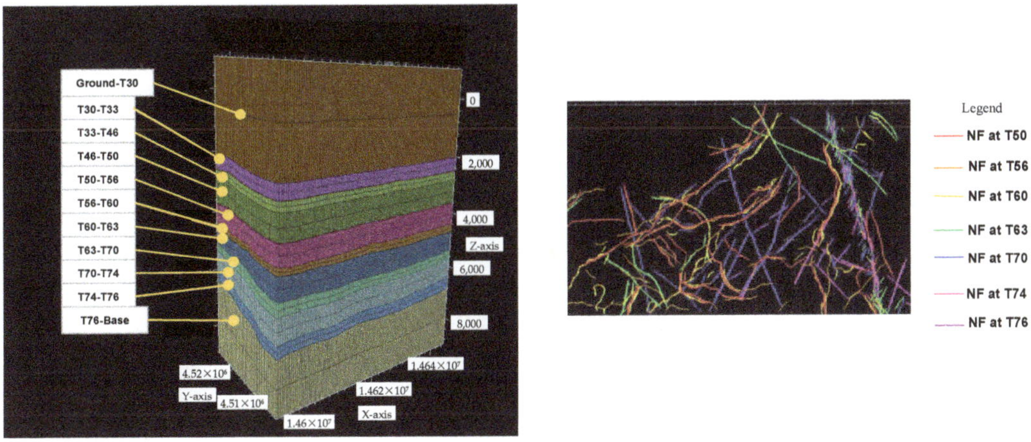

Figure 2. 3D geological model of target formations at Fuman Oilfield (**left**) and the overlapped plane view of natural fractures (**right**) at each interface.

The model consists of 11 layers of rock formations, which are bounded by the geological stratification interface: layer-1 is from the ground surface to the T30 interface; layer-2 is from the T30 to T33 interface; and layer-3 to layer-11 are formed by the following horizons in pairs, respectively: T33–T46 interfaces; T46–T50 interfaces; T50–T56 interfaces; T56–T60 interfaces; T60–T63 interfaces; T63–T70 interfaces; T70–T74 interfaces; T74–T76 interfaces; and T76 interface—bottom plane. There are 11 layers of rock formations in total. Layers from the T30 to T76 interfaces are the layers determined with the interpretation of seismic data. In the target block, F8B, the layer of the Ordovician carbonate rock formation, is located in the T74–T76 layer, with a TVD value of approximately 7000–7600 m.

2.2.2. Analysis of Natural Fractures within Each Layer

The target layer of the Ordovician carbonate rock formation is the one bounded by the T74 to T76 interfaces below 7000 m, as shown in Figure 2. In order to minimize the cost of computation, details of the geological structures from the ground surface to a depth of 3000 m are omitted, as they would introduce trivial inaccuracy. The left-hand section of Figure 2 shows the horizon interfaces from T30 to T76 interpreted with seismic data, and the right-hand section shows the plane view of the distribution of the natural fracture lines of each layer superimposed together. Figure 3 is a separate display of the natural fracture lines on each horizon interface.

Different colors in Figure 3a–g represent 7 sets of natural fractures of formation layers at different depths. The curves shown in Figure 3 are natural fractures obtained with seismic data interpretation. Combined with the azimuth directions of each set of natural fractures and the penetration degree of natural fractures in different rock formations, natural fractures within these carbonate rock formations of the Ordovician reservoir can be classified into five groups as follows:

(1) Natural fractures within the upper layers of formations from the T50 to T60 interfaces form a fracture system with the same sets of values of azimuth angles, as well as the same values of depth of penetration.
(2) Natural fractures of the formation layer from the T60 to T63 interfaces are the second group.

(3) Natural fractures at the T70 interface have some characteristics of both its upper formation and the formation from its lower side. The fault on the far right has not formed a north–south connection at this depth of the T70 interface. Formations around this depth belong to the transitional area between 2 different tectonic movements, and fractures within these formations belong to the third group.

(4) Mudstone of the Sangtamu formation with the T72 interface as its top horizon is a capping layer with fewer natural fractures, which is the fourth group. Distributions of natural fractures are not obvious and thus it is omitted here.

(5) Natural fractures at the T74 and T76 interfaces belong to the same set of natural fracture systems, which are the fifth group. At this depth, these natural fractures are connected vertically from the top to bottom interfaces within this range of depth. The major large faults/natural fractures run from north to south.

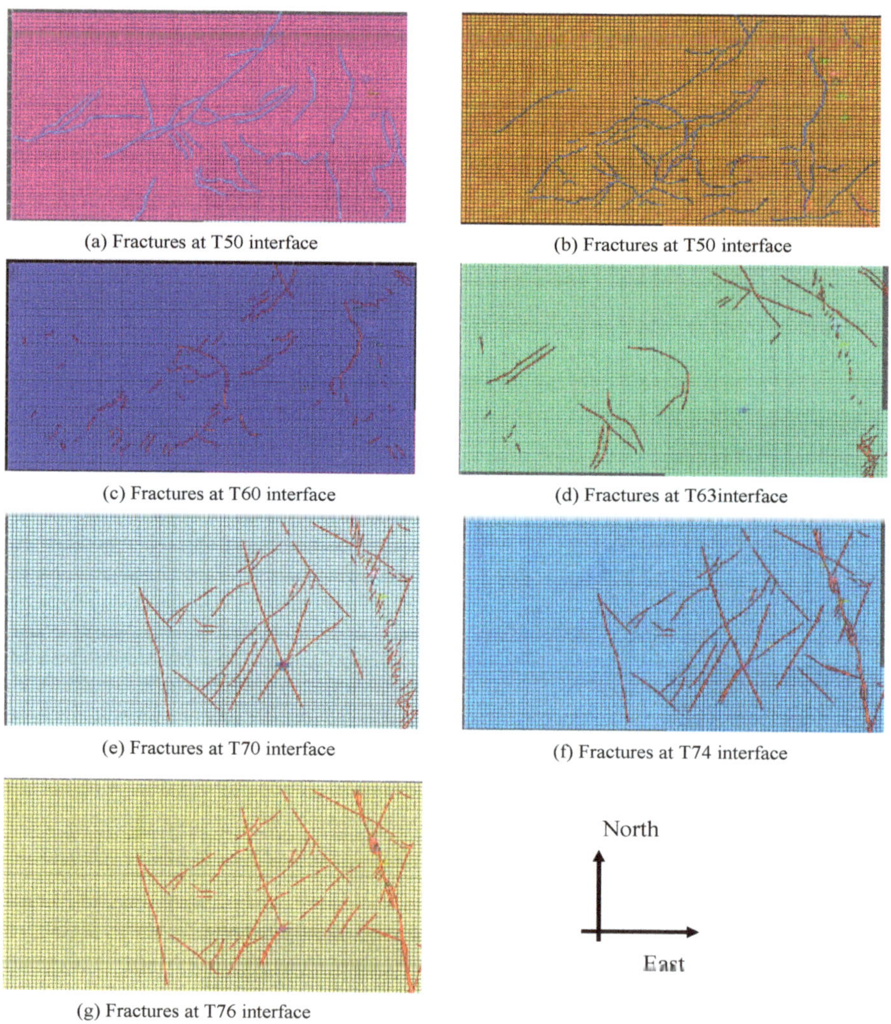

Figure 3. Plane view of distribution of natural fractures in each layer obtained from seismic data interpretation.

These five groups of natural fractures correspond to different tectonic movements. Most of the natural fractures are close to vertical natural fractures, and the values of their dip angles are within the range of 80~90°. At interface T70 in Figure 3e, there are a number of natural fractures/small faults on the right-hand side of the T70 interface, which are a set of oblique lines. They are obviously different from the natural fractures/faults developed in the lower formation's interface, such as those shown in Figure 3g, which form a connected curve from north to south.

Figure 3 shows the differences between natural fractures of neighboring different formations. This is the key for deriving the tectonic loading conditions in the geomechanical model later on. With the information presented in Figure 3, the tectonic loading condition in the geomechanical finite element model obtains support from the information on the geology.

2.3. Mechanical Characteristics of Natural Fractures in the Target Carbonate Rock Formations

The plane view of the natural fractures at the T74 interface in the Ordovician carbonate reservoir of the F8B block is shown in Figure 4. It is seen from Figure 4 that there are three sets of natural fractures/faults with different orientations.

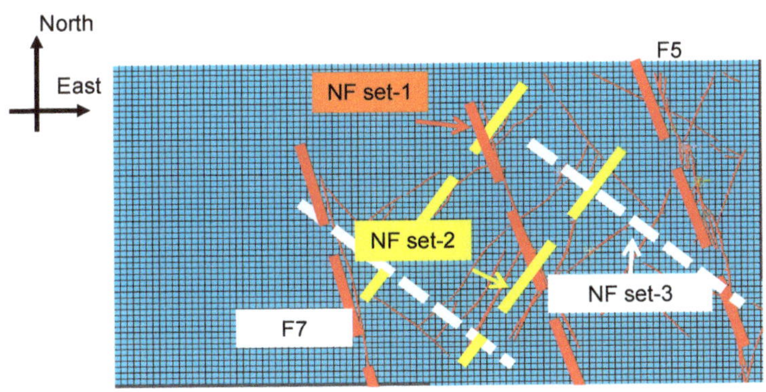

Figure 4. Plane view of natural fractures (NF) at T74 interface and illustration of classification.

The first set of natural fractures are long and penetrating natural fractures, including the No. 5 fault and No. 7 fault, which are represented by red dotted lines.

The second set of natural fractures is intermittently distributed, and their orientation is shown in Figure 4 as a yellow straight line.

The third set of natural fractures is intermittently distributed, and its orientation is shown as a white straight line. The third group of natural fractures and the second group of natural fractures are crossed and appear in pairs.

With reference to the aforementioned classification of natural fractures within the target Ordovician reservoir, it can be concluded that (1) natural fractures of set-1 of the Ordovician reservoir are tensile and shear fractures corresponding to tensile loading from tectonic movement; (2) natural fractures of set-2 and set-3 are fractures produced by compression from tectonic movement, and they correspond to compressive loading conditions.

2.4. Analysis of Mechanism of Natural Fractures Generated by Tectonic Movement in Carbonate Rock Formation

In order to investigate and verify the mechanism of natural fractures generated by tectonic movement in Ordovician carbonate rock formations, we established an Abaqus finite element model on the basis of the 3D geological model given in Figure 1. By jointly using the forward modeling technology of geomechanics for tectonic movement and technology for modeling fractures with continuum damage mechanics, values of parameters such as the orientation and magnitude of the displacement/strain of tectonic movement are determined by the trial-and-error method. Namely, the values of parameters of displacement/strain are regarded as the correct ones if the numerical results of the distribution of natural fractures are in good accordance with the ones interpreted from seismic data, as shown in Figure 4.

(1) The Finite Element model

The Ordovician carbonate rock formation has an age between 485 million years and 443 million years [12]. The sedimentary history of the Ordovician carbonate rock formation between the T74 and T76 interfaces, which is the target formation, is from the middle Caledonian stage I to stage II [5,6]. Tectonic movements experienced by this target Ordovician carbonate rock reservoir are mainly those of the middle Caledonian stage I and II. The subsequent tectonic movements also have an impact on the natural fracture distribution of these rock formations, but they are trivial. As shown in Figure 5, in the finite element model for calculating the natural fractures of the target Ordovician carbonate rock formations between the T74 and T76 interfaces, the T72 interface and its overburden rock formations are omitted in the 3D geomechanical model used here. This model has its top and bottom rock formations added to the target Ordovician carbonate rock formations between the T74 and T76 interfaces for the purpose of convenience of boundary conditions. The finite element mesh used here has 38,885 C3D8R elements and 45,696 nodes in total. Although the Tarim Basin's edge is formed by a set of irregular oblique lines, the plane view of the target block of this finite element model is only a rectangular plane that has regular borders.

Figure 5. The mesh of the finite element model.

Displacement boundary conditions of the model are set as follows: the zero-displacement constraint is applied to the four sides and bottom plane, and the top surface is the surface with traction loads.

Loads of the finite element model include (a) gravity load, including the gravity load of T70–T80 and other rock formations; (b) the traction load on the top surface is set as 50 MPa, which simulates the possible overburden pressure and also plays a role in stabilizing the convergence effect in numerical calculations.

In this calculation, the 'initial strain method' is used to simulate the compressive load of tectonic movements. The initial strain method is used to apply an initial strain tensor ε to each Gaussian integration point of the finite element model. Thus, the strain load borne by each point in the formation is applied. Advantages of this method are as follows: (1) it avoids the serious plastic deformation near the loading point caused by the boundary loading method, and (2) the distribution of the contour of damage variables caused by tectonic movement and their magnitude is reasonable.

The tectonic movement in the period of Caledonian stage I and stage II is the compression from basin edges in the northeast direction [5,6]. The exact values of the orientation angle and the magnitude of displacement are not clearly recorded. The method of phenomenon matching needs to be used to determine the value of displacement/strain of tectonic movement in terms of the coincidence between the distribution of natural fractures in the numerical results obtained and the distribution of natural fractures obtained by the interpretation of seismic data.

The 'initial strain' method is used to model the values of the orientation and magnitude of strain of the rock caused by tectonic movement [13]. This method applies a non-zero initial strain tensor to simulate the loading of tectonic movement applied to the target formations.

In this calculation, in order to determine the direction of the principal component of the displacement/strain of tectonic movement, the orientation angle of the local principal direction 2 of the initial strain tensor is written as α, as shown in Figure 6. A series of five different values are given to α as follows:

$$(a)\ \alpha = 30^\circ;\ (b)\ \alpha = 15^\circ;\ (c)\ \alpha = 0^\circ;\ (d)\ \alpha = -15^\circ;\ (e)\ \alpha = -30^\circ \quad (1)$$

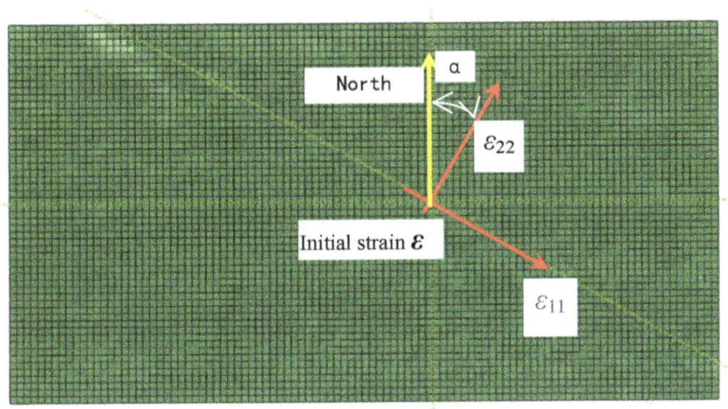

Figure 6. Illustration of initial strain and its principal directions.

The value of the second principal component of initial strain ε_{22} in the local principal direction ($\alpha = 15^\circ$) is taken as a series of different values between 0 and 0.5. The numerical results show that the component ε_{22} of initial strain is given as $\varepsilon_{22} = 0.024$, i.e., 2.4%. Values of the initial strain component in other principal directions are set as zero. With these values of parameters as input data, the numerical solution of natural fractures' distribution obtained with forward modeling of tectonic movement and damage mechanics is in good agreement with the known distribution of natural fractures.

The constitutive model of damage mechanics for carbonate rock that was introduced in reference [9,10,13] is used here to model fractures' initiation and evolution. In this model of damage mechanics, two scalar damage variables, tensile damage D_T and compressive damage D_C, are used to represent the damage initiation and evolution mechanisms of rock formations under tectonic loading conditions. The comprehensive damage variable scalar SDEG is used to comprehensively express the material damage degree caused by D_T and D_C. Theoretical formulation of parameters of this damage mechanics model, including the description of the damage evolution rate and damage initiation criterion, can be seen in the literature [9,10,13]. Values of model parameters are calibrated by the method of 'trial-and-error' and phenomenon matching. After a set of trial calculations, values of parameters of the damage evolution law of the Ordovician carbonate rock reservoir were obtained, and they are listed in Tables 1 and 2. Parameter ε_i represents the value of inelastic strain, and D_c is the value of the compressive damage variable. These two parameters, listed in Table 1, form a diagram of the damage evolution law under the compressive loading condition. Table 2 refers to damage evolution under tensile loading conditions.

Table 1. Values of parameters of damage evolution under compression.

Value of D_c	Values of ε_i
0	0
0.0034	0.00012
0.05	0.003
0.25	0.01
0.27	0.05
0.45	0.055
0.46	0.1
0.5	0.11
0.75	0.5

Table 2. Values of parameters of damage evolution under tension.

Value of D_t	Values of ε_i
0	0
0.1	5×10^{-5}
0.15	0.00015
0.7	0.0005
0.75	0.0015
0.9	0.002
0.905	0.5

3. Results

3.1. Numerical Solution of Contour of Damage Variable SDEG of the Global Model

The analysis performed in the work reported here is the 3D static finite element calculation of the deformation of target formations. With initial strain as the loading of tectonic movement to the target formations, the numerical results of damage initiation and evolution are presented. The contours of damage variables that represent the distribution of natural fractures are given in Figure 7. The finite element submodel and its results for damage contours are given in Figures 8 and 9.

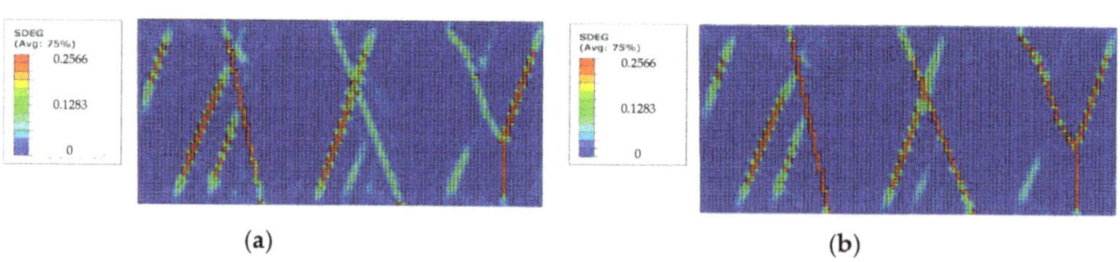

Figure 7. Plane view of contour of damage variable with $\alpha = 15°$ and $\varepsilon22 = 0.024$: (**a**) contour of damage SDEG at T74 interface, and (**b**) contour of SDEG at T76 interface.

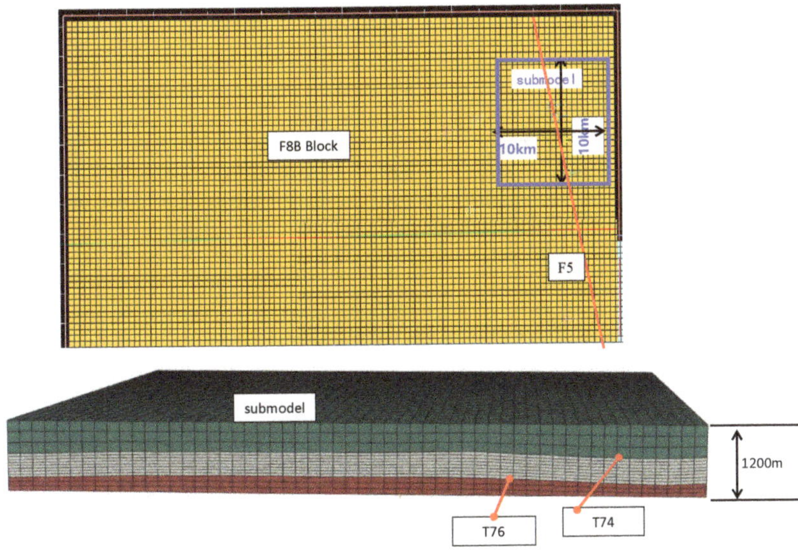

Figure 8. Illustration of relative location of the submodel and its finite element mesh.

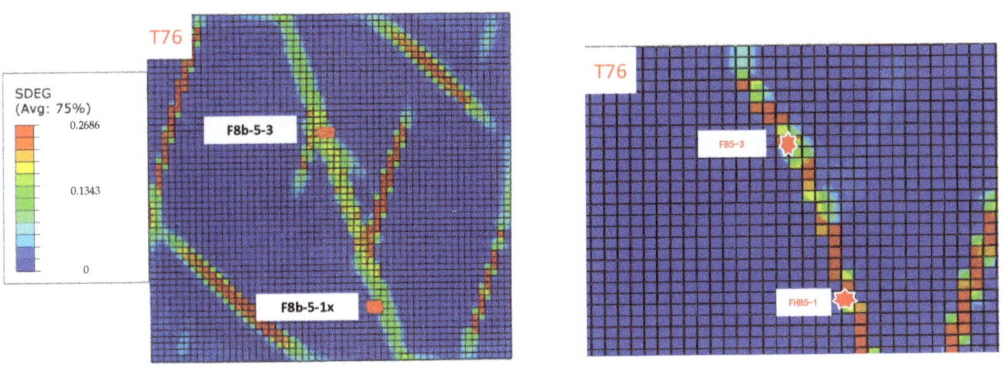

(**a**) (**b**)

Figure 9. Contours of SDEG at T76 interface: (**a**) numerical results of submodel and (**b**) results of global model for the same region.

Figure 7 gives a plane view of the contours of the damage variable SDEG obtained with $\alpha = 15°$ and $\varepsilon_{22} = 0.024$. Figure 7a shows the contour of damage SDEG at the T74 interface, which is the top of the target formation, and Figure 7b shows the contour of SDEG at the T76 interface, which is the bottom of the target formation. It can be seen that the maximum value of the damage variable SDEG is 0.2566, i.e., the degree of fragmentation of the rock is 25.66%. As can be seen by comparing the contours of SDEG in Figure 7a with the one in Figure 7b, the value of the damage variable SDEG on the T76 interface is slightly larger than that on the T74 interface. The damage localization zone shown in Figure 7 represents the distribution of natural fractures in the formation under loading of tectonic movement. The larger the value of the damage variable SDEG at a nodal point within the finite element model, the higher the degree of fragmentation of natural fractures within carbonate rock formations at this location will be.

By comparing the contour of SDEG in Figure 7 with the natural fracture distribution at the T74 and T76 interfaces shown in Figures 3 and 4, it can be concluded that the numerical solution of natural fractures obtained by the forward modeling of tectonic movement is very close to the results of natural fractures obtained with seismic data interpretation: the values of the orientation of major sets of natural fractures are the same and the details of natural fractures of a smaller size are different.

Therefore, according to the numerical solution of the natural fractures' distribution shown in Figure 7, it can be concluded that the mechanical characteristics of the tectonic movement that is applied to the Ordovician carbonate rock formation of the T74–T76 interfaces are as follows: (1) it is a compressive load; (2) the value of the orientation angle of the load is $\alpha = 15°$; and (3) the value of the principal component of the strain tensor generated by tectonic movement is $\varepsilon_{22} = 0.024$, which is 2.4%. Values of other components of this initial strain tensor are set to zero.

3.2. Submodel and Its Numerical Solution of Damage Field for Distribution of Natural Fractures

Submodeling is a type of numerical technology used in finite element calculations to minimize computational costs while maximizing the accuracy of the numerical solution by refining the mesh size only at the key target region [13]. Definition of the submodel of the finite element model that is used here includes mesh and boundary conditions. The material models and loading conditions of the submodel are the same as those of the global model.

Figure 8 shows the location of the submodel of the key target region in the block F8B and its finite element mesh. The red line in Figure 8 shows the location of the known major fault F5.

The submodel is 10 km long and 10 km wide. Its thickness is 1200 m, including the formation of the Ordovician carbonate reservoir between the T74 and T76 interfaces, the top layer, and the bottom layer, with a total of three layers. Values of element size are 200 m in both length and width and 50 m in the thickness direction. In total, 46,000 nodes and 45,000 C3D8R elements are used to discretize the geometry of the submodel.

The boundary conditions of the submodel are given with values of nodal displacement of the numerical solution of the global model of F8B. They are non-zero displacement boundary conditions, which are subtracted from the numerical solution of the displacement field of the global model of the F8B block shown in Figure 5. Values of material parameters such as the damage evolution law, etc., used in the submodel are the same as those in the global model.

Figure 9a,b show the distribution of natural fractures represented by the contour of the damage variable SDEG obtained with the submodel shown in Figure 8 and that with the global model shown in Figure 5, respectively. The contour is given for values of SDEG of nodal points at the T76 interface.

For brevity, only the contour of the synthetic variable SDEG at the T76 interface, which is the bottom of the target formation, is presented in Figure 9a. The contour of SDEG at the T74 interface, which is the top of the target formation, is similar to the case shown in Figure 9a. In Figure 9a, the maximum value of the damage variable SDEG is 0.2686, i.e., the degree of fragmentation of carbonate rock is 26.86%. The width of the damage localization zone is approximately 500 m to 1 km.

In Figure 9b, the contour of the numerical solution of the damage variable SDEG obtained with the global model is shown. By comparing Figure 9a with Figure 9b, it can be seen that the contour of the SDEG of the submodel has more branches than the contour of the SDEG of the global model. This indicates that the results of the natural fractures' distribution obtained by the submodel are more accurate than those obtained with the global model of the F8B block.

3.3. Validation of Numerical Results of Natural Fractures Represented by Contour Damage Variables

The model size is at the level of kilometers, and the carbonate rock formation has a high level of heterogeneity. Due to these characteristics of the engineering problem, validation of the numerical results of natural fractures represented by the contours of the damage variables is not able to be performed as in the cases at the scale of meters: it cannot be validated by direct measurement on values of damage variables. Alternatively, an indirect method can be used for the purpose of validation of the results of the distribution of natural fractures.

The method adopted here is to use information on stress variation within the target reservoir and its overburden seal formation obtained at each well existing in the region of the model. When there is a higher density of natural fractures, values of the strength of the rock will be lower than places where the rock has a lower facture density. Consequently, its magnitude of stress will be lower for the same formation. Therefore, for these wells in the same formation of the target reservoir, the one with a lower stress magnitude is the place where there is a higher density of natural fractures.

Figure 9 shows the contour of the damage variable SDEG, as well as the locations of wells F8b-5-1x and F8b-5-3. Values of damage SDEG at these two wells have the following properties:

(1) As shown in Figure 9, there are two wells drilled in the region of the model, i.e., the wells F8b-5-3 and F8b-5-1x. It is seen in Figure 9 that well F8b-5-3 is located in the center of the zone, where there is a higher value of the damage variable SDG, and this means that this well goes through a formation that has a higher density of natural fractures.
(2) Moreover, in Figure 9, well F8b-5-1x is located at the edge of the zone of natural fractures. The value of the damage variable SDEG at location of F8B-5-1x is less than that at the center of the damage zone.

These two points will be verified by the distribution of the stress gradient at the locations of these two wells.

Figure 10 gives the distribution of the minimum horizontal stress gradient ShG of these two wells. The detailed procedure of analysis and data logging that is used as input to derive these two curves of the stress gradient ShG can be found in [14].

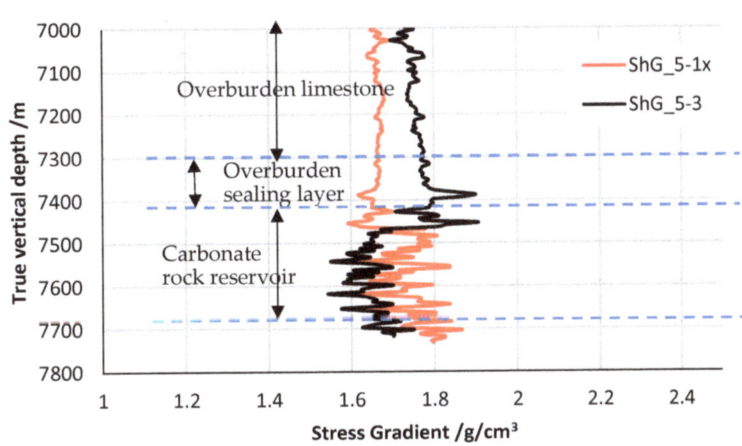

Figure 10. Distribution of minimum horizontal stress gradient ShG at locations of 2 wells.

There are three layers of formations related to the depth interval shown in Figure 10: the carbonate rock reservoir, the overburden sealing layer, and the overburden limestone formation. In the overburden sealing layer, there are very few natural fractures, and its sealing function is well preserved. With an increase in depth, the magnitude of ShG at both wells maintains the same trend as that in the formation of overburden limestone. When these two wells enter the carbonate rock reservoir, it is seen that at the depth interval within the formation of the carbonate rock reservoir, the magnitude of ShG at well F8B-5-3 decreases significantly from 1.8 g/cm^3 to 1.6 g/cm^3. This means that the density of natural fractures has increased significantly.

On the other hand, the magnitude of ShG at well F8B-5-1x does not decrease but undergoes a stepped increase–decrease when the well exits the sealing layer and enters the carbonate rock reservoir. After this stepped increase–decrease, the curve of ShG maintains the same trend as it is in the overburden sealing layer. This indicates that the density of natural fractures at this depth interval of well F8B-5-1x is small.

Therefore, the distribution of ShG at the two wells of F8B-5-1x and F8B-5-3 indicates the distribution of the density of natural fractures in this area. Conclusions derived from Figure 10 are in good accordance with the contour of damage SDEG.

4. Conclusions

Natural fractures within carbonate rock formations were investigated by using forward modeling technology. Continuum damage mechanics and the finite element method were used to numerically model the process of fracture initiation and evolution under the loading of tectonic movement. Based on the geological analysis of natural fractures provided by the interpretation of seismic data and the numerical results of the mechanical analysis of the contour of the damage variable SDEG, the following conclusions can be obtained.

(1) Regarding natural fractures within the Ordovician carbonate rock formation of target block F8B, they are mainly generated by tectonic movement in the Caledonian period stage 1 and stage 2. Consequently, the loading conditions of the model of geomechanics analysis includes only the tectonic movement in the Caledonian period stage 1 and stage 2, and excludes other tectonic movements.
(2) The numerical results of the distribution of natural fractures within target formations are represented by the contour of the damage variable SDEG shown in Figure 7. It indicates that the distribution of natural fractures obtained by the forward modeling of tectonic movement is very close to the results of natural fractures obtained with the interpretation of seismic data. It also indicates that the technology developed here for the forward modeling of natural fractures is an effective tool.
(3) The results of the natural fractures' distribution obtained by the submodel are more accurate than those obtained with the global model. This indicates that the technology of submodeling is a necessary component of forward modeling for the purpose of minimizing computational costs and maximizing the accuracy of the numerical results of the distribution of natural fractures.
(4) The values of parameters of tectonic movement of the Caledonian period stage 1 and stage 2 were calibrated in the process of phenomena matching for natural fractures: its orientation angle can be represented by the directional angle of the principal component of initial strain ε_{22}, which is $\alpha = 15°$, and its magnitude can be represented by $\varepsilon_{22} = 0.024$.
(5) The numerical results of the natural fracture distribution obtained by the global model of the F8B block indicate that the maximum value of SDEG is 0.2686, i.e., the degree of fragmentation of rock within the natural fractures is 26.86%. The width of the damage localization zone, which is the width of the area of the natural fractures, is in the range of 500 m to 1000 m.
(6) The numerical results of the contour of the damage variable SDEG show that the 3D damage mechanics model and finite element numerical analysis can better simulate the 3D spatial distribution of natural fractures caused by tectonic movement than the method of interpretation of seismic data: it not only presents conventional data of natural fractures, such as the azimuth angle and location, but also presents additional

data, which include the fracture width, the degree of fragmentation of the rock, spatial penetration, and so on.

(7) Validation of the results of the distribution of natural fractures has been performed indirectly via distribution of the minimum horizontal stress gradient ShG at two wells, F8b-5-1x and F8b-5-3.

The technology of forward modeling with damage mechanics to identify the distribution of natural fractures is particularly useful for blocks with small natural fractures/small fault displacements, which are difficult to identify by the interpretation of seismic data.

Author Contributions: Conceptualization, M.Y. and X.S.; methodology, X.S.; software, X.S.; validation, L.J.; formal analysis, J.W.; investigation, M.Y.; resources, X.S.; data curation, M.Y.; writing—original draft preparation, M.Y. and X.S.; writing—review and editing, M.Y. and X.S.; visualization, M.Y. and X.S.; supervision, X.S.; project administration, L.J.; funding acquisition, M.Y. and X.S; revising and data interpretation, Z.H.; G.S., software and data processing. All authors have read and agreed to the published version of the manuscript.

Funding: This research was funded jointly by the China National Petroleum Corporation (CNPC), grant number 2020F-46, key project "Integration and test of safe, optimal and fast drilling and completion technology for complex ultra-deep wells", and the National Natural Science Foundation of China, grant number 11272216, Project of "Theoretical and Experimental Investigation on key mechanical problems in unconventional natural gas exploitation".

Institutional Review Board Statement: Not applicable.

Informed Consent Statement: Not applicable.

Conflicts of Interest: The authors declare no conflict of interest.

References

1. Deng, S.; Li, H.L.; Zhang, Z.P.; Wu, X.; Zhang, J.B. Characteristics of differential activities in major strike-slip fault zones and their control on hydrocarbon enrichment in Shunbei area and its surroundings, Tarim Basin. *Oil Gas Geol.* **2018**, *39*, 878–888.
2. Liu, S.; Zhang, Z.; Yun, J.; Li, H.; Gao, S.; Li, J. NE-trending fault belts in Tanggubasi depression of the Tarim Basin: Features, genetic mechanism, and petroleum geological significance. *Oil Gas Geol.* **2018**, *39*, 964–975.
3. Tian, I.; Yang, H.; Zhu, Y.; Deng, X.; Xie, Z.; Zhang, Y.; Li, S.; Cai, Q.; Zhang, Y. Geological conditions for hydrocarbon accumulation and key technologies for exploration and development in Fuman and key technologies for exploration and development in Fuman Oilfield, Tarim Basin. *Acta Pet. Sin.* **2021**, *42*, 971–985.
4. He, D.; Zhang, L.; Wu, S.; Li, D.; Zhen, Y. Tectonic evolution stages and features of the Junggar Basin. *Oil Gas Geol.* **2018**, *39*, 845–861.
5. Wang, W.; Yin, H.; Zhou, P. Deformation characteristics and mechanism of salt-related fold thrust belt in Tarim basin. *Xinjiang Pet. Geol.* **2019**, *40*, 68–73.
6. Lefranc, M.; Hussein, A.M.; Tan, C.P. 3D Structural restoration and geomechanical forward modeling in a visco-plastic medium to natural fracture prediction in a Malay producing field, offshore Malaysia. In Proceedings of the The Offshore Technology Conference-Asia, Kuala-Lumpur, Malaysia, 25–28 March 2014; Paper No.: OTC-24753-MS.
7. Noufal, A.; Sirat, M.; Steiner, S.; Trubienko, O.; Minton, J.; Zhang, X.; Herrera, A. Estimates of in situ stress and faults/fractures in carbonate reservoirs in Onshore Abu Dhabi using geomechanical forward modeling. In Proceedings of the Abu Dhabi International Petroleum Exhibition and Conference, Abu Dhabi, UAE, 9 November 2015. [CrossRef]
8. Roberts, D.T.; Profit, M.L.; Armstrong, J.A.; Yu, J.; Crook, A.J.L. Strategies for forward modeling the evolution of geological structures undergoing large deformation. In Proceedings of the The 49th US Rock Mechanics/Geomechanics Symposium, San Francisco, CA, USA, 28 June–1 July 2015; Paper No.: ARMA 15-348.
9. Shen, X.; Yang, L. *Experiments and Theory of Damage Mechanics for Concrete*, 1st ed.; Science Press: Beijing, China, 2009.
10. Lee, J.; Fenves, G.L. Plastic-damage model for cyclic loading of concrete structures. *J. Eng. Mech.* **1998**, *124*, 892–900. [CrossRef]
11. Liu, J.; Ding, W.; Xiao, Z.; Dai, J. Advances in comprehensive characterization and prediction of reservoir fractures. *Prog. Geophys.* **2019**, *34*, 2283–2300. (In Chinese)
12. Qian, Y.; Sha, X.; Li, H.; Shen, X.; Xia, Y.; Liu, B.; You, D.; Wu, S.; Shi, Q. An approach to Caledonian unconformities and sequence stratigraphic pattern of Ordovician carbonate in the western Tazhong Area, Tarim Basin. *Earth Sci. Front.* **2013**, *20*, 260–274.
13. Dassault Systèmes Simulia Corp. *Abaqus User's Manual*; Volume 2: Analysis, Version 6.10; Dassault Systems: Vélizy-Villacoublay, France, 2010.
14. Xiuping, C.; Xinpu, S.; Jingtao, L.; Guoxiao, S. Refined finite element modeling and analysis on 3D geostress field of a block in Shunbei oil field. *J. Shenyang Univ. Technol.* **2021**, *43*, 1–7. Available online: http://kns.cnki.net/kcms/detail/21.1189.T.20210315.0932.004.html (accessed on 1 October 2021).

energies

MDPI

Article

Analysis of Undergraduates' Environmentally Friendly Behavior: Case Study of Tzu Chi University Environmental Education Program

Angela Yi Jing Tsai [1] and Alex Yong Kwang Tan [2,*]

[1] Center of General Education, Tzu Chi University, Hualien 97004, Taiwan; tsai001@gms.tcu.edu.tw
[2] International College, Tzu Chi University, Hualien 97074, Taiwan
* Correspondence: ttz021878@gms.tcu.edu.tw

Abstract: The Tzu Chi University Environmental Education Program, based on a theory of change, consisted of four weeks of lessons involving environmental and sustainability topics, followed by hands-on sorting of recyclables and four weeks of weekly documenting of environmentally friendly behavior. The Program was analyzed using written thoughts from the Experimental Group, as well as 78 and 116 valid survey responses of the Control and Experimental Groups, respectively. The survey consisted of questions regarding demographics and five constructs: environmental awareness, attitudes, norms, efficacy and behavior. No significant average differences were found between the pre-tests of the Control and Experimental Groups, or between the pre- and post-test of the Control Group. The post-test of the Experimental Group displayed a significantly higher average value when compared to both the pre-test of the Experimental Group and the post-test of the Control Group, as the means of the self-reported environmental awareness, attitudes, norms, efficacy and behavior significantly improved statistically after participating in the Program. Analysis revealed that lessons from the Program increased undergraduates' environmental awareness and attitudes; "hands-on recyclables sorting" and "weekly documentation of environmentally friendly behavior" strengthened undergraduates' environmental norms and efficacy, while their combination resulted in a significant improvement toward environmentally friendly behavior.

Keywords: environmental education; undergraduates; environmentally friendly behavior; awareness; attitude; efficacy; norms

Citation: Tsai, A.Y.J.; Tan, A.Y.K. Analysis of Undergraduates' Environmentally Friendly Behavior: Case Study of Tzu Chi University Environmental Education Program. *Energies* **2022**, *15*, 4853. https://doi.org/10.3390/en15134853

Academic Editor: Dimitrios A. Georgakellos

Received: 1 June 2022
Accepted: 29 June 2022
Published: 1 July 2022

1. Introduction

The widespread deterioration of environmental ecosystems and the convening of the United Nations Conference on the Human Environment in Stockholm, Sweden in 1972 (Stockholm Conference) marked the start of a modern global environmental movement, including the establishment of the United Nations Environment Programme. A decade later, with no significant improvements in various global environmental problems, the Brundtland Commission was founded in 1983, popularizing the term "sustainable development" after the release of its Brundtland Report. Another decade later, after realizing that global environmental challenges needed worldwide cooperation among nations, the United Nations Conference on Environment and Development (Earth Summit) was held in Rio de Janeiro, Brazil in 1992. A significant achievement of the Earth Summit was the formation of the United Nations Framework Convention on Climate Change (UNFCCC) which, in turn, led to the Kyoto Protocol in 1997 and the Paris Agreement in 2015.

The Paris Agreement aimed to maintain the rise in global average temperature at 2 °C above pre-industrial levels, in an attempt to substantially reduce the effects of climate change. In order to achieve the above aim, significant improvements needed to be implemented regarding personal environmental behaviors. Working alongside other environmental stakeholders, environmental education programs served to increase awareness, influence attitudes, strengthen norms and efficacy, as well as improve actual environmentally friendly behavior.

Traditional environmental educators employed the Bloom taxonomy [1] of educational learning whereby environmentally friendly behavior was achieved via the interactions between the cognitive, affective and psychomotor domains of learning, although the taxonomy was criticized for lacking a systematic rationale of construction [2]. Another widely used educational theory is the Hungerford learning method [3], whereby environmental education imparted knowledge and raised awareness regarding environmental protection; in turn, this changed attitudes, leading to better environmentally friendly decisions and behaviors. This "Knowledge-Attitudes-Behavior" theory was widely challenged after more than twenty years of research [4] due to its narrow definition of education. In addition, the ABC theory of emotions [5] used in environmental education—emphasizing that an activating event did not directly cause any consequence, but rather, the belief regarding that event did—was often a focus of criticism due to a lack of conceptually discrete definitions of constructs, as well as the overlapping definitions between activating events, beliefs and consequences [6]. Recently, drawing upon research in environmental economics, psychology and sociology, environmental education programs were positioned as part of broader cultural and social movements that included knowledge and attitudes, as well as norms, identity, efficacy, connections and trust [7].

Traditionally, there were three approaches to environmental education. Education about the environment approached the environment as a scientific topic and aimed to improve awareness, knowledge and understanding of the human–environmental interface. Education about the environment used the outdoor environment as a teaching medium, encouraging awareness and concern through personal interaction with nature. Education about the environment developed a sense of responsibility and active participation in the resolution of environmental issues using an issue-based approach. As all three of the mentioned approaches were unable to fully complete the cycle of awareness, knowledge, understanding, concern, responsibility, action, and back to awareness in the area of environmental education, these approaches were integrated into a threefold approach, forming an education about, in and for the environment [8]; this has been widely adopted in Taiwan's current environmental education [9].

This research aimed to employ a theory-of-change approach to design an environmental education program which incorporated the above-mentioned threefold approach, where in-class lessons sought to improve the awareness and attitudes of university undergraduates. The outdoor environment was used as a teaching medium to improve personal interaction with existing environmental problems, as well as to develop active participation in changing their personal environment-related behaviors in order to contribute toward reducing carbon emission. The undergraduates were separated into an Experimental Group and a Control Group in order to examine whether the designed environmental education program was successful in improving their environmentally friendly behavior.

2. Literature Review

The integrated approach to environmental education has become an important area of research, with the environmental behaviors of undergraduates becoming a promising area of research, as they play an important role in protecting the environment in the future. Teksoz et al. [10] proposed an environmental literacy components model in order to understand the relationships between environmental knowledge, attitudes, responsibility and concern, as well as outdoor activities among Turkish undergraduates. Using structural equation modeling, environmental knowledge was found to significantly predict environmental concern, attitudes, and responsibility, while having significantly indirect relationships with environmental attitudes and responsibility. Vicente-Molina et al. [11] examined the influences of environmental knowledge, education, gender, motivation, attitudes and perceived effectiveness on environmental behavior among undergraduates from America, Spain,

Mexico and Brazil. The survey results showed that motivation and perceived effectiveness were significant factors in influencing environmentally friendly behavior.

Surveying undergraduates from separate universities from Spain, Brazil and the United Arab Emirates, Chuvieco et al. [12] found that undergraduates from environment-related majors had better environmentally friendly behavior, while their country of origin had no significant effects. Liang et al. [13] conducted an environmental literacy survey among undergraduates in Taiwan, showing no significant correlations between knowledge and attitudes, or between knowledge and behavior, although stronger environmental attitudes were significantly correlated with behaviors. Jurdi-Hage et al. [14] examined the environmentally friendly behavior of undergraduates from a Canadian University and found that convenience and habits played significant roles in improving undergraduates' environmentally friendly behavior. Zhao et al. [15] asked undergraduates from Macau, China about their awareness, attitudes, knowledge and behaviors regarding energy saving. More than 90% of students surveyed understood the importance of energy saving. However, less than 10% of students participated in energy-saving activities, while around 20% of students never participated in any energy-saving activities.

Hansmann et al. [16] tried to determine the environmentally friendly behavior of the students and staff of the Swiss Federal Institute of Technology, Lausanne. Results from an online survey showed that gender, age and class standings had a positive correlation with environmentally friendly behavior. Balinska et al. [17] tried to understand the role of eco-friendly mobile applications on the environmentally friendly behavior of undergraduates in Poland. The results showed that applications widely promoted in traditional media gained stronger recognition, while statistically, females understood the usefulness of these applications better then males. Grodek-Szostak et al. [18] surveyed undergraduates from Poland, Ukraine and the Czech Republic in order to understand their awareness and behaviors in energy conservation, where the results showed that roughly 60% of undergraduates followed the principles of energy conservation, although their behaviors varied across countries. Leiva-Brondo et al. [19] attempted to understand the awareness and perception of sustainable development goals (SDGs) of Spanish undergraduates, reporting that only 15.9% of those surveyed had a good understanding of SDGs and the sustainability literacy level was 63%, indicating a lack of knowledge.

The impacts of environmental education programs on environmentally friendly behavior [20–22] are a constant topic of research, with residents as the majority of targeted audiences [23–25], although the impacts of environmental education programs on undergraduates are a growing area of research. Hse [26] conducted an environmental education program during an entire semester on a class of undergraduates. The results showed that the undergraduates had stronger environmentally friendly behaviors after the Program, which were further maintained after two months. By providing accurate and useful environment-related information to the students and staff of Fudan University, Jiang et al. [27] found that combined with supporting low-carbon management, environmental awareness improved, which led to stronger environmentally friendly behavior.

Dupre and Meineri [28] displayed a persuasive message, feedback chart and social comparative feedback chart of recycled weights within three cafeterias in a French university, respectively. The results showed that only the social comparative feedback approach statistically increased recycling behaviors which continued even after the feedback was removed. Godfrey and Feng [29] examined the effectiveness of an environmental education program within a university whereby a communication campaign was designed to showcase the water footprint of food available in the campus dining hall, in an attempt to improve environmentally friendly behavior in food consumption. The results showed that food consumption behaviors did not change significantly, due to the preference for convenience and time pressure over environmental protection.

Cosic et al. [30] found that raising awareness and including an external descriptive social norm successfully improved the recycling rate of plastic coffee cups within a university in Italy. Moreover, by reducing the size of the rubbish bin and maintaining a relatively bigger recycling bin, the recycling rate was successfully "nudged" to almost 98%. Similarly, Poortinga and Whitaker [31] installed environmental awareness posters at twelve universities and business cafeterias to determine their influences on the usage of reusable coffee cups. Together with the charging of disposable cups, the usage of reusable coffee cups increased by 33.7%.

Henkel et al. [32] focus on strengthening undergraduates' environmentally friendly behavior by employing "nudging" in the field of green information systems. The results showed that the Experimental Group undergoing nudging with status quo bias improved their environmentally friendly behavior. Using prompts and support cues, Leoniak and Cwalina [33] found that an injunctive norm successfully induced the energy saving behavior of undergraduates in terms of switching off lights after leaving the restrooms. Telesiene et al. [34] attempted to understand the impact of an intervention course, "Sustainable Development", using the competence–learning–intervention–assessment model. The Environmental Citizenship Questionnaire was used, which includes questions regarding environmental knowledge, attitudes and values, as well as connections with nature. The pre- and post-test results showed a significant improvement in the mean values of students' scores.

From the review of existing literature, the framework for environmental education programs, which lead to improved environmentally friendly behavior in undergraduates, needs to be further developed. This research attempts to construct a theory of change [7,35–39] involving in the improvement of undergraduates' environmental awareness, attitudes, norms and efficacy, as well as an increase in environmentally friendly behavior as the ultimate environmental outcome, in order for stakeholders to learn from experience and continue to challenge existing assumptions; the following hypotheses are proposed:

Hypothesis 1 (H1): The Tzu Chi University Environmental Education Program's lessons involving environmental and sustainability topics will significantly improve the environmental awareness and attitudes of undergraduates from the Experimental Group.

Hypothesis 2 (H2): The Tzu Chi University Environmental Education Program's personal interaction of recyclables sorting at a recycling center, as well as active participation and documentation of personal environmentally friendly behavior, will significantly improve the environmental norms and efficacy of undergraduates from the Experimental Group.

Hypothesis 3 (H3): By significantly improving the environmental awareness, attitudes, norms and efficacy of undergraduates from the Experimental Group, the Tzu Chi University Environmental Education Program will significantly improve their environmentally friendly behavior.

3. Methodology

As shown in Table 1 and Figure 1, Tzu Chi University's Environmental Education Program was designed and based on its theory of change. The Environmental Education Program consisted of nine weeks of compulsory lessons and assignments (known as "activities" within our theory-of-change framework); these involved various environmental and sustainability topics, aiming to improve undergraduates' environmental awareness, attitudes, norms, efficacy (known as "capacity changes" within our theory-of-change framework), which, in turn, aimed to improve environmentally friendly behavior (known as "behavioral changes" within our theory-of-change framework). The first four weeks of lessons was conducted on a weekly basis over a period of two hours, wherein topics included global warming, sustainable development, plant-based diet and climate-change-induced disasters. The first assignment consisted of arranging for undergraduates to visit a local recycling center and participate hands-on in the sorting of recyclables. The next four weeks of assignments consisted of weekly documenting of the efforts and challenges faced while seeking to improve their environmentally friendly behavior.

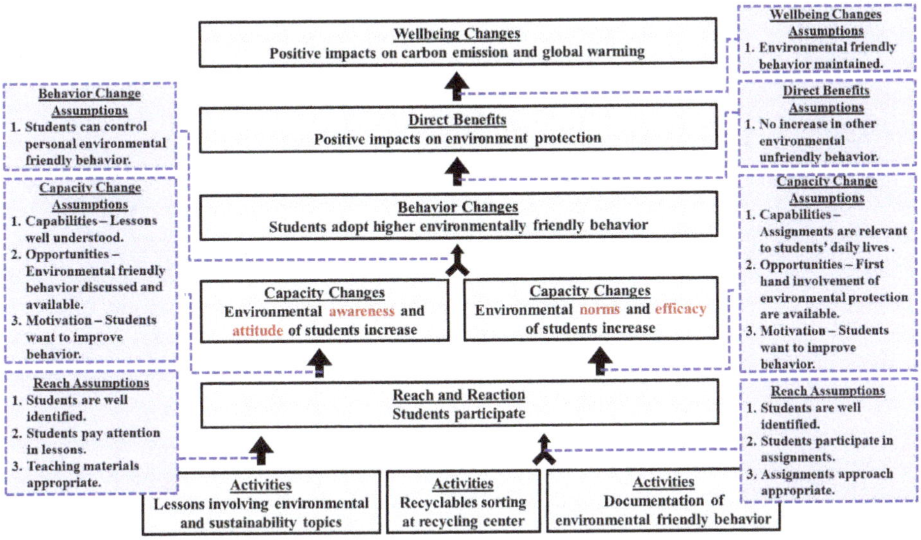

Figure 1. Tzu Chi University Environmental Education Program and its theory of change.

Table 1. Tzu Chi University Environmental Education Program.

Week		Topics
1	Lesson	Introduction to global warming and sustainable development
2	Lesson	Introduction to plant-based diet and its effects on environmental protection
3	Lesson	Climate-change-induced disaster: technology-assisted disaster relief
4	Lesson	Climate-change-induced disaster: case study of cyclone Idai and its effects
5	Assignment	Hands-on recyclables sorting at a recycling center
6–9	Assignment	Weekly documentation of individual environmentally friendly behavior

All freshmen of Tzu Chi University were required to enroll in two freshmen courses: "Education for Life" during the first semester and "Tzu Chi Humanities and Service Learning" for the second semester; each course was further separated into 15 classes. The Control Group consisted of 87 freshman undergraduates, predominately from the department of Nursing (two classes) while the Experimental Group (known as "reach and reaction" within our theory-of-change framework) consisted of 142 freshman undergraduates from the departments of molecular biology and human genetics, human development and psychology, communication studies, English language and literature, as well as international service industry management (three classes). Undergraduates were informed beforehand about the Environmental Education Program, and were free to switch to another 10 classes not involved in this research. Undergraduates from the Experimental Group underwent the entire Environmental Education Program while undergraduates from the Control Group only participated in hands-on recyclables sorting at a recycling center.

This research employed the parallel mixed methods of research, whereby qualitative and quantitative data were collected and analyzed concurrently [40]. Qualitative data were written thoughts submitted by the undergraduates at the end of each lesson and assignment, while quantitative data were survey findings. As a quasi-experimental design, the respondents of the Control Group and Experimental Group were not randomly selected, and hence, there was a possibility that undergraduates from the Experimental Group were more open to environmental protection. Furthermore, as direct or indirect observations were not carried out due to their heavy demands on manpower, another limitation of this research was that only self-reported feedback and survey responses were collected.

In total, 289 written thoughts regarding the first four weeks of lessons were collected from the Experimental Group. At the same time, 127 written thoughts regarding the assignment of "hands-on recyclables sorting at a recycling center" and 121 written thoughts regarding the assignment of "weekly documentation of individual environmentally friendly behavior" were collected from the Experimental Group.

The criteria for the theory-of-change analysis involved well-defined and measured results following a logical sequence within a plausible timeframe [35–39]. The Environmental Education Program was administrated within the plausible timeframe of the academic year 2021/2022, where pre-tests and post-tests using the survey were conducted. Undergraduates from those 5 classes made up around 35% of the entire cohort of freshmen and voluntarily participated in answering the surveys. For the Control Group, 78 valid pre- and post-tests result (90% response rate) were obtained, while for the Experimental Group, 116 valid pre- and post-test results (82% response rate) were obtained, with both groups fulfilling the 5% margin of error at a 95% confidence level.

The pre-tests and post-tests were identical surveys (shown in Appendix A) performed online, thus ensuring a logical sequence. SurveyCake, which is frequently used as a survey tool in Taiwan, had "required questions" built in to ensure the quality of answers. Since all the questions were required questions in the online surveys, there were no missing data. The design of the survey underwent a pilot-test of 20 personnel to assess its validity and identify unresolved ambiguities, and the content of the survey was later modified based on feedback obtained.

The survey consisted of two sections, with a total of 30 questions. The first section had a total of five questions, collecting basic demographic data (as shown in Table 2) which included gender, national identification number (last four digits), age, department of studies, and nationality. The last four digits of the national identification number, which were not known to those involved in this research, and hence, did not compromise the anonymity of the questionnaire participant, were used to match undergraduates' pre- and post-test responses. The second section had a total of 25 questions and was further divided into five constructs, attempting to understand the environmental awareness, attitudes, norms, efficacy and behavior of undergraduates. Each construct, as shown in Table 3, was rated based on five-point Likert scale questions, whereby 1 indicated "strongly disagree" or "never" and 5 indicated "strongly agree" or "always", producing well-defined and measured results. The measures of norms and efficacy were modified from previous studies [41–45], while the measures of environmental awareness, attitudes and behavior were created based on previous studies [43–47] and discussions with university personnel involved with environment protection policies.

Table 2. Basic demographic of survey samples.

Variables	Control Group		Experimental Group	
	n = 78	%	n = 116	%
Gender				
Male	13	16.6%	38	32.8%
Female	64	82.1%	78	67.2%
Declined to disclose	1	1.3%	-	-
Age				
19	38	48.7%	54	46.6%
20	29	37.2%	48	41.4%
21	6	7.7%	7	6.0%
22	2	2.6%	4	3.4%
≥23	3	3.8%	3	2.6%

Table 2. *Cont.*

Variables	Control Group		Experimental Group	
	n = 78	%	n = 116	%
Department of				
Nursing	75	96.2%	-	-
Molecular biology and human genetics	-	-	28	24.1%
Human development and psychology	-	-	28	24.1%
Communication studies,	-	-	27	23.3%
English language and literature	-	-	12	10.3%
International service industry management	-	-	10	8.6%
Others	3	3.8%	11	9.5%
Nationality				
Republic of China (Taiwan)	75	96.2%	107	92.2%
Others	3	3.8%	9	7.8%

Table 3. Mean and standard deviation of survey items.

Item	Control Group						Experimental Group					
	Pre-Test			Post-Test			Pre-Test			Post-Test		
	Mean	S.D.	r	Mean	S.D.	r	Mean	S.D.	r	Mean	S.D.	r
Awareness												
A1.	4.54	0.935	0.709	4.41	0.859	0.781	4.61	0.669	0.656	4.69	0.501	0.703
A2.	4.37	0.775	0.696	4.35	0.770	0.800	4.54	0.638	0.806	4.57	0.515	0.812
A3.	4.32	0.655	0.732	4.28	0.643	0.859	4.41	0.710	0.721	4.55	0.565	0.779
A4.	4.35	0.680	0.706	4.29	0.723	0.730	4.40	0.696	0.776	4.55	0.623	0.812
A5.	4.29	0.723	0.728	4.15	0.774	0.752	4.37	0.679	0.740	4.48	0.639	0.809
Attitude												
T6.	4.53	0.528	0.850	4.37	0.584	0.762	4.45	0.637	0.888	4.56	0.579	0.860
T7.	4.27	0.617	0.807	4.27	0.678	0.895	4.41	0.633	0.891	4.50	0.611	0.883
T8.	3.90	0.731	0.788	3.95	0.754	0.812	4.08	0.759	0.863	4.27	0.773	0.887
Norms												
N9.	4.18	0.679	0.728	4.10	0.815	0.675	4.10	0.848	0.731	4.30	0.760	0.777
N10.	3.83	0.763	0.764	3.99	0.730	0.756	3.95	0.903	0.826	4.24	0.787	0.793
N11.	3.45	0.878	0.790	3.63	0.955	0.776	3.66	0.845	0.808	3.94	0.887	0.818
N12.	3.81	0.722	0.835	3.87	0.812	0.822	3.89	0.810	0.878	4.11	0.755	0.905
Efficacy												
E13.	2.86	1.003	0.758	3.09	1.153	0.863	3.16	1.046	0.837	3.55	1.074	0.821
E14.	3.64	0.882	0.827	3.88	0.853	0.763	3.88	0.846	0.823	4.24	0.809	0.812
E15.	3.83	0.932	0.814	3.94	0.873	0.817	4.03	0.849	0.824	4.24	0.753	0.824
Behavior												
B16.	3.60	1.049	0.494	3.56	1.076	0.469	3.53	1.067	0.433	4.02	0.769	0.484
B17.	4.17	0.918	0.678	4.14	0.922	0.751	4.14	0.932	0.567	4.55	0.609	0.503
B18.	3.78	0.878	0.608	3.72	0.938	0.681	3.91	0.965	0.645	4.28	0.800	0.648
B19.	4.46	0.863	0.563	4.19	1.033	0.613	4.48	0.818	0.518	4.57	0.688	0.493
B20.	3.79	0.985	0.574	3.76	1.095	0.664	3.72	1.124	0.513	3.93	1.069	0.650
B21.	3.32	1.087	0.474	3.38	1.154	0.681	3.46	1.091	0.633	3.89	0.949	0.752
B22.	3.14	1.102	0.529	3.44	1.088	0.753	3.29	1.072	0.752	3.76	0.929	0.819
B23.	4.04	0.986	0.562	4.05	0.938	0.776	4.21	0.860	0.554	4.27	0.858	0.657
B24.	3.81	0.757	0.615	3.55	1.065	0.780	4.12	0.876	0.632	4.10	0.784	0.742
B25.	2.76	1.164	0.517	2.58	1.212	0.587	2.78	1.259	0.534	3.31	1.058	0.525

Note: r refers to Pearson correlation value, where the corresponding critical Pearson correlation value for the Control Group and Experimental Group are $r_c = 0.223$ (degrees of freedom = 76) and $r_c = 0.182$ (degrees of freedom = 114), respectively, for a significance level of 0.05 (two-tailed). All values of r are significant at *** $p < 0.001$.

To analyze the collected data, descriptive statistics were carried out using the software tool SPSS 25.0 [48]. As the five-point Likert scale was of ordinal scale, non-parametric statistical testing was performed. Wilcoxon rank-sum test (between the Control Group and Experimental Group) and Wilcoxon signed-rank test (between pre- and post-test), which were identified according to the last four digits of the national identification numbers of the undergraduates, were performed at a 95% confidence interval, with the null hypotheses stating that the difference between the population means was equal to zero.

4. Findings and Discussion

4.1. Theory of Change for the Environmental Education Program

In order for our theory of change to be feasible, the assumptions shown in Figure 1 needed to be well defined, justified, realizable and measurable [35–39]. Within the "reach" assumption, undergraduates from the Experimental Group were well defined, realizable and measurable, as they were registered for the courses throughout the semesters. Furthermore, the rates of attendance to lessons and assignments handed in were above 90%. Regarding the "capacity changes" assumption, more than 70% of undergraduate feedback stated "very satisfied" regarding the topic and content of the lessons conducted, according to the five-point Likert scale.

Opportunities for recyclables sorting within the local recycling center were also constantly available due to the high number of recyclables sent in daily, as shown in Figure 2. The undergraduates performed hands-on recyclables sorting within the semi-enclosed spaces of the recycling center, and remarked, in their written thoughts, that most of the plastic bottles gave off a pungent smell as they were unwashed before recycling. Furthermore, most of the recycled plastic bottles were bottled water instead of other beverages, and undergraduates realized that using water bottles filled with plain water would have significantly reduced the production of plastic bottles.

Figure 2. Undergraduates performing hands-on recyclables sorting within the recycling center.

The "behavior change" assumption that undergraduates could control their personal environmentally friendly behavior was found to be justified, realizable and sustainable, as more than 90% of undergraduates from Tzu Chi University stayed in university dormitories with freshmen from the same major assigned among each other as roommates. Hence, away from senior family members and together with roommates participating in the Environmental Education Program, undergraduates from the Experimental Group could independently control their personal environmentally friendly behavior. The "direct benefits" assumption was also found to be well defined, justified and realizable, as the ten environmentally friendly behaviors used during the weekly assignment and survey were common behaviors within the local setting and, after discussion, with undergraduates and various stakeholders. Finally, the "wellbeing changes" assumption was well defined, justified and measurable, as the post-test survey was conducted two weeks after the end of the Environmental Education Program.

4.2. Statistical Analysis of Surveyed Results

With respect to the results of the pre- and post-test, the mean, standard deviation and Pearson correlation values of the survey items are presented in Table 3, while the Cronbach's alpha, mean and standard deviation of each construct are presented in Table 4. The Cronbach's alpha values across the various groups and tests were greater than 0.7 [47] and ranged from 0.713 to 0.879, indicating excellent internal consistency reliability of the scales. At the same time, the Pearson correlation values across the various groups and tests were greater than the critical Pearson correlation values r_c of 0.223 (degrees of freedom = 76 for the Control Group) and 0.182 (degrees of freedom = 114 for the Experimental Group), respectively, indicating strong validity.

Table 4. Cronbach's alpha, mean and standard deviation of survey constructs.

Item	Control Group						Experimental Group					
	Pre-Test			Post-Test			Pre-Test			Post-Test		
	α	Mean	S.D.	α	Mean	S.D.	α	Mean	S.D.	α	Mean	S.D.
Awareness	0.751	4.37	0.538	0.838	4.30	0.590	0.792	4.47	0.502	0.841	4.57	0.447
Attitudes	0.728	4.23	0.508	0.758	4.20	0.554	0.849	4.31	0.595	0.840	4.44	0.574
Norms	0.780	3.82	0.593	0.747	3.90	0.627	0.825	3.90	0.690	0.839	4.15	0.656
Efficacy	0.713	3.44	0.750	0.740	3.64	0.786	0.764	3.69	0.757	0.735	4.01	0.720
Behavior	0.747	3.69	0.544	0.863	3.64	0.707	0.772	3.76	0.581	0.828	4.07	0.541

For the environmental awareness construct, the average scores for all items were above 4.15, with standard deviations ranging from 0.501 to 0.935, showing that undergraduates have high environmental awareness. The pre- and post-test average scores for the Control Group for environmental awareness were 4.37 and 4.30 (as shown in Table 4), with the Wilcoxon signed-rank test showing no significant average difference ($z_{76} = -1.123$, $p = 0.261$, as shown in Table 5) at the 95% confidence interval. However, the pre- and post-test average scores for the Experimental Group for environmental awareness were 4.47 and 4.57 (as shown in Table 4), with the Wilcoxon signed-rank test showing a significant average difference at the 95% confidence interval ($z_{114} = -2.758$, $p < 0.01$, as shown in Table 6).

Table 5. Wilcoxon signed-rank tests between pre- and post-test for the Control Group.

Construct	Test	Mean	S.D.	z-Value	p-Value
Awareness	Pre-test	4.37	0.538	−1.123	0.261 (NS)
	Post-test	4.30	0.590		
Attitudes	Pre-test	4.23	0.508	−0.915	0.360 (NS)
	Post-test	4.20	0.554		
Norms	Pre-test	3.82	0.593	−1.484	0.138 (NS)
	Post-test	3.90	0.627		
Efficacy	Pre-test	3.50	0.657	−2.138	0.033 *
	Post-test	3.75	0.709		
Behavior	Pre-test	3.69	0.544	−0.371	0.710 (NS)
	Post-test	3.64	0.707		

Note: * $p < 0.05$ (two-tailed), and NS means non-significant.

Regarding the environmental attitude construct, the average scores for all items were above 3.90, with standard deviations ranging from 0.528 to 0.773, showing that undergraduates have high and consistent environmental attitudes. The pre- and post-test average scores for the Control Group for environmental attitude were 4.23 and 4.20 (as shown in Table 4), with the Wilcoxon signed-rank test showing no significant average difference ($z_{76} = -0.915$, $p = 0.360$, as shown in Table 5) at the 95% confidence interval. At the same time, the pre- and post-test average scores for the Experimental Group for environmental attitude were 4.31 and 4.44 (as shown in Table 4), with the Wilcoxon signed-rank test showing a significant average difference ($z_{114} = -2.785$, $p < 0.01$, as shown in Table 6) at the 95% confidence interval.

Table 6. Wilcoxon signed-rank tests between pre- and post-test for the Experimental Group.

Construct	Test	Mean	S.D.	*z*-Value	*p*-Value
Awareness	Pre-test	4.47	0.502	−2.758	0.006 **
	Post-test	4.57	0.447		
Attitudes	Pre-test	4.31	0.595	−2.785	0.005 **
	Post-test	4.44	0.574		
Norms	Pre-test	3.90	0.690	−4.016	0.000 ***
	Post-test	4.15	0.656		
Efficacy	Pre-test	3.76	0.739	−4.755	0.000 ***
	Post-test	4.07	0.691		
Behavior	Pre-test	3.76	0.581	−4.939	0.000 ***
	Post-test	4.07	0.541		

Note: ** $p < 0.01$ and *** $p < 0.001$, respectively (two-tailed).

The average scores for all items belonging to the environmental norms construct ranged from 3.45 to 4.30, with standard deviations ranging from 0.679 to 0.955, where undergraduates had slightly diverse views on environmental norms. The pre- and post-test average scores for the Control Group for environmental norms were 3.82 and 3.90 (as shown in Table 4), with the Wilcoxon signed-rank test showing no significant average difference ($z_{76} = -1.484$, $p = 0.138$, as shown in Table 5) at the 95% confidence interval. Subjected to the Environmental Education Program, the pre- and post-test average scores for the Experimental Group for environmental norms were 3.90 and 4.15 (as shown in Table 4), with the Wilcoxon signed-rank test showing a significant average difference ($z_{114} = -4.016$, $p < 0.001$, as shown in Table 6) at the 95% confidence interval.

Next, the average scores for all items from the environmental efficacy construct ranged from 2.86 to 4.24, with standard deviations ranging from 0.753 to 1.153, giving the insight that undergraduates were divided on the topic of environmental efficacy. The pre- and post-test average scores for the Control Group for environmental efficacy were 3.50 and 3.75 (as shown in Table 4), with the Wilcoxon signed-rank test showing a significant average difference ($z_{76} = -2.138$, $p < 0.05$, as shown in Table 5) at the 95% confidence interval. As undergraduates from the Control Group only participated in hands-on recyclables sorting at a recycling center, this exposure prompted undergraduates to realize that they had the capacity to contribute to environmental protection. Similarly, the pre- and post-test average scores for the Experimental Group for environmental efficacy were 3.76 and 4.07 (as shown in Table 4), with the Wilcoxon signed-rank test showing a significant average difference ($z_{114} = -4.016$, $p < 0.001$, as shown in Table 6) at the 95% confidence interval.

Finally, the average scores for all items from the ultimate outcome of environmentally friendly behavior construct ranged from 2.58 to 4.57, with standard deviations ranging from 0.609 to 1.259, giving the insight that undergraduates had a wide range of environmentally friendly behavior. The pre- and post-test average scores for the Control Group for environmentally friendly behavior were 3.69 and 3.64 (as shown in Table 4), with the Wilcoxon signed-rank test showing no significant average difference ($z_{76} = -0.371$, $p = 0.710$, as shown in Table 5) at the 95% confidence interval. However, the pre- and post-test average scores for the Experimental Group for environmentally friendly behavior were 3.76 and 4.07 (as shown in Table 4), with the Wilcoxon signed-rank test showing a significant average difference at the 95% confidence interval ($z_{114} = -4.939$, $p < 0.001$, as shown in Table 6).

In addition, when the Control and Experimental Groups, together with pre- and post-test, are compared based on the average of all constructs within the surveys, the results indicate meaningful and significant differences, as seen in Figure 3 and Table 7. The Wilcoxon rank-sum test at the 95% confidence interval revealed that no significant average differences were found between the pre-tests of the Control Group and Experimental Group ($z = -1.345$, $p = 0.179$). Likewise, no significant average differences were found in the Wilcoxon signed-rank test between pre- and post-test of the Control Group ($z_{76} = -0.141$, $p = 0.888$). However, the post-test of the Experimental Group displayed a significantly higher average value when compared to both the pre-test of the Experimental Group ($z_{114} = -5.623$, $p < 0.001$) and the post-test of the Control Group ($z = -4.708$, $p < 0.001$).

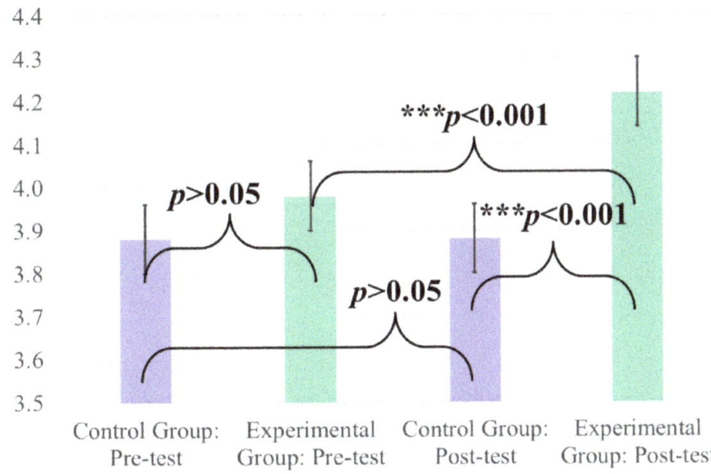

Figure 3. Mean values for Control and Experimental Groups and their significance of mean differences.

Table 7. Wilcoxon rank-sum tests and Wilcoxon signed-rank tests between Control and Experimental Groups.

Construct	Test	Mean	S.D.	*z*-Value	*p*-Value
Control	Pre-test	3.88	0.411	−1.345	0.179 (NS)
Experimental	Pre-test	3.98	0.450		
Control	Pre-test	3.88	0.411	−0.141	0.888 (NS)
Control	Post-test	3.88	0.493		
Experimental	Pre-test	3.98	0.450	−5.623	0.000 ***
Experimental	Post-test	4.22	0.448		
Control	Post-test	3.88	0.493	−4.708	0.000 ***
Experimental	Post-test	4.22	0.448		

Note: *** $p < 0.001$, respectively (two-tailed), and NS means non-significant.

4.3. Correlations between Constructs

Statistically, the relationship between two variables was generally considered high when the magnitude of the Pearson correlation coefficient (r) was greater than 0.7; the correlation was moderate when r was between 0.5 and 0.7; the correlation was low when r was between 0.3 and 0.5; and there was no significant correlation when r was less than 0.3 [48]. As shown in Tables 8 and 9, the Pearson correlation analysis was conducted to understand the relationship between environmental awareness, attitudes, norms, efficacy and behaviors among undergraduates from Tzu Chi University. For the pre-test, 65.1% of the variability in environmental attitudes was explained by the variability in environmental awareness, while 61.7% of the variability in environmental efficacy was explained by the variability in environmental norms. Similarly, 22.7%, 38.5%, 35.6% and 32.1% of the variability in environmentally friendly behavior was explained by the variability in environmental awareness, attitudes, norms and efficacy, respectively. For the post-test, 81.6% of the variability in environmental attitudes was explained by the variability in environmental awareness, while 73.3% of the variability in environmental efficacy was explained by the variability in environmental norms. Furthermore, 38.3%, 49.1%, 44.1% and 29.7% of the variability in environmentally friendly behavior was explained by the variability in environmental awareness, attitudes, norms and efficacy, respectively.

Table 8. Correlations among constructs for Experimental Group (pre-test).

Pearson, r	Awareness	Attitudes	Norms	Efficacy	Behavior
Awareness	1				
Attitudes	0.651 ***	1			
Norms	0.485 ***	0.646 ***	1		
Efficacy	0.368 ***	0.599 ***	0.647 ***	1	
Behavior	0.227 *	0.385 ***	0.356 ***	0.321 ***	1

Note: * $p < 0.05$ and *** $p < 0.001$, respectively (two-tailed).

Table 9. Correlations among constructs for Experimental Group (post-test).

Pearson, r	Awareness	Attitudes	Norms	Efficacy	Behavior
Awareness	1				
Attitudes	0.816 ***	1			
Norms	0.614 ***	0.756 ***	1		
Efficacy	0.537 ***	0.661 ***	0.733 ***	1	
Behavior	0.383 ***	0.491 ***	0.441 ***	0.297 **	1

Note: ** $p < 0.01$ and *** $p < 0.001$, respectively (two-tailed).

4.4. Discussion

By interacting with undergraduates from both the Control and Experimental Groups, as well as examining the written thoughts from the Experimental Group, a better understanding emerged. "Introduction to global warming and sustainable development" and other lessons allowed undergraduates a greater in-depth understanding of the impacts of global warming internationally, and on their future lifestyles. The trip to the "recycling center together with hands-on recyclables sorting", where they were active participants, gave undergraduates a visual insight into daily environmentally unfriendly behaviors; several undergraduates reflected upon the countless food boxes with food residuals giving off a pungent smell, and realized that it was what they did daily. This led to an increase in environmental efficacy, accounting for the statistically improvement in efficacy between the pre- and post-test of the Control Group. However, this sole improvement in efficacy in the Control Group did not continue to have an effect on environmentally friendly behavior.

In addition, undergraduates reflected that the inertia to adopt more environmentally friendly behavior was still very high; the habits of ordering takeout for meals meant that paper boxes, wooden chopsticks and plastic bags were disposed almost daily. The success of improving environmentally friendly behavior in using reusable coffee cups from Cosic et al. [30] and Poortinga and Whitaker [31], as well as the lack of improvement in the consumption of meals with a low water footprint from Godfrey and Feng [29], showed the importance of convenience, with similar results also reported by Jurdi-Hage et al. [14]. For example, the environmentally friendly behavior of "B24. During a week, I use reusable eating utensils instead of disposable eating utensils" did not improve statistically (mean of 4.12 and 4.10 for the pre- and post-test of the Experimental Group, respectively) after the Environmental Education Program, wherein written thoughts by undergraduates showed that the inconvenience of bringing a bulky reusable container for meals hampered undergraduates' willingness to be more environmentally friendly. At the same time, the environmentally friendly behavior of "B20. During a week, I walk or cycle (non-electric) for distances less than a kilometer or requiring less than 10 min" did not improve statistically (mean of 3.72 and 3.93 for the pre- and post-test of the Experimental Group, respectively) after the Environmental Education Program. As Hualien, the location of Tzu Chi University, did not have an existing public transport system, undergraduates commonly travelled using their personal motorcycles. In addition, under the hot and humid summer of Hualien, undergraduates would not walk under the hot sun for more than 500 m or more than 5 min. These insights would be useful for future modifications to the Environmental Education Program, wherein the importance of convenience and habits regarding undergraduates' environmentally friendly behavior needs to be further incorporated.

Furthermore, forgetfulness led to the behavior of not switching off lights and other electrical appliances when not in use, and laziness led to the behavior of taking the elevator instead of stairs, even when only going up or down less than three floors. However, significant changes were observed during and after the four weeks of "weekly documentation of individual environmentally friendly behavior", wherein undergraduates realized that eating a plant-based diet instead of meat- and plant-based diet was not as unappetizing as initially thought, and improves skin complexion; constant encouragement among friends in taking the stairs instead of the elevator when only going up or down less than three floors actually resulted in a shorter commuting time; mutual reminders among roommates to switch off lights and other electrical appliances when not in use strengthened this environmentally friendly behavior. Hence, environmental efficacy was strengthened. Furthermore, the widespread usages of social media (Instagram and Facebook) by undergraduates to document their individual environmentally friendly behavior strengthened their personal and social norms, as seen from the various photos within their written thoughts (Figure 4). These constant reminders from friends and roommates, who were also participating as undergraduates from the Experimental Group, successfully fulfilled the roles of nudges and social influences, which were similar to the results in the published literature carried out by Henkel et al. [32], as well as Leoniak and Cwalina [33].

Figure 4. Photos within written thoughts of undergraduates from Experimental Group documenting their environmentally friendly behavior: (clockwise from top left) cycle for distance less than a kilometer or requiring less than 10 min; walk for distance less than a kilometer or requiring less than 10 min; take the stairs instead of the elevator when walking up/down less than three floors; use reusable bags instead of disposable bags; turn off the tap while brushing teeth; switch off electrical appliances when not in use; drink plain water instead of bottled beverages; use reusable instead of disposable eating utensils; eat plant-based diet instead of meat and plant-based diet.

Moreover, undergraduates, indeed, sought to improve their environmentally friendly behavior after realizing the additional benefits of better skin complexion and a healthier body. As suggested in the mitochondrial free-radical theory of aging, mitochondrial free radicals are by-products of metabolism and result in oxidative damage to cells which, in turn, is one of the major causes of aging [49,50]. Research has shown that plants are enriched with antioxidative compounds, and a plant-based diet could induce more antioxidative enzyme production in the body and in cells. Thus, a plant-based diet was able to lower oxidative stress and, subsequently, reduce the aging of skin [51,52].

The environmental education program conducted by Hsu [26] was carried out during the academic year of 1998/1999, which was more than 20 years ago, although both National Dong Hwa University and Tzu Chi University were located in Hualien, Taiwan. Furthermore, the environmentally friendly behavior reported by Hsu [26] covered the five aspects of eco-management, consumer action, persuasion, political action and legal action, while the current research only concentrated on personal individual behaviors. Hence, the basis for comparison was limited. However, the environmental education program performed by Jiang et al. [27] was implemented in the year 2010. Measures included raising environmental awareness, providing information to highlight the impact of individual changes in environmentally friendly behavior, and personal pledges to overcome personal habitual barriers, which were similar to the approach of this research. However, Jiang et al. [27] went further and included measures to overcome university-level barriers for greater environmentally friendly behavior, as well as long-term systematic commitment towards environmental protection. For instance, the open display of the consumption of energy and water in various buildings within the campus, as well as monetary discounts for energy and water savings, could be implemented within Tzu Chi University to complement the current Environmental Education Program.

5. Conclusions

In the context of environmental education, researchers are going beyond the narrow "Knowledge-Attitudes-Behavior" theory, and instead, are emphasizing the importance of connecting the influences of environmental education programs and their wider impacts on environmental behavioral changes. This study provided a response to this direction of study by investigating the influence of a unique Environmental Education Program developed by Tzu Chi University, to target the characteristics of environmentally friendly behavior using a holistic and multidisciplinary approach, as well as implementing various teaching methods associated with developing environmental awareness, attitudes, efficacy and norms.

The results showed that the self-reported environmental awareness, attitudes, norms, efficacy and behavior of undergraduates significantly improved statistically after participating in the Environmental Education Program, as seen from the increase in the mean values among all constructs between the pre- and post-test of the Experimental Group. Statistical analysis and written thoughts from undergraduates revealed that lessons from the Environmental Education Program, showcasing real-life instances of the destructive impacts of global warming and climate change from all over the world, significantly increased the environmental awareness and attitudes of undergraduates from the Experimental Group, thus allowing us to accept Hypothesis 1. Together with encouragement and reminders from friends within the Environmental Education Program and via social media, the trip to the "recycling center together with hands-on recyclables sorting" and four weeks of "weekly documentation of individual environmentally friendly behavior" significantly strengthened the environmental efficacy and norms of undergraduates from the Experimental Group, thus allowing us to accept Hypothesis 2. Regarding Hypothesis 3, although low correlations were observed between environmental awareness, attitudes, norms, efficacy and environmentally friendly behavior among undergraduates from the Experimental Group, the Environmental Education Program significantly improved their awareness of the importance of environmental protection, changed their environmental attitudes, and increased their environmental norms and efficacy towards physical participation; finally, there was also a significant improvement in their environmentally friendly behavior.

However, it is important to note that self-reporting of surveys ran the risk of undergraduates having different interpretations of the five-point Likert scale or not being honest regarding their environmentally friendly behavior. Another limitation of the study lies in its execution during the COVID-19 pandemic, where many eateries stopped the usage of reusable eating utensils, and instead, provided disposable eating utensils out of public health concerns. Indeed, out of hygiene concerns, undergraduates also preferred to use those disposable eating utensils provided by the eateries, or use personal reusable utensils if they carry them with them. This overarching concern due to the COVID-19 pandemic did not enable undergraduates to fully realize their potential in improving their environmentally friendly behavior. Furthermore, the lack of direct or indirect observations meant that undergraduates not paying attention during lessons or assignments might have introduced uncertainties into the applied analysis.

Moreover, it was important to note that the unique arrangement of undergraduates staying together in the university dormitories and the compulsory nature of the Environmental Education Program fostered an ambience that encouraged and maintained the assumptions of our theory-of-change framework. In view of the aim of connecting undergraduates' environmentally friendly behavior and the Tzu Chi University's Environmental Education Program, it is important to repeat this study with improved and modified assumptions. Future research could expand the sample sizes to the majority of the freshman population, further randomizing the selection of freshmen for the Experimental Group, as well as arrange interviews and direct or indirect observations for selected participants in order to strengthen the process of data collection. Lastly, a delayed post-test could be arranged in order to determine whether the undergraduates' environmentally friendly behavior was retained after two months of the completion of the Environmental Education Program. This might allow clearer identification of the influences of the specific constructs of the Program.

Author Contributions: Conceptualization, methodology, survey design, software analysis and writing—review and editing, A.Y.J.T.; Funding acquisition, conceptualization, methodology, survey design, software analysis, investigation and writing—original draft, A.Y.K.T. All authors have read and agreed to the published version of the manuscript.

Funding: Funding for this research was provided by The Yin Shun and Cheng Yen Education Foundation and Tzu Chi University (funding number: TCU108400A).

Informed Consent Statement: All subjects gave their informed consent for inclusion before they participated in the study.

Data Availability Statement: The dataset will be provided upon request.

Conflicts of Interest: The authors declare no conflict of interest.

Appendix A Survey Items

Section 1: Basic Demographics

1. Gender
2. Last four digits of the national identification number
3. Age
4 Department of studies and level
5. Nationality

Section 2: Constructs

A1. Global warming is happening
A2. Global warming caused climate change
A3. Environmental protection allows human to co-exist with the Earth
A4. Sorting of rubbish is beneficious for the environment
A5. Green products benefit the environment
T6. I believe that environmental protection is very important
T7. I will protect our Earth's environment
T8. I am glad to adopt environmental protection behaviors
N9. Environmental protection is a moral issue
N10. My friends and family supported me in concerning about environmental protection
N11. Environmental protection allowed me to find extra meaning in life
N12. My friends and family supported me in adopting environmental protection behaviors
E13. I have sufficient money to protect the environment
E14. I have limitless potential in protecting the environment

E15. It is mostly up to me whether I adopt environmental protection behaviors
B16. During a week, I drink plain water instead of bottled beverages
B17. During a week, I eat in moderation and do not waste food
B18. During a week, I switch off lights and other electrical appliances when not in use
B19. During a week, I turn off the tap while brushing my teeth
B20. During a week, I walk or cycle (non-electric) for distances less than a kilometer or requiring less than 10 min
B21. During a week, I take the stairs instead of using the elevator when walking up/down less than three floors
B22. During a week, I use reusable bags instead of disposable bags
B23. During a week, I sort my rubbish according to regulations
B24. During a week, I use reusable eating utensils instead of disposable eating utensils
B25. During a week, I eat a plant-based diet instead of meat and plant-based diet

References

1. Bloom, B.S.; Engelhart, M.D.; Furst, E.J.; Hill, W.H.; Krathwohl, D.R. Taxonomy of educational objectives: The classification of educational goals. In *Handbook I: Cognitive Domain*; David McKay Company: New York, NY, USA, 1956.
2. Morshead, R.W. On taxonomy of educational objectives. Handbook II: Affective domain. *Stud. Philos. Educ.* **1965**, *4*, 164–170. [CrossRef]
3. Hungerford, H.R.; Volk, T.L. Changing learner behavior through environmental education. *J. Environ. Educ.* **1990**, *21*, 8–22. [CrossRef]
4. Heimlich, J.E. Environmental education evaluation: Reinterpreting education as a strategy for meeting mission. *Eval. Program. Plan.* **2010**, *33*, 180–185. [CrossRef] [PubMed]
5. Ellis, A. Rational psychotherapy and individual psychology. *J. Individ. Psychol.* **1957**, *13*, 38–44.
6. Muran, J.C. A reformation of the ABC model in cognitive psychotherapies: Implications for assessment and treatment. *Clin. Psychol. Rev.* **1991**, *11*, 399–418. [CrossRef]
7. Krasny, M.E. *Advancing Environmental Education Practice*; Cornell University Press: New York, NY, USA, 2020.
8. Tilbury, D. Environmental education for sustainability: Defining the new focus of environmental education in the 1990s. *Environ. Educ. Res.* **1995**, *1*, 195–212. [CrossRef]
9. Fang, W.T. *Environmental Education*; Wu-Nan Book Inc.: Taipei, Taiwan, 2019.
10. Teksoz, G.; Sahin, E.; Tekkaya-Oztekin, C. Modeling environmental literacy of university students. *J. Sci. Educ. Technol.* **2011**, *21*, 157–166. [CrossRef]
11. Vicente-Molina, M.A.; Fernandez-Sainz, A.; Izagirre-Olaizola, J. Environmental knowledge and other variables affecting pro-environmental behaviour: Comparison of university students from emerging and advanced countries. *J. Clean. Prod.* **2013**, *61*, 130–138. [CrossRef]
12. Chuvieco, E.; Burgui, M.B.; Da Silva, E.V.; Hussein, K.; Alkaabi, K. Factors affecting environmental sustainability habits of university students: Intercomparison analysis in three countries (Spain, Brazil and UAE). *J. Clean. Prod.* **2018**, *198*, 1372–1380. [CrossRef]
13. Liang, S.; Fang, W.; Yeh, S.; Liu, S.; Tsai, H.; Chou, J.; Ng, E. A nationwide survey evaluating the environmental literacy of undergraduate students in Taiwan. *Sustainability* **2018**, *10*, 1730. [CrossRef]
14. Jurdi-Hage, R.; Hage, H.S.; Chow, H.P.H. Cognitive and behavioral environmental concern among university students in a Canadian city: Implications for institutional interventions. *Aust. J. Environ. Educ.* **2019**, *35*, 28–61. [CrossRef]
15. Zhao, S.; Song, Q.; Wang, C. Characterizing the energy-saving behaviors, attitudes and awareness of university students in Macau. *Sustainability* **2019**, *11*, 6341. [CrossRef]
16. Hansmann, R.; Laurenti, R.; Mehdi, T.; Binder, C. Determinants of pro-environmental behavior: A comparsion of university students and staff from diverse faculties at a Swiss University. *J. Clean. Prod.* **2020**, *268*, 121864. [CrossRef]
17. Balinska, A.; Jaska, E.; Werenowska, A. The role of eco-apps in encouraging pro-environmental behavior of young people studying in Poland. *Energies* **2021**, *14*, 4946. [CrossRef]
18. Grodek-Szostak, Z.; Malinowski, M.; Suder, M.; Kwiecien, K.; Bodziacki, S.; Vaverkova, M.D.; Maxianova, A.; Krakowiak-Bal, A.; Ziemianczyk, U.; Uskij, H.; et al. Energy conservation behaviors and awareness of Polish, Czech and Ukrainian students: A case study. *Energies* **2021**, *14*, 5599. [CrossRef]
19. Leiva-Brondo, M.; Lajara-Camilleri, N.; Vidal-Melo, A.; Atares, A.; Lull, C. Spanish university students' awareness and perception of sustainable development goals and sustainability literacy. *Sustainability* **2022**, *14*, 4552. [CrossRef]
20. Bergquist, M.; Nilsson, A.; Schultz, W.P. A meta-analysis of field-experiments using social norms to promote pro-environmental behaviors. *Glob. Environ. Change* **2019**, *59*, 101941. [CrossRef]
21. Chatzigeorgiou, I.M.; Andreou, G.T. A systematic review on feedback research for residential energy behavior change through mobile and web interfaces. *Renew. Sustain. Energy Rev.* **2021**, *135*, 110187. [CrossRef]
22. Grilli, G.; Curtis, J. Encouraging pro-environmental behaviors: A review of methods and approaches. *Renew. Sustain. Energy Rev.* **2021**, *135*, 110039. [CrossRef]

23. Lede, E.; Meleady, R.; Seger, C.R. Optimizing the influence of social norms interventions: Apply social identity insights to motivate resident water conservation. *J. Environ. Psychol.* **2019**, *62*, 105–114. [CrossRef]
24. Mi, L.; Ding, C.; Yang, J.; Yu, X.; Cong, J.; Zhu, H.; Liu, Q. Chinese urban families to save electricity actively—A randomized controlled field trial. *J. Clean. Prod.* **2019**, *226*, 443–453. [CrossRef]
25. Van der Werff, E.; Taufik, D.; Venhoeven, L. Pull the plug: How private commitment strategies can strengthen personal norms and promote energy-saving in the Netherlands. *Energy Res. Soc. Sci.* **2019**, *54*, 26–33. [CrossRef]
26. Hsu, S. The effects of an environmental education program on responsible environmental behavior and associated environmental literacy variables on Taiwanese college students. *J. Environ. Educ.* **2004**, *35*, 37–48. [CrossRef]
27. Jiang, P.; Chen, Y.; Xu, B.; Dong, W.; Kennedy, E. Building low carbon communities in China: The role of individual's behavior change and enagement. *Energy Policy* **2013**, *60*, 611–620. [CrossRef]
28. Dupre, M.; Meineri, S. Increasing recycling through displaying feedback and social comparative feedback. *J. Environ. Psychol.* **2016**, *48*, 101–107. [CrossRef]
29. Godfrey, M.; Feng, P. Communicating sustainability: Student perceptions of a behavior change campaign. *Int. J. Sustain. High. Educ.* **2017**, *18*, 2–22. [CrossRef]
30. Cosic, A.; Cosic, H.; Ille, S. Can nudges affect students' green behaviors? A field experiment. *J. Behav. Econ. Policy* **2018**, *2*, 107–111.
31. Poortinga, W.; Whitaker, L. Promoting the use of reusable coffee cups through environmental messaging, the provision of alternatives and financial incentives. *Sustainability* **2018**, *10*, 873. [CrossRef]
32. Henkel, C.; Seidler, A.; Kranz, J.; Fiedler, M. How to nudge pro-environmental behaviour: An experimental study. In Proceedings of the 27th European Conference on Information Systems (ECIS), Stockholm & Uppsala, Sweden, 8–14 June 2019.
33. Leoniak, K.; Cwalina, W. The role of normative prompts and norm support cues in promoting light-switching behavior: A field study. *J. Environ. Psychol.* **2019**, *64*, 1–11. [CrossRef]
34. Telesiene, A.; Boeve-de Pauw, J.; Goldman, D.; Hansmann, R. Evaluating an educational intervention designed to foster environmental citizenship among undergraduate university students. *Sustainability* **2021**, *13*, 8219. [CrossRef]
35. Dyson, A.; Todd, L. Dealing with complexity: Theory of change evaluation and the full service extended schools initiative. *Int. J. Res. Method Educ.* **2010**, *33*, 119–134. [CrossRef]
36. Jackson, E.T. Interrogating the theory of change: Evaluating impact investing where it matters most. *J. Sustain. Financ. Investig.* **2013**, *3*, 95–110. [CrossRef]
37. De Silva, M.J.; Breuer, E.; Lee, L.; Asher, L.; Chowdhary, N.; Lund, C.; Patel, V. Theory of change: A theory-driven approach to enhance the medical research council's framework for complex interventions. *Trials* **2014**, *15*, 267–279. [CrossRef]
38. Mayne, J. Theory of change analysis: Building robust theories of change. *Can. J. Program Eval.* **2017**, *32*, 155–173. [CrossRef]
39. Lam, S. Toward learning from change pathways: Reviewing theory of change and its discontents. *Can. J. Program Eval.* **2020**, *35*, 188–203. [CrossRef]
40. Shorten, A.; Smith, J. Mixed methods research: Expanding the evidence base. *Evid. Based Nurs.* **2017**, *20*, 74–75. [CrossRef] [PubMed]
41. García Mejías, M.V.; Arroyo-Arcos, L.; Segrado-Pavón, R.G.; De Marez, L.; Ponnet, K. Factors influencing the intention to adopt a pro-environmental behavior by tourist operators of a Mexican national marine park. *Tur. Soc.* **2021**, *29*, 221–238. Available online: https://ssrn.com/abstract=3875874 (accessed on 5 May 2022). [CrossRef]
42. Hsu, C.; Chen, M.; Lin, Y. Information technology adoption for sustainable development: Green e-books as an example. *Inf. Technol. Dev.* **2017**, *23*, 261–280. [CrossRef]
43. Arli, D.; Tan, L.P.; Tjiptono, F.; Yang, L. Exploring consumers' purchase intention towards green products in an emerging market: The role of consumers' perceived readiness. *Int. J. Consum. Stud.* **2018**, *42*, 389–401. [CrossRef]
44. Du, J.; Pan, W. Examining energy saving behaviors in student dormitories using an expanded theory of planned behavior. *Habitat Int.* **2021**, *107*, 102308. [CrossRef]
45. Jang, S.O.; Chung, J.Y.; Kim, Y.G. Effects of environmentally friendly perceptions on customers' intentions to visit environmentally friendly restaurants: An extended theory of planned behavior. *Asia Pac. J. Tour. Res.* **2015**, *20*, 599–618. [CrossRef]
46. Nie, H.; Vasseur, V.; Fan, Y.; Xu, J. Exploring reasons behind careful-use, energy-saving behaviors in residential sector based on the theory of planned behavior: Evidence from Changchun, China. *J. Clean. Prod.* **2019**, *230*, 29–37. [CrossRef]
47. Wang, S.; Lin, S.; Li, J. Exploring the effects of non-cognitive and emotional factors on household electricity saving behavior. *Energy Policy* **2018**, *115*, 171–180. [CrossRef]
48. Field, A. *Discovering Statistics Using IBM SPSS Statistics, 5th ed*; SAGE Publications: London, UK, 2017.
49. Sanz, A.; Stefanatos, R.K.A. The mitochondrial free radical theory of aging: A critical view. *Curr. Aging Sci.* **2008**, *1*, 10–21. [CrossRef] [PubMed]
50. Ziada, A.S.; Smith, M.; Cote, H.C.F. Updating the free radical theory of aging. *Front. Cell Dev. Biol.* **2020**, *8*, 575645. [CrossRef] [PubMed]
51. Thaler, R.; Karlic, H.; Rust, P.; Haslberger, A.G. Epigenetic regulation of human buccal mucosa mitochondrial superoxide dismutase gene expression by diet. *Br. J. Nutr.* **2009**, *101*, 743–749. [CrossRef]
52. Wachtel-Galor, S.; Siu, P.M.; Benzie, I.F.F. Antioxidants, vegetarian diets and aging. In *Aging: Oxidative Stress and Dietary Antioxidants*; Preedy, V.R., Patel, V.B., Eds.; Academic Press: Cambridge, MA, USA, 2014; pp. 81–91.

energies

MDPI

Article

Rapid Quantitation of Coal Proximate Analysis by Using Laser-Induced Breakdown Spectroscopy

Yulin Liu, Dongming Wang and Xiaohan Ren *

Institute of Thermal Science and Technology, Shandong University, Jinan 250061, China; 202014243@mail.sdu.edu.cn (Y.L.); 202134238@mail.sdu.edu.cn (D.W.)
* Correspondence: xiaohan09126@gmail.com

Abstract: Proximate analysis of coal is of great significance to ensure the safe and economic operation of coal-fired and biomass-fired power generation units. Laser-induced breakdown spectroscopy (LIBS) assisted by chemometric methods could realize the prediction of coal proximate analysis rapidly, which makes up for the shortcomings of the traditional method. In this paper, three quantitative models were proposed to predict the proximate analysis of coal, including principal component regression (PCR), artificial neural networks (ANNs), and principal component analysis coupled with ANN (PCA-ANN). Three model evaluation indicators, such as the coefficient of determination (R^2), root-mean-square error of cross-validation (RMSECV), and mean square error (MSE), were applied to measure the accuracy and stability of the models. The most accurate and stable prediction of coal proximate analysis was achieved by PCR, of which the average R^2, RMSECV, and MSE values were 0.9944, 0.39%, and 0.21, respectively. Although the R^2 values of ANN and PCA-ANN were greater than 0.9, the higher RMSECV and MSE values indicated that ANN and PCA-ANN were inferior to PCR. Compared with the other two models, PCR could not only achieve accurate prediction, but also shorten the modeling time.

Keywords: LIBS; coal; proximate analysis; PCR; ANN

Citation: Liu, Y.; Wang, D.; Ren, X. Rapid Quantitation of Coal Proximate Analysis by Using Laser-Induced Breakdown Spectroscopy. *Energies* **2022**, *15*, 2728. https://doi.org/10.3390/en15082728

Academic Editor: Dimitrios A. Georgakellos

Received: 22 February 2022
Accepted: 31 March 2022
Published: 8 April 2022

1. Introduction

In the next decades, coal is expected to remain an important energy source and comprise a large proportion of energy consumption worldwide; for instance, more than half of the electric power is supplied by coal-fired power plants in China [1]. Coal-fired burners face an urgent problem in that the flame combustion stability needs to be settled, which is not conducive to ignition, combustion efficiency, extinction, and pollutant control. If the physical and chemical properties of coal could be determined in time, it could lead to extensive advances in combustion control and optimization. Proximate analysis is a key parameter that could preliminarily distinguish the types of coal and assess the coal quality, contributing well to the knowledge of the changes in fuel characteristics in boilers [2]. Traditional chemical measurement is time-consuming and labor-intensive, as characterizing coal requires involved sample preparation and cannot obtain multi-component information simultaneously, which is unfavorable for real-time boiler combustion optimization.

Many rapid sample analysis techniques exist, such as prompt gamma neutron activation analysis (PGNAA) [3], X-ray fluorescence (XRF) [4], and inductively coupled plasma-optical emission spectroscopy (ICP-OES) [5]. However, the neutron source of PGNAA has potential radiation hazards; XRF cannot analyze C, H, and other low-atomic-number elements; and ICP–OES has several weaknesses; for example, the costly analysis is derived from a large amount of argon consumption and the occurrence of several matrix effects is difficult to avoid or mitigate. As many limitations exist in practical applications, searching for an efficient and convenient method to assess coal quality is necessary. Laser-induced breakdown spectroscopy (LIBS) is an atomic emission analysis approach that is based on plasma techniques [6]. Under the action of an intense pulsed laser, the ablation point of the sample is ionized instantaneously to form high-temperature plasma. The plasma radiates different characteristic spectral emission lines during the cooling process, providing conditions for qualitative and quantitative analysis. Less than 10 min is needed to complete the whole process, including sample preparation and the prediction of chemometrics. Owing to its simple pretreatment process [7], rapid response, non-destructive testing, and simultaneous multi-element measurement [8], LIBS has unique value in some rapid detection fields, including coal analysis [9], soil testing [10], and food safety [11].

After LIBS spectra are obtained, a chemometric process is needed to further predict the properties of coal and ash samples [12]. The existing chemometric methods, including principal component analysis (PCA) [13], partial least squares (PLS) [14], support vector machine (SVM) [15], and ANNs [16], could improve the accuracy of consequence. Zhang et al. [16] investigated the coal fusion characteristics directly from coal instead of coal ash by using a LIBS experimental setup that was assisted by PLS, SVM, and an ANN. The ash content, volatiles, fixed carbon, and moisture were integrated into the model to improve the prediction accuracy of coal ash fusion temperatures (AFTs). The RMSECV values were 4.88 °C and 9.11 °C, and the corresponding root-mean-square error of prediction (RMSEP) values were 8.15 °C and 11.3 °C, respectively. The relationships between the AFTs and the intensities of the spectral lines of the LIBS spectra were further explored, which demonstrated the variation trend of AFTs with elemental content. In addition to the single chemometric method, some applications of hybrid methods, such as independent component analysis–wavelet neural network (ICA-WNN) and genetic algorithm-ANN (GA-ANN), have been widely used in industrial fields. Zhang et al. [17] employed PCA to extract the LIBS spectra of coal ash, and then independent component analysis (ICA) was used to optimize the input variables for the wavelet neural network (WNN). LIBS coupled with ICA-WNN, with 100% classification accuracy, was confirmed to classify coal ash by principal components. Lu et al. [18] developed LIBS assisted by GA-ANN to estimate the gross calorific value of continuous coal particle flows. As a result, the mean standard deviation (MSD) and the mean absolute error (MAE) of the gross calorific value for the prediction set were 0.38 and 0.39 MJ/kg, which could meet the Chinese national standard (GB/T 29161-2012) in the neutron activation method of coal analysis.

This work aimed at offering a more suitable quantitative method that was based on LIBS to predict the proximate analysis of coal expediently and precisely. Pulverized coal with added KBr binder was mixed and then pressed into a tablet, and sixteen samples were obtained in total. The LIBS experimental setup was established to investigate the atomic and molecular emissions related to the proximate analysis. The PCR, ANN, and PCA-ANN models, namely linear, nonlinear, and a combination of these two, were applied to predict coal proximate analysis, in which their quantitative accuracy and stability were analyzed and compared.

2. Materials and Methods

2.1. Coal Samples

Coal is a significant basic energy source around the world, with abound reserves, low cost, and high heating values. Two types of coal were investigated in this work, including bituminous coal and anthracite from Hebei Province, China. The raw coal was ground into finely ground powders, with a particle size of less than 75 μm, because coal is a heterogeneous mixture with a complex chemical composition. The powders mixed with KBr binder were placed in an agate mortar to make the sample uniform and eliminate the matrix effect of coal. Then, the mixture was pressed into tablets using a tablet press with a diameter of 20 mm and a thickness of 10 mm under 15 MPa pressure. The function of KBr was to enhance the intensities of the spectra and bond coal powders closely [19]. Bromide could not only inhibit the gas-phase radicals to control the chemical reaction, but could also consume H/OH radicals in the reaction zone to contribute to a stable condition [20].

According to the literature [21], the appropriate mass fraction of KBr was determined as 90 wt. %.

On the basis of the Chinese national standard (GB/T 212-2008), coal proximate analysis was carried out in an electric oven (5E-DHG) and muffle furnace (5E-MF6000). The heating values of the coal were measured with a calorimeter (5E-AC/PL) in accordance with standard GB/T 213-2008. All the instruments of coal proximate analysis mentioned above were produced by Changsha Kaiyuan Instrument, Ltd., Changsha, China. The proximate analyses were detailed in Ref. [22], and the results are shown in Table 1.

Table 1. Proximate analysis of 16 coal samples (air-dried basis).

No.	Moisture	Volatiles	Ash	Fixed Carbon [1]	Heating Value	Coal Type [2]
	(wt. %)	(wt. %)	(wt. %)	(wt. %)	(MJ/kg)	
1	9.15	29.15	9.31	52.39	29.42	Bituminous coal
2	5.61	31.13	6.35	56.91	29.65	Bituminous coal
3	3.71	30.49	6.81	58.99	30.18	Bituminous coal
4	2.12	31.33	9.00	57.55	29.84	Bituminous coal
5	3.74	30.94	3.84	61.48	30.55	Bituminous coal
6	3.95	29.86	5.55	60.64	30.24	Bituminous coal
7	8.18	32.25	3.23	56.34	29.75	Bituminous coal
8	4.50	31.28	4.97	59.25	30.00	Bituminous coal
9	3.71	30.35	6.87	59.07	30.09	Bituminous coal
10	4.04	29.68	5.59	60.69	30.42	Bituminous coal
11	0.14	17.63	10.54	71.69	32.16	Bituminous coal
12	2.11	7.07	17.61	73.21	27.19	Anthracite
13	2.25	7.38	11.83	78.54	30.23	Anthracite
14	0.59	25.65	8.72	65.04	31.93	Bituminous coal
15	0.80	12.52	27.01	59.67	23.77	Bituminous coal
16	1.64	6.51	21.87	69.98	25.48	Anthracite

[1] Determined by difference; [2] Classified by Chinese National Standard (GB5751-86).

2.2. Experimental Setup

The coal samples were assessed using a mobile LIBS system with a coaxial structure, as shown in Figure 1. A Q-switched Nd: YAG laser was used for a laser source of 1064 nm in wavelength, 1–20 Hz in repetition rate, and 7 ns in pulse width. The parameters of the linearly polarized pulses were optimized to 1 Hz in repetition rate and 10 mJ/pulse in energy (less than 1%). The LIBS signal, collected by two plano-convex lenses with 50 mm focal length, was accessed by a 50 μm single-channel spectrometer (AvaSpec-ULS4096-EVO, Avantes). The single-channel spectrometer could record the plasma emission in the wavelength range of 200–1100 nm with a resolution of 0.05–0.24 nm. Under the circumstance of a high signal-to-noise ratio (SNR), the delay time and integration time were set to 1.2 and 30 μs, respectively. The coal sample was placed on a 2D stage, which could be shifted continuously in a serpentine (S-shaped) path relative to the laser, to increase the reproducibility of the measurement. High-purity air (21% O_2 and 79% N_2) at a flow rate of 1 L/min was applied to purge the plasma region, which increased the intensity of the emission lines and decreased the limit of detection [23].

Some differences exist in the intensities of the spectral emission lines between ablation points. Every coal tablet was divided into 16 parts (4 × 4), and each part was placed at 16 ablation points (4 × 4) with a spacing of 500 μm. The spectral intensities of the 256 ablation points were averaged to represent the spectrum of each coal sample.

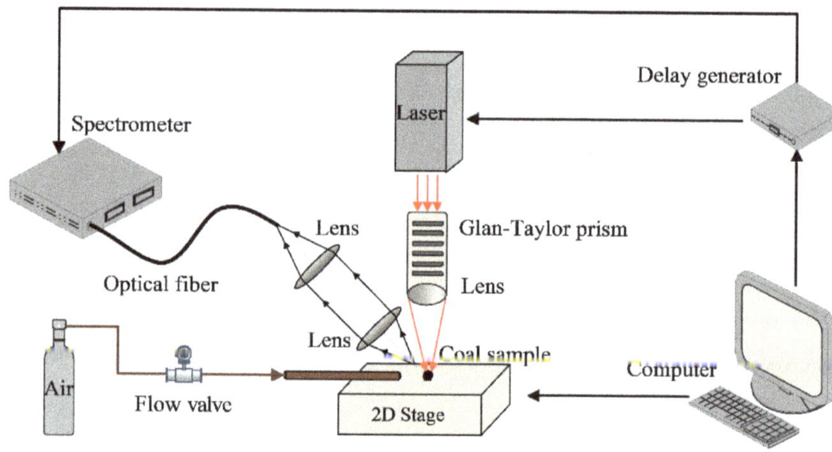

Figure 1. Schematic of the LIBS experimental setup.

2.3. *Spectral Pretreatment*

Taking the average spectrum of 256 spectra could enhance the repeatability of the LIBS signal and improve the accuracy of quantitative analysis. For example, the calculated relative standard deviation (RSD) of Fe 461.88 nm emission (one of 256 spectra selected in Section 3.1) was 0.09. However, the baseline of the average spectrum was prone to drift. Removing the baseline has a significant effect on increasing the SNR. The baseline was removed by outlier elimination and the first-order derivative [24], which could enhance the implicit peaks and separate the overlapped peaks [25].

$$I = I_{raw} - I_{baseline} \tag{1}$$

where I is the intensity of the corrected spectrum, I_{raw} is the intensity of the raw spectrum, and $I_{baseline}$ is the intensity of the estimated baseline.

2.4. *Chemometric Methods*

2.4.1. PCR

PCA combines the independent variables into a few principal components that could fully reflect the overall information through linear transformation to avoid the collinearity problem between variables. PCA could not only distinguish and remove the redundant spectral emission lines, but also extract the most related information [26]. Then, the principal components are used for the multiple linear regression (MLR) of dependent variables, namely, PCR.

Herein, the independent variables are the intensities of the selected spectra, referring to the characteristic emission lines in Section 3.1, and the dependent variable is the proximate analysis. The cumulative explained variance of the selected principal components was set to 90% or 95% [27]. In accordance with the component matrix generated by PCA, the principal components were expressed by the original variables for linear regression. Finally, MLR was used to map the selected principal components to the coal proximate analysis. Compared with MLR, PCR reduces the number of input variables and improves the stability and fitness of the regression. The key to the accuracy and stability of the PCR model is the number of principal components.

2.4.2. ANNs

Originating from biology, ANNs combine mathematics with physics to abstract the human brain's neural network from the perspective of information processing. ANNs are usually composed of an input layer, a hidden layer, and an output layer, with basic units called neurons. The input data are inputted into the activation function in the hidden layer after weighted summation to calculate the output value. Meanwhile, the value of the loss function is calculated to update the weight. ANNs have strong input-output nonlinear mapping, self-adaptability, and learning ability. However, ANNs have a low study rate and local minimum solution, leading to the inaccuracy of predictive performance. ANNs cannot easily study the contribution of input variables to the output [28]. Though traditional ANNs could predict quantitative analysis [29], a large number of input variables could increase the difficulty of the training model and reduce the predictive precision [30].

Trained with a backpropagation algorithm, the ANN model consists of sigmoid hidden layer neurons and linear output neurons. The more neurons in the hidden layer, the higher the prediction accuracy. If the network is too complex, it may not converge. Therefore, the trade-offs between network complexity and training efficiency are worthy of attention. The number of hidden layer neurons is roughly determined by the following equation [31]:

$$N^{(h)} = \sqrt{N^{(i)} \cdot N^{(o)}} \tag{2}$$

where $N^{(h)}$ is the number of hidden layer neurons, $N^{(i)}$ is the number of input neurons, and $N^{(o)}$ is the number of output neurons.

2.4.3. PCA–ANN

The selected principal components with large variances were first used to replace the original variables to eliminate information redundancy in the raw data. An ANN assisted by PCA was used to reduce the input variables and improve the efficiency of training models. Compared with the ANN model, the PCA-ANN model simplifies the network structure and computational process.

The principal component values calculated by the component matrix and the coal proximate analysis were treated as independent and dependent variables for training the ANN model, respectively. When the principal component scores of a coal sample are inputted into the trained ANN model, a predicted proximate analysis result is outputted. The feature extraction of the LIBS spectrum performed by PCA-ANN can be found in previous research [32].

2.5. Model Evaluation Indicators

Leave-one-out cross-validation (LOO-CV) was adopted to verify the generalization effect of these three models. Three indicators were used to evaluate the accuracy and stability of the models. R^2 represents the fitting degree of a model (a value equal to one is a perfect fit). RMSECV describes the deviation between the predicted and real values (a value equal to zero is a perfect fit). MSE enlarges the value with a large deviation and compares the stability of different models (a value equal to zero is a perfect fit).

$$R^2 = 1 - \frac{\sum_i^n (y_i - \hat{y}_i)^2}{\sum_i^n (y_i - \overline{y}_i)^2} \tag{3}$$

$$RMSECV = \sqrt{\frac{\sum_i^n (y_i - \hat{y}_i)^2}{n-1}} \tag{4}$$

$$MSE = \frac{1}{n}\sum_i^n (y_i - \hat{y}_i)^2 \tag{5}$$

where y_i, $\overline{y}_i$, and $\hat{y}_i$ are the real, mean, and predicted values from the training set, respectively; and n is the number of samples in the training set.

3. Results and Discussion

3.1. Selection of Characteristic Emission Lines

Five indices of proximate analysis, which consisted of moisture, ash, volatiles, fixed carbon, and heating value, were utilized to assess the coal quality in this context. The organic emission lines (e.g., C, H, and O) associated with these five indices were suitable to integrate into noise; thus, setting up a database alone was difficult [15]. The molecular spectra of CN and C_2 were chosen to complement the reduction in atomic C emissions, because the collisions of O and N with C in the air enhanced the formation of molecular species, such as CN, C_2, and CO [33]. Several mineral elements correlated with volatiles and ash were also taken into consideration to cover all elements in coal as much as possible. Therefore, the database was composed of 63 emissions, including C, H, O, N, Li, Na, Mg, Al, Si, Ca, Ti, Fe, C_2, and CN [34], as shown in Table 2. Owing to the addition of 90 wt. % KBr, the intensity of the K emission line (766.49 nm) was the highest in all wavelengths, as shown in Figure 2. Thus, the K emission lines were eliminated to prevent the high concentration from affecting the prediction accuracy. The removal of the baseline not only retained the original appearance of spectral peaks, but also improved the SNR, which contributed to the selection of characteristic emission lines. The basic considerations for selecting the emission lines were as follows [35]:

- High intensity to obtain a high SNR;
- High probability of excitation to ensure the repeatability of the experiment;
- Interference-free spectrum to exclude the influence of spectral line overlap.

Table 2. Emission lines related to coal proximate analysis.

Type	Name	Emission Lines [1] (nm)
Organic atom	C	247.86 (I [2])
	H	656.29 (I)
	O	777.19 (I), 844.68 (I)
	N	744.23 (I), 746.83 (I)
Inorganic atom	Si	220.80 (I), 221.67 (I), 250.69 (I), 251.43 (I), 251.61 (I), 251.92 (I), 252.85 (I), 288.16 (I)
	Li	670.78 (I)
	Na	589.00 (I), 589.59 (I)
	Mg	279.55 (II [3]), 280.27 (II), 285.21 (I)
	Al	226.91 (I), 237.31 (I), 308.22 (I), 309.27 (I), 394.40 (I), 396.15 (I)
	Ca	315.89 (II) 317.93 (II), 393.37 (II), 396.85 (II), 422.67 (I), 443.50 (I), 445.48 (I), 610.27 (I), 612.22 (I), 616.13 (I), 643.91 (I), 646.26 (I), 854.21 (II), 866.21 (II)
	Ti	334.94 (II), 336.12 (II), 398.18 (I), 398.96 (I), 399.90 (I)
	Fe	234.35 (II), 238.18 (I), 239.33 (II), 240.48 (II), 258.55 (I), 260.65 (II), 261.11 (II), 273.92 (I), 274.89 (II), 275.55 (I), 404.58 (I), 461.88 (I)
Molecule	C_2	469.76, 471.52, 473.71
	CN	386.19, 387.14, 388.34

[1] Data from NIST; [2] Atomic spectral line; [3] Ion spectral line.

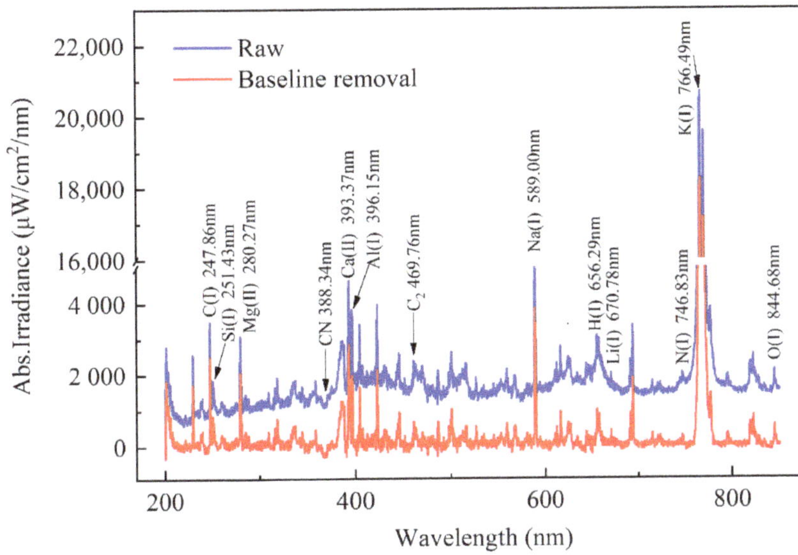

Figure 2. Average emission spectrum of coal sample 1.

3.2. PCR

Several uncorrelated principal components were obtained by dimensional reduction of PCA. According to Figure 3a, four principal components could account for 92.68% of the cumulative explained variance. Even if the cumulative explained variance was over 90%, the model evaluation indicators, such as R^2 and RMSECV, as shown in Figure 3b, were still essential. When the number of principal components was selected to be 14, the RMSECV value was the smallest, at only 0.53%, and the cumulative explained variance met the convention of 90% cumulative explained variance [27]. Then, 14 principal components of each coal sample were determined and mapped to the coal proximate analysis by MLR.

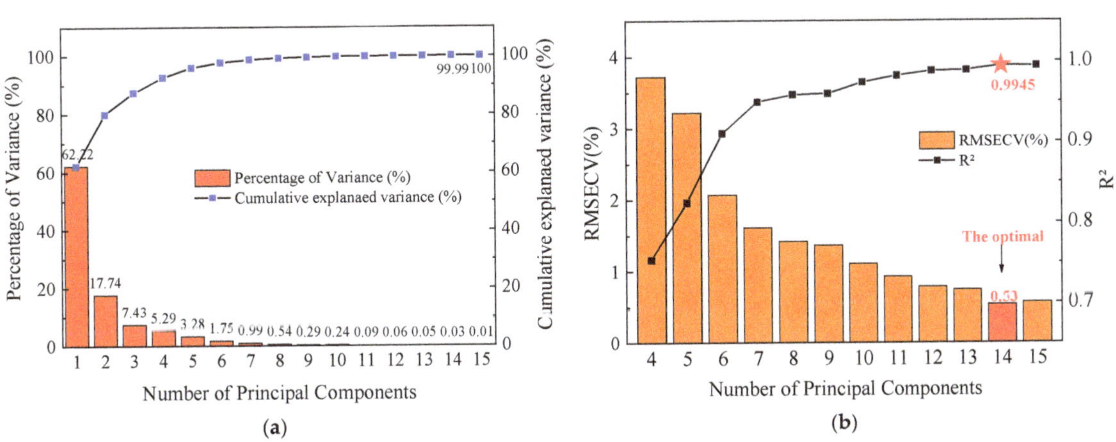

Figure 3. Scree plot (**a**) and RMSECV and R^2 values (**b**) of PCR with different numbers of principal components.

The correlation between the input emission lines and the first two principal components was shown in Figure 4. Principal component 1 was positively correlated with the intensities of the C, O, Li, Na, Mg, Ca, Fe, C_2, and CN emission lines. As principal components 1 and 2 were orthogonal, some emission lines were negatively correlated with principal component 2, such as O, Na, Mg, Ca, and CN. The scoring trends of the C, Si, Li, Fe, and C_2 emission lines were generally similar, whereas those of the H, O, Na, Mg, Ca, and CN emission lines were the opposite. Although each principal component had no clear physical meaning, the loadings of the first two principal components were devoted to determining which potential emission lines were vital to select the emission lines [36].

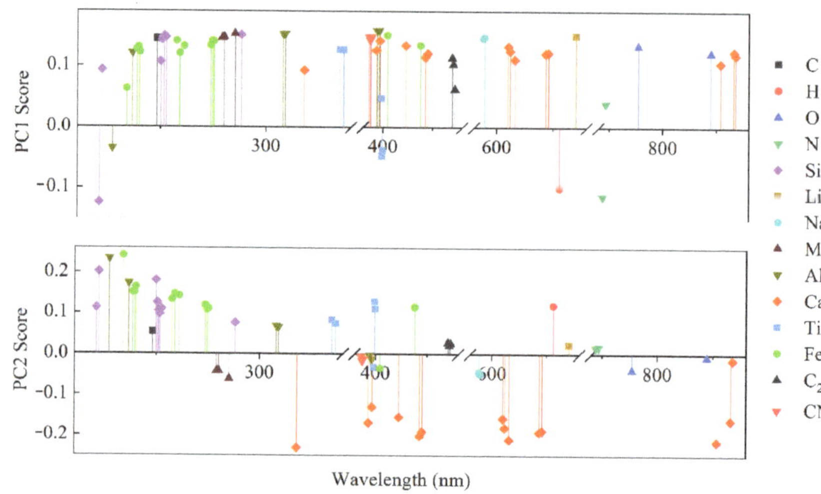

Figure 4. Loading plot of spectral emission lines on the first two principal components.

A clustering phenomenon could be observed in Figure 5, and the first two principal component scores of each coal sample corresponded to the types of coal, especially in bituminous coal. The scores of principal components calculated by PCA were widely used in sample classification, implying that similar samples could obtain close scores of principal components [37]. Thus, PCA could be applied to roughly classify a large number of coal samples into different types in accordance with the scores. Furthermore, the predictive model could be trained by different groups to predict the coal proximate analysis accurately.

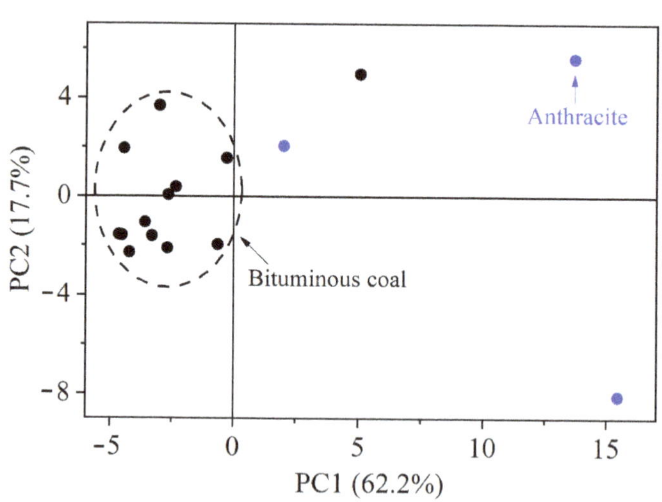

Figure 5. Score plot of 16 coal samples on the first two principal components.

3.3. ANNs

ANNs are devoted to the mapping of multi-inputs and multi-outputs; thus, they are suitable for solving nonlinear problems with complex internal mechanisms. However, an overfitted situation, where the ANN model performs well in training but poorly in prediction, may occur. The most direct method to avoid overfitting is to add the number of spectra for each coal sample [38]. Thus, the 256 spectra of each coal sample were used to cover more changes and enlarge the training set.

The intensities of the 63 emission lines selected in Table 2 were inputted as independent variables, and the proximate analysis was inputted as the dependent variable to train the model. In this work, 12 coal samples were selected as the training set, two samples were regarded as the validation set, and the remaining two were used as the prediction set. The proportions of the training set, validation set, and prediction set were approximately 70%, 15%, and 15%, respectively. The most suitable number of hidden layer neurons was 11, fixed by the minimum RMSECV and the maximum R^2, as shown in Figure 6. When the number of hidden layer neurons was 13, the ANN model was overfitted, which indicated that ANNs may need to conduct data regularization for enhanced stability [38].

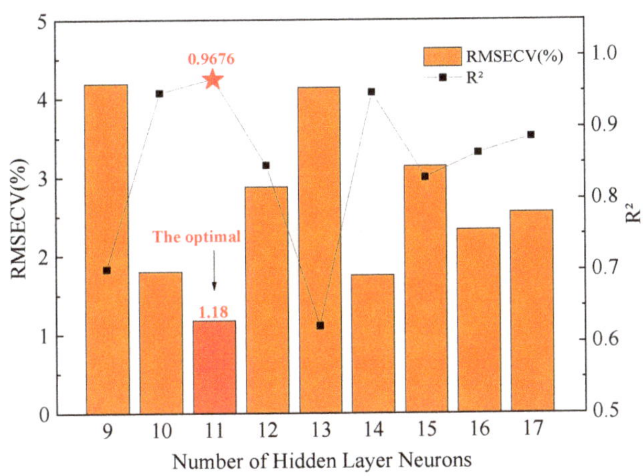

Figure 6. The RMSECV and R^2 values of ANN with different numbers of hidden layer neurons.

3.4. PCA-ANN

The PCA-ANN model contained the nonlinear regression of ANN. Thus, the problem of overfitting in Section 3.3 also existed when the number of hidden layer neurons was nine, as shown in Figure 7. If a sufficient number of principal components was used, the accuracy of classification could be improved [39], indicating the importance of selecting principal component input for ANN. According to Figure 3a, almost 100% of the variance could be explained by the first 15 principal components, which were inputted from large to small by eigenvalue (EV-PCA-ANN) to build an ANN model. The maximum R^2 and minimum RMSECV optimized the number of hidden layer neurons to five, at which point the performance of the PCA-ANN model was the best.

3.5. Comparison and Analysis

Due to the complex and diverse composition of coal, three chemometric methods were proposed to predict the coal proximate analysis after spectral pretreatment (See Table 3). PCR is regarded as a linear algorithm, and ANN is a nonlinear algorithm. PCA-ANN first performed linear dimensionality reduction (PCA) and then mapped the principal component scores to coal proximate analysis through nonlinear transformation (ANN). The optimal parameters of each algorithm were determined by cross-validation. The comparison of predicted and real values of coal proximate analysis by three different chemometric methods was shown in Figure 8.

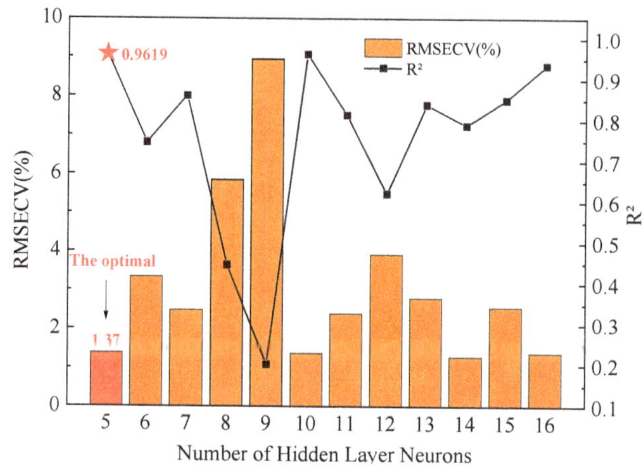

Figure 7. The RMSECV and R^2 values of PCA-ANN with different numbers of hidden layer neurons.

Table 3. Quantitative analysis of the three chemometric methods.

Proximate Analysis	Methods	R^2	RMSECV (%)	MSE	Modeling Time (s)
Moisture	PCR	0.9904	0.2458	0.0565	10
	ANN	0.9816	0.3386	0.1082	89
	PCA-ANN	0.9513	0.5555	0.2887	28
Ash	PCR	0.9986	0.2455	0.0564	10
	ANN	0.9770	1.0060	0.9496	85
	PCA-ANN	0.9179	1.8955	11.6299	25
Fixed carbon	PCR	0.9891	0.6620	0.4097	10
	ANN	0.9330	1.6334	2.5081	90
	PCA-ANN	0.9770	0.5765	1.7799	30
Volatiles	PCR	0.9958	0.6857	0.4399	10
	ANN	0.9928	0.8973	0.7579	79
	PCA-ANN	0.9917	1.6452	24.5579	22
Heating value	PCR	0.9979	0.0980	0.0900	10
	ANN	0.9342	0.5482	0.2837	83
	PCA-ANN	0.9438	0.5120	0.2452	24

A slight difference was found among the three predictions of moisture. Among them, the PCR model performed best, and its R^2, RMSECV, and MSE values were 0.9904, 0.2458%, and 0.0565, respectively. For ash, the prediction of PCR was the most accurate and stable. The R^2 values of ash between the predicted and real values, as calculated by PCR, ANN, and PCA-ANN, were 0.9986, 0.9770, and 0.9179, respectively; the corresponding RMSECV values obtained by LOO-CV were 0.2455%, 1.0060%, and 1.8955%, respectively. The MSE values of PCR and ANN were 0.0564 and 0.9496, while the MSE value of PCA-ANN was 11.6299, indicating that the prediction model was not as stable as the other two. In terms of fixed carbon, the best prediction model was PCR. The R^2 and RMSECV values of the three models were close. The MSE value of PCR was far lower than that of the other two, indicating that PCR was more robust. The PCR model had the best performance for the quantification of volatiles. Though the R^2 values of all models were greater than 0.99, the robustness of the model still needed to be considered. The RMSECV values of PCR, ANN, and PCA-ANN were 0.6857%, 0.8973%, and 1.6452%, respectively. Compared with PCR and ANN, the MSE value of PCA-ANN reached 24.5579, which showed that PCA–ANN was not suitable for the prediction of volatiles. When predicting the heating value, the highest R^2, and the lowest RMSECV and MSE demonstrated that the PCR was still the prime model for prediction.

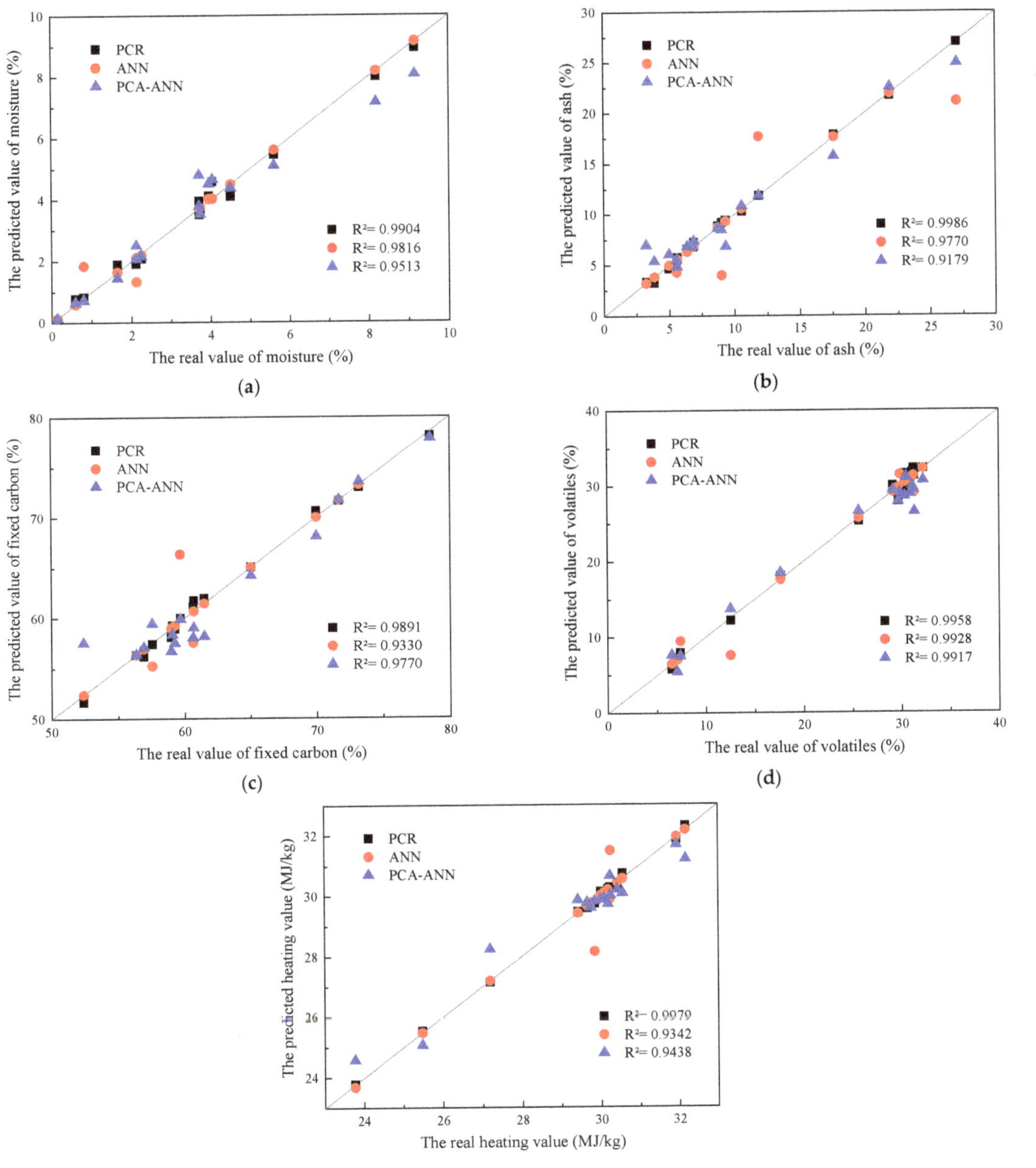

Figure 8. Comparison of predicted and real values of moisture (**a**), ash (**b**), fixed carbon (**c**), volatiles (**d**), and heating value (**e**) on an air-dries basis by three different chemometric methods.

The modeling time was considered as an important index of the modeling efficiency. When the same algorithm used different parameters, its modeling time was similar, which was consistent with Zhang's work [15]. The modeling time of different algorithm models varied greatly; for instance, the modeling times of PCR, ANN, and PCA-ANN to predict moisture were 10, 89, and 28 s, respectively. The modeling time of PCR was generally the shortest, PCA-ANN ranked second, and the modeling efficiency of ANN was the lowest. The prediction results of the cross-validation and modeling time were shown in Table 3.

As a result, the PCR should be considered first when coal proximate analysis was predicted under similar experimental conditions. After dimensional reduction and linear regression, PCR had good accuracy and stability, and its training time was the shortest. The intensities of the selected emission lines had a great linear correlation with the coal proximate analysis. First, the principal components were extracted directly from the 63 emission lines, non-reabsorbed and non-overlapping, which provided the possibility of linear correlation. With the addition of KBr binder, the generation and disappearance of shock waves were at the same level during the ablation process of coal [40], which suppressed the matrix effect of coal. Meanwhile, the laser-induced ablation pyrolysis of volatiles may have been strongly inhibited to provide a good environment for forming plasmas with stable emissions. Thus, a strong corresponding relationship existed between the component concentration and spectral intensity in coal.

Due to the large number of variables to be inputted, the training of the ANN was time consuming, while its prediction accuracy and stability were slightly lower than those of PCR. Some of the selected emission lines may not have mattered, which wasted computing resources and resulted in inaccurate results. Although the correlation between the component concentration and elemental concentration in coal was nonlinear, the selected emission lines weakened some nonlinear factors, to an extent. Another reasonable explanation was that the temporal and spatial uniformity of the plasma, provided by the KBr, improved the linear relation, causing the ANN to not be as applicable as PCR.

In contrast, PCA-ANN was not suitable for the prediction of coal proximate analysis. The reason why PCA-ANN lacked accuracy and stability was that the principal components generated by PCA and the input data for ANN were unrelated, consistent with the results of Drezga et al. [41]. The principal components obtained by PCA were usually sorted by EV–PCA-ANN for training ANN, similar to the present work. The results revealed that the principal components, obtained by PCA and then selected by GA, for training ANN (PCA-GA-ANN) had a better performance than the EV-PCA-ANN in literature [32].

4. Conclusions

In this work, the spectra of 16 coal samples with added KBr binder were investigated by a LIBS experimental setup. Three types of chemometric methods were applied to predict the coal proximate analysis: PCR, ANN, and PCA-ANN, representing linear, nonlinear, and a combination of linear and nonlinear methods, respectively. Three indicators that expressed the accuracy and stability of the model were employed to fix the optimal parameters and evaluate the performance of the model. With the highest R^2 and the lowest RMSECV and MSE values, the PCR method was deemed the most effective approach to quantitative proximate analysis. Although the R^2 values of ANN and PCA-ANN were both above 0.9, their RMSECV and MSE values were much higher than those of PCR. The training time of ANN was the longest, and a gap was found between the prediction and training ability. PCA reduced the number of input variables of ANN, whereas the nonlinear fitting of the unrelated variables reduced the robustness of PCA-ANN.

On the basis of the advantages of multi-element rapid measurement, simple sample preparation, and remote measurement, LIBS could be able to provide in-situ and online measurement and analysis for coal-fired boilers in the future. The three chemometric methods could be used to predict coal proximate analysis to provide an expedient method for obtaining the proximate analysis of coal burning in power plant boilers.

Author Contributions: Conceptualization, X.R. and Y.L.; methodology, Y.L.; software, Y.L.; validation, D.W.; formal analysis, Y.L.; investigation, D.W.; resources, X.R.; data curation, D.W.; writing—original draft preparation, Y.L.; writing—review and editing, X.R.; supervision, X.R.; project administration, X.R. All authors have read and agreed to the published version of the manuscript.

Funding: Scientific research program of Hebei administration for market regulation (No. 2022ZD09).

Institutional Review Board Statement: Not applicable.

Informed Consent Statement: Not applicable.

Data Availability Statement: Not applicable.

Acknowledgments: The authors are grateful to the anonymous reviewers whose comments have helped to improve the manuscript.

Conflicts of Interest: The authors declare no conflict of interest.

References

1. Statistical Review of World Energy 2021. Available online: https://www.bp.com/content/dam/bp/business-sites/en/global/corporate/pdfs/energy-economics/statistical-review/bp-stats-review-2021-full-report.pdf (accessed on 14 March 2022).
2. Ghetti, P. DTG combustion behaviour of coal: Correlations with proximate and ultimate analysis data. *Fuel* **1986**, *65*, 636–639. [CrossRef]
3. Charbucinski, J.; Nichols, W. Application of spectrometric nuclear borehole logging for reserves estimation and mine planning at Callide coalfields open-cut mine. *Appl. Energy* **2003**, *74*, 313–322. [CrossRef]
4. Parus, J.; Kierzek, J.; Malozewska-Bucko, B. Determination of the carbon content in coal and ash by XRF. *X-ray Spectrom. Int. J.* **2000**, *29*, 192–195. [CrossRef]
5. Ctvrtnickova, T.; Mateo, M.P.; Yañez, A.; Nicolas, G. Application of LIBS and TMA for the determination of combustion predictive indices of coals and coal blends. *Appl. Surf. Sci.* **2011**, *257*, 5447–5451. [CrossRef]
6. Carranza, J.E.; Hahn, D.W. Sampling statistics and considerations for single-shot analysis using laser-induced breakdown spectroscopy. *Spectrochim. Acta Part B At. Spectrosc.* **2002**, *57*, 779–790. [CrossRef]
7. Singh, J.P.; Almirall, J.R.; Sabsabi, M.; Miziolek, A.W. Laser-induced breakdown spectroscopy (LIBS). *Anal. Bioanal. Chem.* **2011**, *400*, 3191–3192. [CrossRef]
8. Rosenwasser, S.; Asimellis, G.; Bromley, B.; Hazlett, R.; Martin, J.; Pearce, T.; Zigler, A. Development of a method for automated quantitative analysis of ores using LIBS. *Spectrochim. Acta Part B At. Spectrosc.* **2001**, *56*, 707–714. [CrossRef]
9. Zhang, Y.; Dong, M.; Cheng, L.; Wei, L.; Cai, J.; Lu, J. Improved measurement in quantitative analysis of coal properties using laser induced breakdown spectroscopy. *J. Anal. At. Spectrom.* **2020**, *35*, 810–818. [CrossRef]
10. Senesi, G.S.; Senesi, N. Laser-induced breakdown spectroscopy (LIBS) to measure quantitatively soil carbon with emphasis on soil organic carbon. A review. *Anal. Chim. Acta* **2016**, *938*, 7–17. [CrossRef]
11. Sezer, B.; Bilge, G.; Boyaci, I.H. Capabilities and limitations of LIBS in food analysis. *Trends Analyt. Chem.* **2017**, *97*, 345–353. [CrossRef]
12. Hahn, D.W.; Omenetto, N. Laser-Induced Breakdown Spectroscopy (LIBS), Part I: Review of Basic Diagnostics and Plasma-Particle Interactions: Still-Challenging Issues within the Analytical Plasma Community. *Appl. Spectrosc.* **2010**, *64*, 335A–366A. [CrossRef]
13. Yang, Y.; Li, C.; Liu, S.; Min, H.; Yan, C.; Yang, M.; Yu, J. Classification and identification of brands of iron ores using laser-induced breakdown spectroscopy combined with principal component analysis and artificial neural networks. *Anal. Methods* **2020**, *12*, 1316–1323. [CrossRef]
14. Zhang, W.; Zhuo, Z.; Lu, P.; Tang, J.; Tang, H.; Lu, J.; Xing, T.; Wang, Y. LIBS analysis of the ash content, volatile matter, and calorific value in coal by partial least squares regression based on ash classification. *J. Anal. At. Spectrom.* **2020**, *35*, 1621–1631. [CrossRef]
15. Zhang, Y.; Xiong, Z.; Ma, Y.; Zhu, C.; Zhou, R.; Li, X.; Li, Q.; Zeng, Q. Quantitative analysis of coal quality by laser-induced breakdown spectroscopy assisted with different chemometric methods. *Anal. Methods* **2020**, *12*, 3530–3536. [CrossRef] [PubMed]
16. Zhang, W.; Zhuo, Z.; Lu, P.; Lu, J.; Sun, T.; Tang, J.; Tang, H.; Zhou, T.; Li, L. Laser induced breakdown spectroscopy for quantitative and qualitative analysis of the ash fusion temperatures of coal in power plants. *J. Anal. At. Spectrom.* **2021**, *36*, 576–589. [CrossRef]
17. Zhang, T.; Yan, C.; Qi, J.; Tang, H.; Li, H. Classification and discrimination of coal ash by laser-induced breakdown spectroscopy (LIBS) coupled with advanced chemometric methods. *J. Anal. At. Spectrom.* **2017**, *32*, 1960–1965. [CrossRef]
18. Lu, Z.; Mo, J.; Yao, S.; Zhao, J.; Lu, J. Rapid Determination of the Gross Calorific Value of Coal Using Laser-Induced Breakdown Spectroscopy Coupled with Artificial Neural Networks and Genetic Algorithm. *Energy Fuels* **2017**, *31*, 3849–3855. [CrossRef]
19. Yao, S.; Zhao, J.; Xu, J.; Lu, Z.; Lu, J. Optimizing the binder percentage to reduce matrix effects for the LIBS analysis of carbon in coal. *J. Anal. At. Spectrom.* **2017**, *32*, 766–772. [CrossRef]
20. Dixon-Lewis, G.; Marshall, P.; Ruscic, B.; Burcat, A.; Goos, E.; Cuoci, A.; Frassoldati, A.; Faravelli, T.; Glarborg, P. Inhibition of hydrogen oxidation by HBr and Br2. *Combust. Flame* **2012**, *159*, 528–540. [CrossRef]

21. Zhu, W.; Li, X.; Sun, R.; Zhang, Y.; Yan, Y.; Yu, X.; Ren, X. Investigation of mineral-element migration upon pyrolysis and quantitative prediction of volatiles in coal using laser-induced breakdown spectroscopy. *J. Anal. At. Spectrom.* **2021**, *36*, 1399–1409. [CrossRef]
22. Ren, X.; Rokni, E.; Sun, R.; Meng, X.; Levendis, Y.A. Evolution of Chlorine-Bearing Gases during Corn Straw Torrefaction at Different Temperatures. *Energy Fuels* **2017**, *31*, 13713–13723. [CrossRef]
23. Dong, M.; Mao, X.; Gonzalez, J.J.; Lu, J.; Russo, R.E. Time-resolved LIBS of atomic and molecular carbon from coal in air, argon and helium. *J. Anal. At. Spectrom.* **2012**, *27*, 2066–2075. [CrossRef]
24. Liu, H.; Shah, S.; Jiang, W. On-line outlier detection and data cleaning. *Comput. Chem. Eng.* **2004**, *28*, 1635–1647. [CrossRef]
25. Chen, D.; Huang, Z.; Wang, T.; Ma, Y.; Zhang, Y.; Wang, G.; Zhang, P. High-throughput analysis of single particles by micro laser induced breakdown spectroscopy. *Anal. Chim. Acta* **2020**, *1095*, 14–19. [CrossRef]
26. Yu, H.; Xu, Y.; Lv, N.; Chen, H.; Chen, S. Arc spectral processing technique with its application to wire feed monitoring in Al–Mg alloy pulsed gas tungsten arc welding. *J. Mater. Process Technol.* **2013**, *213*, 707–716. [CrossRef]
27. Jolliffe, I. Principal Component Analysis. In *Encyclopedia of Statistics in Behavioral Science*; Springer Press: New York, NY, USA, 2002; ISBN 9783642048975.
28. Sjöberg, J.; Zhang, Q.; Ljung, L.; Benveniste, A.; Delyon, B.; Glorennec, P.Y.; Hjalmarsson, H.; Juditsky, A. Nonlinear black-box modeling in system identification: A unified overview. *Automatica* **1995**, *31*, 1691–1724. [CrossRef]
29. El Haddad, J.; Bruyère, D.; Ismaël, A.; Gallou, G.; Laperche, V.; Michel, K.; Canioni, L.; Bousquet, B. Application of a series of artificial neural networks to on-site quantitative analysis of lead into real soil samples by laser induced breakdown spectroscopy. *Spectrochim. Acta Part B At. Spectrosc.* **2014**, *97*, 57–64. [CrossRef]
30. Nourani, V.; Komasi, M.; Alami, M.T. Hybrid Wavelet–Genetic Programming Approach to Optimize ANN Modeling of Rainfall–Runoff Process. *J. Hydrol. Eng.* **2012**, *17*, 724–741. [CrossRef]
31. Shibata, K.; Yusuke, I. Effect of number of hidden neurons on learning in large-scale layered neural networks. In Proceedings of the ICROS-SICE International Joint Conference 2009, Fukuoka, Japan, 18–21 August 2009; IEEE: New York, NY, USA, 2009; pp. 5008–5013.
32. Hemmateenejad, B. Correlation ranking procedure for factor selection in PC-ANN modeling and application to ADMETox evaluation. *Chemometr. Intell. Lab. Syst.* **2005**, *75*, 231–245. [CrossRef]
33. Sheta, S.; Afgan, M.S.; Hou, Z.; Yao, S.C.; Zhang, L.; Li, Z.; Wang, Z. Coal analysis by laser-induced breakdown spectroscopy: A tutorial review. *J. Anal. At. Spectrom.* **2019**, *34*, 1047–1082. [CrossRef]
34. Liu, K.; He, C.; Zhu, C.; Chen, J.; Zhan, K.; Li, X. A review of laser-induced breakdown spectroscopy for coal analysis. *Trends Analyt. Chem.* **2021**, *143*, 327–334. [CrossRef]
35. Ferreira, E.C.; Anzano, J.M.; Milori, D.M.; Ferreira, E.J.; Lasheras, R.J.; Bonilla, B.; Montull-Ibor, B.; Casas, J.; Martin Neto, L. Multiple response optimization of laser-induced breakdown spectroscopy parameters for multi-element analysis of soil samples. *Appl. Spectrosc.* **2009**, *63*, 1081–1088. [CrossRef] [PubMed]
36. Pořízka, P.; Klus, J.; Képeš, E.; Prochazka, D.; Hahn, D.W.; Kaiser, J. On the utilization of principal component analysis in laser-induced breakdown spectroscopy data analysis, a review. *Spectrochim. Acta Part B At. Spectrosc.* **2018**, *148*, 65–82. [CrossRef]
37. Chatterjee, S.; Singh, M.; Biswal, B.P.; Sinha, U.K.; Patbhaje, S.; Sarkar, A. Application of laser-induced breakdown spectroscopy (LIBS) coupled with PCA for rapid classification of soil samples in geothermal areas. *Anal. Bioanal. Chem.* **2019**, *411*, 2855–2866. [CrossRef]
38. Li, L.N.; Liu, X.F.; Yang, F.; Xu, W.M.; Wang, J.Y.; Shu, R. A review of artificial neural network based chemometrics applied in laser-induced breakdown spectroscopy analysis. *Spectrochim. Acta Part B At. Spectrosc.* **2021**, *180*, 106183. [CrossRef]
39. Pokrajac, D.; Lazarevic, A.; Kecman, V.; Marcano, A.; Markushin, Y.; Vance, T.; Reljin, N.; McDaniel, S.; Melikechi, N. Automatic classification of laser-induced breakdown spectroscopy (LIBS) data of protein biomarker solutions. *Appl. Spectrosc.* **2014**, *68*, 1067–1075. [CrossRef]
40. Fu, Y.T.; Gu, W.L.; Hou, Z.Y.; Muhammed, S.A.; Li, T.Q.; Wang, Y.; Wang, Z. Mechanism of signal uncertainty generation for laser-induced breakdown spectroscopy. *Front. Phys.* **2020**, *16*, 22502. [CrossRef]
41. Drezga, I.; Rahman, S. Input variable selection for ANN-based short-term load forecasting. *IEEE Trans. Power Syst.* **1998**, *13*, 1238–1244. [CrossRef]

energies

MDPI

Article

Assessment of Interchangeability of Fuels Used in the Process of Heat Production and Comparison of Their Selected Characteristics: A Case Study

Martina Hlatká [1], Vieroslav Molnár [2], Gabriel Fedorko [3,*], Beáta Stehlíková [3] and Gabriela Bogdanovská [3]

1 The Institute of Technology and Business in České Budějovice, Okružní 517/10, 370 01 Ceske Budejovice, Czech Republic; hlatka@mail.vstecb.cz
2 Faculty of Manufacturing Technologies, Technical University of Košice with a Seat in Prešov, Bayerova 1, 080 01 Presov, Slovakia; vieroslav.molnar@tuke.sk
3 Faculty of Mining, Ecology, Process Control and Geotechnologies, Technical University of Košice, Park Komenského 14, 040 01 Kosice, Slovakia; beata.stehlikova@tuke.sk (B.S.); gabriela.bogdanovska@tuke.sk (G.B.)
* Correspondence: gabriel.fedorko@tuke.sk

Abstract: Exchangeability means the possibility of the fuel changing, with conservation of the required energy and environmental criteria. The assessment of fuel exchangeability should be realized by a suitable method, which must reliably present the possibility of the exchangeability of fuels, or reject it. In the presented paper, research on the exchangeability of solid fuels in the field of heating production is surveyed by the case study. Based on the available published knowledge from previous studies on fuel exchangeability, the statistical method was chosen for evaluation. The application of this method is useful. For example, by evaluating the exchangeability of natural gas, the manuscript will describe its application for the field of solid fuels in heat production. The research evaluated and analyzed the sample of 12 fuels. For each fuel sample, 35 gas attributes were measured, which were classified into separate flue gas attribute groups: ash content, combustion heat, heating capacity, sulfur content, combustible content, water content, emission factor, carbon content, hydrogen content, and oxygen content. Attributes of flues were evaluated and grouped according to the fuel properties—ash content and combustion heat, sulphur content, water content, emission factor, carbon content, hydrogen content, nitrogen content, oxygen content, and combustible content.

Keywords: heat production; interchangeability of fuel; assessment; flue gases; combustion

Citation: Hlatká, M.; Molnár, V.; Fedorko, G.; Stehlíková, B.; Bogdanovská, G. Assessment of Interchangeability of Fuels Used in the Process of Heat Production and Comparison of Their Selected Characteristics: A Case Study. *Energies* **2022**, *15*, 124. https://doi.org/10.3390/en15010124

Academic Editor: Dimitrios A. Georgakellos

Received: 16 November 2021
Accepted: 22 December 2021
Published: 24 December 2021

1. Introduction

The issue of fuel interchangeability is more and more topical in various fields of human activity. Interest in this issue is mainly associated with fields such as transport, ecology, economy, or energy [1].

The interchangeability of fuels means the possibility of fuel changing, by conservation of required energy and environmental criteria. The interchangeability of fuels can be assessed from a general point of view based on different parameters, such as burning rate, flame stability, combustion perfection, the temperature of combustion, amount, and composition of flue gas, most of which cannot be met at the same time [2]. The interchangeability of fuels can be also realized based on criteria that are specific only for the specific type of fuel. This approach was applied by Ferguson et al. [3].

Wobbe and Weaver applied different historical parameters and specific indexes for the determination of fuel interchangeability. Park et al. [4] used another approach. They used the model of the combustion plant for an evaluation of interchangeability of different types of fossil fuels, including synthesis gas from the coal gasification, steel gases, and blast furnace gas (BFG). The realized evaluation was based on the criteria such as the emission trend (NOx), exhaust gas temperature distribution according to the combustor pressure, syngas composition, and dilution ratio at the fuel side.

Research on the interchangeability of fuels is now more and more actual for the field of heat production. One of the reasons is the trend of changeover from fossil energy to renewable energy; this fact causes the question of fuel interchangeability to be a necessity [5].

The heat production process in heating plants is based on technologies that mostly use different categories of solid fuels as the primary energy source. Coal, wood, or biomass are most often used as fuels.

Fossil fuels are currently one of the most widely used energy sources in heating plants. During their combustion, a large amount of gaseous emissions and solid waste are generated. Current trends in fossil fuel combustion are aimed at making the combustion process as efficient as possible so that thermal energy is obtained most efficiently. At the same time, the main intention is to minimize the negative impact of the combustion process on the environment. In particular, the waste generated by fossil fuel combustion contains various types of substances, which can be toxic in certain cases.

The process of fossil fuel combustion is often influenced by different factors, so the combustion of the same type of fuel in different thermal power plants with different technologies is not the same. This issue, in terms of trace element behaviour, was researched by Lopez-Anton et al. [6] as well as Klein et al. [7]. Trace elements are an important factor in the fossil fuel combustion process and it is very important to recognize research about this problem, as presented by Bool and Helble [8] or Querol et al. [9]. Trace elements created during the combustion of fossil fuels can also be effectively searched in laboratory conditions. Helble [10] searched trace elements using an isothermal laminar flow of tube furnace and sorted the resulting ash samples according to their size. The study of trace elements is also presented in [11].

Different methods and approaches can be used for the evaluation of fuel interchangeability. Keramiotis et al. [12] used an infrared thermography and gas chromatograph for the research of biogas interchangeability. Xiang et al. [13] used a simulation approach with the use of software simulation tools. Pujihatma et al. [14] used a similar approach for the evaluation of the interchangeability of the optimized model. Martinez et al. [15] researched the interchangeability of natural gas based on statistical methods. Thus, it is possible to assume the suitability of statistical methods applied for the evaluation of interchangeability of other types of fuels, also fossil fuels. Based on this fact, the utilization of this method for the field of fuel interchangeability evaluation was searched in the field of heat production. The paper will present in more detail the presented approach, with the evaluation of 12 types of fuels; each fuel was evaluated based on 36 categories of parameters obtained from the ash after their combustion.

2. Materials and Methods

The investigation of fuel properties, according to Figure 1, consisted of two steps. In the first step, the research focused on the analysis of the properties of individual fuels. Based on the processing of the set of flue gas attributes, descriptive statistics were first calculated with verification of normality for all monitored flue gas attributes.

According to the values of the flue gas attributes of individual fuels, in the future, it will be possible to evaluate its position and the percentile within each flue gas attribute for fuel after adding new fuel.

Percentile is a statistical parameter that determines the percentage within the population, where the measured value is from minimum to maximum [16]. Subsequently, the research in the first step continued with the creation of flue gas attribute groups, with the aim to reduce the number of attributes to be monitored (lines in bold in Table 1). An analysis of the dependencies and relations between these attributes has been carried out for the bold attributes of flue gases, from Table 1.

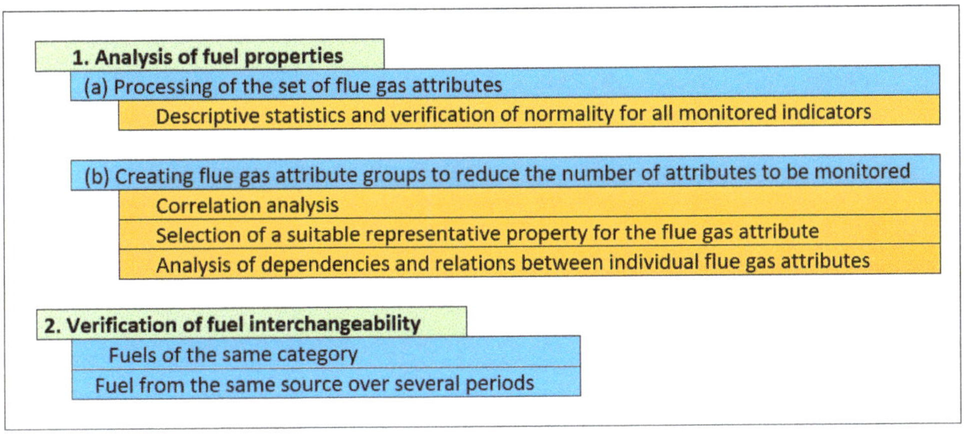

Figure 1. Scheme of the fuel properties research.

Table 1. Descriptive statistics and results of the Shapiro–Wilk flue gas attribute test.

Attribute	Description	Mean	SD	Median	Min	Max	Range	Skew	Kurtosis	Testing Criteria Shapiro–Wilk Test	p_{value}
Aa	Ash content in the analytical sample (%)	20.44	10.25	22.89	2.03	38.91	36.88	−0.42	0.00	0.92	0.26
Ad	Ash content in anhydrous sample (%)	22.12	11.04	24.60	2.05	41.41	39.36	−0.50	−0.01	0.91	0.24
Ar	**Ash content in the original sample (%)**	16.33	8.25	17.40	1.91	29.22	27.31	−0.51	−0.51	0.93	0.38
Qa_s	Combustion heat in analytical sample (MJ/kg)	19.51	2.40	19.50	15.37	23.41	8.04	0.07	−0.85	0.98	0.96
Qd_s	Combustion heat in anhydrous sample (MJ/kg)	20.99	3.03	19.95	16.35	25.43	9.08	0.11	−1.27	0.93	0.41
$Q_s$daf	Combustion heat of pure combustible (MJ/kg)	27.25	4.20	28.61	20.09	31.56	11.47	−0.67	−1.05	0.86	0.05
Qsr	**Combustion heat of original sample (MJ/kg)**	15.62	1.92	15.36	11.54	18.31	6.77	−0.41	−0.17	0.94	0.54
Q_ir	Original calorific value (MJ/kg)	14.22	1.92	13.98	10.28	16.98	6.70	−0.32	−0.39	0.95	0.60
S_ta	Sulphur content of the analytical sample (%)	0.98	0.68	0.97	0.03	2.66	2.63	0.94	1.45	0.88	0.09
S_td	Sulphur content of anhydrous sample (%)	1.06	0.73	1.08	0.03	2.83	2.80	0.81	1.15	0.89	0.12
$S_t$daf	Sulphur content of combustible material (%)	1.50	1.23	1.44	0.04	4.83	4.79	1.57	2.80	0.81	0.01
Smr	Specific sulphur saturation (g/MJ)	0.58	0.49	0.53	0.02	1.94	1.92	1.81	3.47	0.76	0.00
Sr	**Total sulphur content of the original sample (%)**	0.77	0.52	0.72	0.02	2.00	1.98	0.80	1.07	0.91	0.20
Vr	Flammable content (%)	57.12	14.34	55.04	40.84	91.13	50.29	1.07	0.63	0.90	0.18
Wa	Water content in the analytical sample (%)	6.73	3.88	8.07	0.98	11.84	10.86	−0.27	−1.36	0.91	0.19
Wex	Coarse water content (%)	20.66	8.86	23.57	6.24	33.30	27.06	−0.37	−1.08	0.90	0.14
Wh	Residual water content (%)	5.36	2.76	6.39	0.77	8.24	7.47	−0.65	−1.17	0.85	0.03
Wtr	**Total water content (%)**	24.82	9.76	28.25	6.96	38.33	31.37	−0.48	−0.83	0.92	0.29
fem	**Emission factor (t_{CO_2}/TJ)**	97.92	2.64	97.75	94.31	104.33	10.02	1.04	0.98	0.91	0.24

Table 1. *Cont.*

Attribute	Description	Mean	SD	Median	Min	Max	Range	Skew	Kurtosis	Testing Criteria Shapiro–Wilk Test	p_{value}
C_ta	Carbon content in the analytical sample (%)	47.44	5.95	48.61	36.63	56.77	20.14	−0.21	−0.78	0.98	0.98
C_td	Carbon content of the anhydrous sample (%)	51.02	7.34	50.28	38.98	61.66	22.68	−0.12	−1.10	0.96	0.72
$C_t$daf	Carbon content in combustible (%)	66.15	9.64	69.80	49.82	75.78	25.96	−0.62	−1.25	0.84	0.03
Cr	**Carbon content in the original sample (%)**	38.01	4.93	37.64	27.51	45.40	17.89	−0.44	−0.15	0.96	0.81
H_ta	Hydrogen content of the analytical sample (%)	5.21	0.73	5.32	3.73	6.30	2.57	−0.60	−0.21	0.95	0.70
H_td	Hydrogen content of the anhydrous sample (%)	4.74	0.93	4.61	3.25	6.26	3.01	0.09	−0.95	0.97	0.85
$H_t$daf	**Hydrogen content in combustible (%)**	6.09	0.72	5.94	4.88	7.47	2.59	0.69	0.18	0.88	0.09
Hr	Hydrogen content of the original sample (%)	4.21	0.83	3.98	2.80	5.92	3.12	0.57	0.04	0.93	0.39
N_ta	Nitrogen content of the analytical sample (%)	0.97	0.73	0.74	0.23	3.07	2.84	2.10	3.74	0.70	0.00
N_td	Nitrogen content of the anhydrous sample (%)	1.03	0.73	0.80	0.24	3.10	2.86	2.07	3.72	0.71	0.00
$N_t$daf	Nitrogen content of combustible material (%)	1.30	0.74	1.05	0.24	3.17	2.93	1.35	1.67	0.83	0.02
Nr	**Nitrogen content in the original sample (%)**	0.82	0.72	0.57	0.18	2.89	2.71	2.16	3.87	0.69	0.00
O_da	Oxygen content in the analytical sample (%)	18.11	10.97	12.29	10.21	39.65	29.44	1.24	−0.02	0.73	0.00
O_dd	Oxygen content of the anhydrous sample (%)	19.85	10.25	14.68	12.27	40.42	28.15	1.29	0.14	0.73	0.00
$O_d$daf	Oxygen content in combustible (%)	24.79	9.37	21.49	15.46	41.61	26.15	0.80	−0.85	0.84	0.03
Odr	**Oxygen content in the original sample (%)**	14.95	9.89	9.53	7.31	36.85	29.54	1.19	0.07	0.78	0.01

In the second step, the interchangeability of the fuels in the same category was verified based on the flue gas attributes.

This methodology was verified within the research and was performed in a real heating plant, which was a suitable object for such verification. The given heating plant was chosen because its fuel consumption was approximately uniform and it was repeated cyclically. This fact is also confirmed by a record of the four-year cycle of the heating plant fuel consumption, which is shown in Figure 2.

This consumption monitoring was carried out before the research of interchangeability of different fuel types.

The following research, concerning properties of the selected fuels, was realized on the basis of measured ash attributes in the heating plant during the 5 years. Before starting the research of the fuel interchangeability in the given heating plant, the fuel used so far was analyzed based on the residues that remained after the combustion process. The main characteristics of the fuel are presented in Figure 3. During the 3rd year it was installed, the desulphurization technology in the heating plant influenced the obtained parameters.

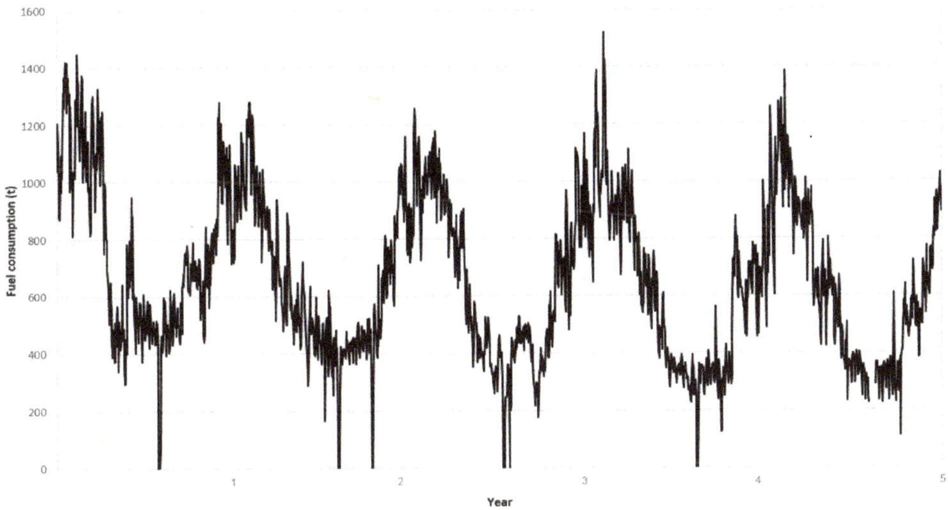

Figure 2. Overview of fuel consumption.

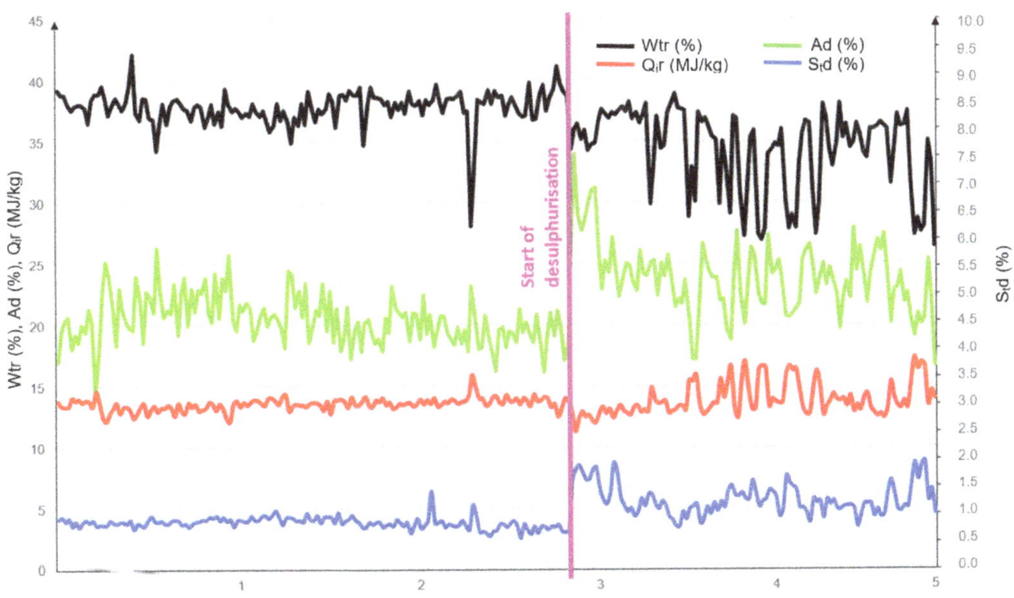

Figure 3. Fuel parameters.

The methodology developed for verifying the fuel interchangeability was tested on the research sample of 12 fuels, which were applied in the given heating plant for 5 years.

The group of 12 samples of examined fuels consisted of: brown coal from several suppliers, wood chips, wood pellets, non-wood pellets, and two solid alternative fuels.

For each of the 12 fuel samples, 35 flue gas attributes were measured and assigned to separate flue gas attribute groups: ash content, the heat of combustion, calorific value, sulphur content, combustible matter content, water content, emission factor, carbon content, hydrogen content, nitrogen content, and oxygen content (Table 1).

The following were analyzed:

- Solid alternative fuels, designated as TAP1 and TAP2;
- Biomass: wood chips, wood pellets, and non-wood pellets.

Furthermore, the interchangeability of the fuel-brown coal was verified. Coal from the same source was evaluated over several periods from the same supplier over several years, based on the attributes of their flue gases.

3. Results

A total of 35 flue attributes from 12 fuel samples were evaluated. The flue gas attributes were evaluated and grouped according to fuel properties—ash content, the heat of combustion and calorific value, sulphur content, water content, emission factor, carbon content, hydrogen content, nitrogen content, oxygen content, and combustible content.

3.1. Descriptive Fuel Statistics

Descriptive statistics and results of the Shapiro–Wilk test of flue gas attributes of compliance with the normal distribution are given in Table 1. The tested null Hypothesis H0 was: the population is normally distributed. The highlighted *p*-value means a rejection of the null hypothesis and the use of non-parametric tests for the corresponding data.

The following section lists the attributes for which the null hypothesis H0 on the normal distribution was rejected. This means that statistical parametric tests cannot be used for these attributes.

The population's attributes of the flue gases that were not normally distributed are: $Q_s$daf-Combustion heat of pure combustible (MJ/kg), $S_t$daf-Sulphur in combustible (%), Smr-Specific sulphur content (g/MJ), Wh-Residual water (%), $C_t$daf-Carbon in combustible (%), N_ta-Nitrogen in the analytical sample (%), N_td-Nitrogen in the anhydrous sample (%), $N_t$daf-Nitrogen in combustible (%), Nr-Nitrogen in the original sample (%), O_da-Oxygen in the analytical sample (%), O_dd-Oxygen in the anhydrous sample (%), $O_d$daf-Oxygen in the combustible (%), and Oder-Oxygen in the original sample (%).

The flue gas attributes, for which a normal distribution can be considered, are suitable for estimating the relationships between the individual flue gas attributes. The results from Table 1 are discussed in detail in the discussion chapter.

3.2. Creating Groups of Flue Gas Attributes with the Aim to Reduce the Number of Monitored Attributes

For all of the flue gas attributes of the same property, their correlation was examined and one attribute was selected to be used in further analyses.

For flue gas attributes where the hypothesis the population is normally distributed was not rejected, the Pearson's correlation coefficient (with the critical value of 0.532) was used to verify the correlation of the flue gas attribute pairs.

For the flue gas attribute pairs, at least one of which cannot be considered hypothesis, the population is normally distributed, the Spearman correlation coefficient (with the critical value of 0.587) was calculated.

Based on the mutual correlation of the flue gas attributes of the same property and the availability of the flue gas property attribute values, one attribute was chosen, which is used as a representative of the property in subsequent calculations.

3.3. Research of Relations between Flue Gas Attributes

The existence of mutual relations was investigated for the selected flue gas attributes (given in the previous step).

(a) Model: regression models for flue gas attribute pairs (linear or polynomial model of the maximum third degree) were used to analyze the relationships between the flue gas attribute values. The mathematical expression of the models is:

- The linear model has the shape $Y = a_o + a_1x$;
- Degree of a polynomial 2 is expressed by the equation $Y = a_o + a_1x + a_2x^2$;

- Degree of a polynomial 3, the member $+a_3x^3$ is added to the degree of a polynomial 2.

(b) Model evaluation: two model evaluation statistics—the F test of the model regression function and the Adjusted R-squared R^2 were used to evaluate model suitability. The F model regression function test [17].

The null hypothesis of the statistical test is $H_0: \alpha_0 = \gamma; \alpha_1 = \alpha_2 = \ldots = \alpha_p = 0$ This means that all estimated parameters α_i of the regression model, except the parameter α_0, are statistically insignificant. α_0 is a constant, p is the number of flue gas model attributes, except α_0.

The alternative hypothesis is H_1: at least one parameter $\alpha_i \neq 0$ is not statistically insignificant where $i = 1.2 \cdots p$.

The test criterion is calculated according to the Equation (1):

$$F = \frac{\sum_{i=1}^{n}(Y_i - \overline{y})^2}{\sum_{i=1}^{n}(Y_i - y_i)^2} \frac{(n-p-1)}{p} \quad (1)$$

where:

y_i are the measured values (empirical),
Y_i are the modelled values (theoretical),
$\overline{y}$ is the average of the measured values,
n is the number of values from which the estimate of the flue gas attributes α_0 to α_p of the model was calculated.

The hypothesis H0 is rejected if the test criterion F is greater than the critical value for the test. The critical value of the Fisher distribution for significance level α and the number of degrees of freedom p and $(n-p-1)$. $F_{\alpha.p.(n-p-1)}$.

If the hypothesis H0 is not rejected, the mathematical equation function is not suitable. In this case, we should change the function type. The significance level $\alpha = 0.05$ was chosen, it is the probability that the null hypothesis is valid and will be rejected.

The test result can also be evaluated using the p_{value} test, determined as the probability of rejecting the null hypothesis if it is correct.

The value p_{value} is compared with the value α. The condition of acceptance of the model according to the F test result is that the value of p_{value} is less than the significance level $\alpha = 0.05$.

The other evaluation statistics is Adjusted R − squared R^2 calculated according to Equation (2):

$$R^2 = \frac{\sum_{i=1}^{n}(Y_i - \overline{y})^2}{\sum_{i=1}^{n}(y_i - \overline{y})^2} \frac{(n-1)}{(n-p-1)} \quad (2)$$

The calculated value is the ratio of the variability of empirical values.

The components of the calculation are explained in Equation (1). To accept the model, the value R^2 was required to be greater than 0.6.

Residues $e_i = Y_i - y_i$ were calculated for each model and the proportions of residues and model values $rat_i = \frac{e_i}{Y_i}$.

For the acceptance of the model, a condition has been established for the residues that the proportion of the model and actual value difference must be less than 0.25 for all values, which means that the model deviation percentage is less than 25%. In Section 4.1.2, only models that meet the specified requirements are listed. For linear models, the correlation coefficient was calculated instead of Adjusted R-squared R^2.

3.4. *Verification of Fuel Interchangeability*

The Wilcoxon signed-rank sum test [17] was used when data on brown coal from the same source were available for two periods, respectively, two alternatives of the same fuel category. The verified statistical hypothesis was:

H0: the difference between the pairs follows a symmetric distribution around zero,

H1: the difference between the pairs does not follow a symmetric distribution around zero.

The Friedman rank-sum test was used when data on brown coal from the same source were available for three or more periods, respectively three or more alternatives of the same fuel category.

Zero and alternative hypothesis:

Hypothesis 0 (H0). *Differences in flue gas attribute values are random.*

Hypothesis 1 (H1). *Differences in flue gas attribute values are not random.*

The evaluation of the test results was at the significance level $\alpha = 0.05$ using the value p_{value}.

4. Discussion

The results obtained by the performed mathematical-statistical analysis will be presented and commented on in the next chapter. Besides individual descriptive statistics, the possibility of the interchangeability of individual fuels was also verified.

4.1. Processing of a Set of Flue Gas Attributes

This section of paper reviews the descriptive statistics for the sets of attributes of the same quality and the correlation examined among these. In case of significant correlation, it is enough to examine only one attribute, as the others can be obtained by derivation.

The results of processing the set of flue gas attributes, which contains descriptive statistics and verification of normality for all monitored indicators (Table 1), are described in detail in the following subchapters.

4.1.1. Descriptive Statistics of the Flue Gas Ash Content Attributes

Aa Ash in the analytical sample (%)
Ad Ash in the anhydrous sample (%)
Ar Ash in the original sample (%)

The maximum values of all the flue gas attributes of the ash content were measured for brown coal from Severočeské doly a.s. of 2017. The maximum ash content in the analytical sample Aa was 38.91%, and was Ad 41.44% in the anhydrous sample. The lowest values of ash content were for wood pellets in the original sample Aa 1.91% and in the anhydrous sample Ad 2.05%. According to the statistics for skew (−0.42 to −0.51) and kurtosis (0 to −0.51), it can be stated that the flue gas attributes of the ash are symmetrical and have adequate kurtosis. This is also indicated by the position of the median and diameter, which are close to each other (for Ad, the average is 22.12, median 24.60). The Shapiro–Wilk test confirmed that the sets of values came from a population with a normal distribution. The Pearson's correlation coefficient (Table 2) was calculated to verify linear dependence. All of the flue gas attributes of the ash were strongly correlated with each other, as shown by the high correlation coefficient value.

Table 2. The Pearson's correlation coefficient for attributes of the flue ash content.

	Aa	Ad	Ar
Aa	1.00	1.00	0.98
Ad	1.00	1.00	0.97
Ar	0.98	0.97	1.00

Such correlation values allow, if necessary, to observe only one of the flue gas attributes of the ash content and to calculate the others according to a linear regression model. The Ad attribute will be used for the analysis of dependencies and relations between individual flue gas attributes.

4.1.2. Descriptive Statistics of the Flue Gas Attributes for Gross Heat and Calorific Value

Qa_s Combustion heat in the analytical sample (MJ/kg)
Qd_s Combustion heat in the anhydrous sample (MJ/kg)
Q_ir Net calorific value (MJ/kg)
Q_sdaf Combustion heat of pure combustible (MJ/kg)
Qsr Combustion heat of the original sample (MJ/kg)

The combustion heat of pure Q_sdaf cannot be regarded as normally distributed (Table 1). The values of the Pearson's correlation coefficient for the other combustion gas attributes and calorific values with normal distribution are given in Table 3.

Table 3. The Pearson's correlation coefficient for attributes of flue gases of gross heat and calorific value.

	Qa_s	Qd_s	Q_ir	Qsr
Qa_s	1.00	0.97	0.65	0.69
Qd_s	0.97	1.00	0.52	0.55
Q_ir	0.65	0.52	1.00	1.00
Qsr	0.69	0.55	1.00	1.00

It is obvious that the attribute Qa_s is correlated with the attributes of the flue gas of the combustion heat and the calorific value Qd_s, Q_ir, and Qsr. The correlation was not confirmed between the Q_ir and Qd_s flue gas attributes. For each pair of flue gas attributes, except for Qir and Qd_s, it is sufficient to observe only one of the flue gas attributes.

The Spearman correlation coefficient in Table 4 was determined to track the relationship between the Q_sdaf attribute and other ones.

Table 4. The Spearman correlation coefficient for attributes of flue gases of gross heat and calorific value.

	Qa_s	Qd_s	Q_ir	Q_sdaf	Qsr
Q_sdaf	0.62	0.74	0.04	1.00	0.01

The results show that the Q_sdaf attribute is not correlated with the Q_ir and Qsr attributes. The Q_ir attribute will thus be used to analyze dependencies and relations between individual flue gas attributes. For the Q_ir attribute, the minimum value of 10.28 MJ/kg was measured for brown coal PS2 Severočeské doly for 2017. The maximum value of 16.98 MJ/kg was measured for biomass. For the other two types of lignite, calorific values above 16 MJ/kg were found. The average value is 14.22 MJ/kg and the median is 13.98 MJ/kg, which are close values. The distribution is symmetrical, skew is −0.32 and adequately sharp, and the kurtosis coefficient is −0.39. The range is 6.7 MJ/kg.

4.1.3. Descriptive Statistics of the Flue Gas Attributes of the Sulphur Content

S_ta Sulphur in the analytical sample (%)
S_td Sulphur in the anhydrous sample (%)
S_tdaf Sulphur in combustible (%)
Sr All sulphur in original sample (%)
Smr Specific sulphur content (g/MJ)

The attributes of the flue gas sulphur content S_tdaf and Smr do not come from a population with a normal distribution. The values of the Pearson's coefficient (Table 5) show a very strong correlation between the flue gas attributes of sulphur content S_ta, S_td, and Sr. All values are greater than 0.98.

Table 5. The Pearson's correlation coefficient for attributes of flue gases of sulphur content.

	S_ta	S_td	Sr
S_ta	1.00	1.00	1.00
S_td	1.00	1.00	0.99
Sr	0.99	0.99	1.00

For the flue gas attributes of the sulphur content, which cannot be considered as normally distributed—S_tdaf and Smr (Table 1), the Spearman correlation coefficient (Table 6) was calculated, of which all values are above 0.95. This indicates a strong correlation between the flue gas attributes of the sulphur content.

Table 6. The Spearman correlation coefficient for attributes of flue gases of sulphur content.

	S_ta	S_td	S_tdaf	Sr	Smr
S_tdaf	0.97	0.97	1.00	0.98	0.997
Smr	0.95	0.95	0.99	0.96	1.00

The S_td attribute will thus be used to analyze dependencies and relations between individual flue gas attributes. The descriptive statistics for the S_td attribute are:

- A minimum value of 0.03% for wood chips and a maximum value of 2.83% for brown coal PS2 Severočeské doly for 2017.
- The average value is 1.06% and the median is 1.08%.
- A symmetrical distribution is assumed, but the value of the skew coefficient (skew) is 0.81, which is a slight asymmetry.
- The kurtosis coefficient is 1.15, which is a sharper distribution.

4.1.4. Descriptive Statistics of the Flue Gas Attributes for Water Content

Wa Water in analytical sample (%)
Wex Water coarse (%)
Wh Residual water (%)
Wtr All water (%)

The Pearson's correlation coefficient values for attributes of flue gases (Table 7) water content show a linear dependence between attributes.

Table 7. The Pearson's correlation coefficient for the flue gases of the water content.

	Wa	Wex	Wtr
Wa	1	0.68	0.80
Wex	0.68	1.00	0.98
Wtr	0.80	0.98	1.00

The flue gas water content attribute Wh does not come from a population with a normal distribution, so the Spearman correlation coefficient (Table 8) was used to verify the correlation with the other flue gas water content attributes.

Table 8. The Spearman correlation coefficient for attributes of flue gases of water content.

	Wa	Wex	Wh	Wtr
Wh	0.80	0.28	1.00	0.62

Except for the pair of flue gas attributes Wh and Wex, there is a linear dependence between the flue gas attributes of the water content. For the analysis of dependencies and relations between individual flue gas attributes, the attribute Wtr will be used, which can be used to describe all other flue gas content attributes using linear regression.

For the Wtr attribute, descriptive statistics are as follows:

- A minimum value of 6.96% for wood pellets and a maximum value of 38.33% for brown coal from Sokolovská uhelná.
- The range of values is 31.37.
- The average value is 24.82 and the median is 28.25. This indicates a negative skew of −0.48. The kurtosis coefficient is −0.81, so the probability distribution is a flatter normal distribution.

4.1.5. Descriptive Statistics of the Flue Gas Attributes for Carbon Content

C_ta Carbon in the analytical sample (%)
C_td Carbon in the anhydrous sample (%)
$C_t$daf Carbon in the combustible (%)
Cr Carbon in original sample (%)

The values of the Pearson's correlation coefficient (Table 9) confirm the linear relation between the flue gas attributes C_ta, C_td, and Cr.

Table 9. The Pearson's correlation coefficient for attributes of flue gases of carbon content.

	C_ta	C_td	Cr
C_ta	1.00	0.96	0.70
C_td	0.96	1.00	0.56
Cr	0.70	0.56	1.00

The $C_t$daf indicator does not come from a population with a normal distribution (Table 1), so the Spearman correlation coefficient (Table 10) was used to investigate the relationship between it and the other flue gas carbon content attributes. The results indicate that $C_t$daf does not correlate with Cr.

Table 10. The Spearman correlation coefficient for attributes of flue gases of carbon content.

	C_ta	C_td	$C_t$daf	Cr
$C_t$daf	0.63	0.68	1.00	0.06

For the analysis of dependencies and relations between individual flue gas attributes, the attribute Cr will be used. For the attribute Cr, the minimum value is 27.51% for PS2 Severočeské doly, and the maximum value is 45.40% for wood pellets. The average value is 38.1 and the median is 37.64. This indicates that the distribution is symmetrical. The kurtosis coefficient is −0.15 and the distribution is adequately sharp.

4.1.6. Descriptive Statistics of the Flue Gas Attributes for Hydrogen Content

H_ta Hydrogen in the analytical sample (%);
H_td Hydrogen in anhydrous sample (%);
$H_t$daf Hydrogen in combustible (%);
Hr Hydrogen in the original sample (%).

All of the flue gas attributes of the hydrogen content originate from a population with a normal distribution. The values of Pearson's correlation coefficient are displayed in Table 11.

Table 11. The Pearson's correlation coefficient for attributes of flue gases of hydrogen content.

	H_ta	H_td	H_tdaf	Hr
H_ta	1.00	0.96	0.65	0.86
H_td	0.96	1.00	0.71	0.91
H_tdaf	0.65	0.71	1.00	0.69
Hr	0.86	0.91	0.69	1.00

All of the flue gas attributes of the hydrogen content are correlated with each other and it is sufficient to observe only one indicator. Other indicators can be derived using a linear regression model. The attribute Hr will thus be used for the analysis of dependencies and relations between individual flue gas attributes. The values of the attribute Hr range from 2.80% for PS2 Severočeské doly brown coal to 5.92% for wood pellets. The average value is 4.21% and the median is 3.98%. The coefficient of asymmetry is 0.57 and the distribution is suitably symmetrical. The kurtosis coefficient is 0.04, so the distribution is as sharp as the normal distribution.

4.1.7. Descriptive Statistics of the Flue Gas Attributes for Nitrogen Content

N_ta Nitrogen in the analytical sample (%);
N_td Nitrogen in the anhydrous sample (%);
N_tdaf Nitrogen in combustible (%);
Nr Nitrogen in the original sample (%).

The null hypothesis of a normal population distribution was rejected for all nitrogen-containing flue gas attributes. Mutual relations between attributes were observed using the Spearman correlation coefficient. The results show that all flue gas attributes are correlated with each other (Table 12).

Table 12. The Spearman correlation coefficient for attributes of flue gases of nitrogen content.

	N_ta	N_td	N_tdaf	Nr
N_ta	1.00	0.99	0.93	0.98
N_td	0.99	1.00	0.89	0.97
N_tdaf	0.93	0.89	1.00	0.95
Nr	0.98	0.97	0.95	1.00

For the analysis of dependencies and relations between individual flue gas attributes, the attribute Nr will be used. The attribute values range from 0.18% in wood chips to 2.89% in wood pellets. The average value is 0.82% and the median is 0.57%. These values are relatively close to each other, but are closer to the minimum than to the maximum. This determines the value of the asymmetry coefficient of 2.16, thus, more values are smaller than the average. The kurtosis coefficient 3.87 testifies to the flat distribution.

4.1.8. Descriptive Statistics of the Flue Gas Attributes for Oxygen Content

O_da Oxygen in the analytical sample (%);
O_dd Oxygen in the anhydrous sample (%);
O_ddaf Oxygen in combustible (%);
Odr Oxygen in the original sample (%).

For all of the flue gas attributes of the oxygen content, the null hypothesis of a normal population distribution was rejected (Table 1), so the relative relations between them were monitored using the Spearman correlation coefficient, which results in all flue gas attributes correlating with each other (Table 13).

Table 13. The Spearman correlation coefficient for attributes of flue gases of oxygen content.

	O_da	O_dd	$O_d$daf	Odr
O_da	1.00	0.96	0.96	0.96
O_dd	0.96	1.00	0.92	0.89
$O_d$daf	0.96	0.92	1.00	0.91
Odr	0.96	0.89	0.91	1.00

The attribute Odr will be used for the analysis of dependencies and relations between individual flue gas attributes. Th values of the indicator range from 7.31% in brown coal from Sokolovská uhelná to 36.85% in wood pellets. The average value is 14.95% and the median is 9.53%. The variation between the median and the diameter is due to the composition of the fuel selection analyzed, as 8 out of 12 fuels are brown coal and the rest are biomass and alternative fuels. The proportion of oxygen in brown coal is about 7% to 12%. In other types of analyzed fuels, it is 19% to 29%. The asymmetry coefficient 1.19 corresponds to the described fact that more values are smaller than the average. The kurtosis coefficient 0.07 indicates a slightly flat distribution.

4.1.9. Descriptive Statistics of Other Flue Gas Attributes

Vr Combustible content (%).

The values of the attribute Vr range from 40.84% in wood chips to 91.13% in wood pellets. The average value is 57.12% and the median is 55.04%. These values are close to each other, but are closer to the minimum than to the maximum. This determines the value of the asymmetry coefficient of 1.07, thus, more values are smaller than the average. The kurtosis coefficient of 0.63 shows an adequate sharp distribution.

fem Emission factor (t_{CO2}/TJ).

The values of the fem attribute range from 94.31% in one of the solid alternative fuels, to 104.33% in the wood chip. The average value is 97.92% and the median is 97.75%. These values are close to each other but are closer to the minimum than to the maximum. This determines the value of the asymmetric coefficient of 1.04, thus, more values are smaller than the average. The kurtosis coefficient 0.98 gives an adequate sharp distribution.

These flue gas attributes come from a population with a normal distribution. Both Vr and fem will be used to analyze dependencies and relations between individual flue gas attributes.

4.1.10. The Analysis of Dependencies and Relations between Individual Flue Gas Attributes

This chapter presents selected attributes for all components under observation, of which mutual relations regression models were calculated. The analysis was performed for all pairs created from the flue gas attributes selected in the previous step (Sections 4.1.1–4.1.9):

Ad Ash in the anhydrous sample (%);
Qir Net calorific value (MJ/kg);
S_td Sulphur in the anhydrous sample (%);
Wtr All water (%);
Cr Carbon in the original sample (%);
Hr Hydrogen in the original sample (%);
Nr Nitrogen in the original sample (%);
Odr Oxygen in the original sample (%);
Vr Combustible matter (%);
fem Emission factor (t_{CO2}/TJ).

Following the established model acceptance guidelines, the following model relations, as shown in Table 14, have been accepted.

Table 14. Accepted model relations.

Relation	Model	p_{value} of $F-$ test	R^2	rat_i	r
$Q_ir \sim Ad$	$Q_ir = 12.495 + 1.516\ Ad - 0.089\ Ad^2 + 0.001\ Ad^3$	0.0016	0.632	0.206	-
$Q_ir \sim Cr$	$Q_ir = -0.292 + 0.382\ Cr$	1.29×10^{-8}	0.961	0.060	0.982
$Cr \sim Q_ir$	$Cr = 2.064 + 2.528Q_ir$	1.29×10^{-8}	0.962	0.059	0.982
$Vr \sim Ad$	$Vr = 85.554 - 1.22\ Ad$	3.136×10^{-5}	0.702	0.234	–0.849
$Hr \sim S_td$	$Hr = 5.218 - 0.952S_td$	0.00056	0.682	0.149	–0.843
$Cr \sim Vr$	$Cr = -171.8 + 9.7\ Vr - 0.146\ Vr^2 + 0.0007\ Vr^3$	0.01371	0.612	0.177	-

Among the flue gas attributes, $Q_ir \sim Ad$, the accepted relation was described by the degree of a polynomial 3.

Among the flue gas attributes, $Q_ir \sim Cr$ and $Cr \sim Q_ir$, the accepted relation was described by a linear dependence. The correlation coefficient r = 0.982 confirms a strong linear dependence of the calorific value and the amount of carbon.

Among the flue gas attributes, $Vr \sim Ad$, the accepted relation was described by a linear dependence. The correlation coefficient r = –0.849 confirms a strong negative correlation, i.e., fuels with a lower amount of ash contain a higher amount of combustible material.

Among the flue gas attributes, $Hr \sim S_td$, the accepted relation was described by a linear dependence. The correlation coefficient r = –0.843 confirms a strong negative correlation, i.e., fuels with a lower amount of sulphur in the sample without water contain a higher amount of hydrogen in the original sample.

Among the flue gas attributes, $Cr \sim Vr$, the accepted relation was described by the degree of a polynomial 3.

4.2. *Verification of Fuel Interchangeability*

This verification was carried out for the preliminary evaluation of fuel interchangeability for:

- Fuels of the same category:
  - Solid alternative fuels: TAP1 and TAP2,
  - Biomass: wood chips, wood pellets, and non-wood pellets;
- Fuel (brown coal) from three different sources for several periods:
  - Sokolovská uhelná for the years, 2017 and 2018,
  - Severočeské doly a.s., for the years, 2017 and 2018,
  - Severní energetická a.s., for the years, 2016, 2017, and 2018.

4.2.1. Comparison of the Results of Flue Gas Attributes of the Same Category Fuels

The Wilcoxon signed-rank test was used to test the flue gas attribute compliance for Solid Alternative Fuels: TAP1 and TAP2. The test criterion value is V = 307 and the $p_{value} = 0.9035$.

The null hypothesis cannot be rejected. The flue gas attributes of TAP1 and TAP2 are not statistically significantly different.

The comparison of TAP1 and TAP2 results implies their mutual interchangeability.

The Friedman rank-sum test was used to verify the hypothesis of the conformity of the flue gas attributes of the biomass, wood chips, wood pellets, and non-wood pellets. The test criterion value is $X^2 = 6.6232$ and the value $p_{value} = 0.03646$.

The null hypothesis is rejected in favor of the alternative one. Flue gas attributes of at least one fuel pair of wood chips, wood pellets, and non-wood pellets differ statistically significantly.

The Wilcoxon signed-rank test was subsequently used to verify the hypothesis of the conformity of the flue gas attributes for wood chips and wood pellets. The test criterion value is V = 328 and $p_{value} = 0.8378$. Thus, the null hypothesis cannot be rejected. The attributes of flue gases of wood chips and wood pellets do not differ significantly.

The result of the comparison suggests that non-wood pellets differ significantly from the wood chips and wood pellets in the flue gas attributes.

Comparison of biomass results: wood chips, wood pellets and non-wood pellets implies that non-wood pellets are of different attributes; however, wood pellets and wood chips are mutually interchangeable.

4.2.2. Comparison of the Results of Flue Gas Attributes of the Fuel—Brown Coal from the Same Source for Several Periods

To verify the hypothesis of conformity of brown coal flue gas attributes from Sokolovská uhelná, the Wilcoxon signed-rank test was used for 2017 and 2018. The test criterion value is V = 414 and the value p_{value} =0.04733.

The null hypothesis is rejected in favor of the alternative one. In the values of brown coal flue gas values in 2017 and 2018 from Sokolovská uhelná non-conformity, i.e., brown coal supplied between 2017 and 2018, is not interchangeable.

To verify the hypothesis of conformity of brown coal flue gas attributes from Severočeské doly a.s., the Wilcoxon signed-rank test was used for 2017 and 2018. The test criterion value is V = 226 and the value $p_{value} = 0.147$.

The null hypothesis cannot be rejected. In the values of brown coal flue gas attributes in the years 2017 and 2018 from Severočeské doly a.s. is a match, i.e., brown coal supplied between 2017 and 2018 is interchangeable.

To verify the hypothesis of conformity of brown coal flue gas attributes from Severní energetická a.s. for the years 2016, 2017, and 2018, the Friedman rank sum test was used. The test criterion value is $X^2 = 1.338$ and the value $p_{value} = 0.512$.

The null hypothesis cannot be rejected. In the values of brown coal flue gas attributes in the years 2016, 2017, and 2018 from Severní energetická a.s. is a match, i.e., brown coal supplied between 2016, 2017, and 2018 is interchangeable.

The results of the comparison show that two suppliers, Severní energetická a.s. and Severočeské doly a.s., respected the values of the flue gas attributes of the supplied brown coal for 2016 to 2018. Sokolovská uhelná differed in 2016–2018 in terms of attributes.

5. Conclusions

The analysis of fuel interchangeability in the conditions of heat production is realized with the intention of efficiency of the complex technological process. In practice, there are several approaches for the evaluation of fuel interchangeability. In the presented case study, the researched approach was based on the application of mathematical-statistical methods. Specifically, 12 types of solid fuels were evaluated, and also 35 different attributes.

Due to the extent of the monitored attributes, the attributes were classified to simplify the way of their monitoring. Specifically, for all attributes of flue gas with the same property, was searched a cross-correlation and one attribute was selected, which was used in further analyses. The analysis of cross-correlation among attributes was used as regression models for the pairs of attributes.

Evaluation of fuel interchangeability its comparison in terms of the same category and also the same source during several periods was realized by Wilcoxon signed-rank sum test and Friedman rank-sum test, according to the condition two or more fuels in the group.

The result of the realized analysis and preparatory calculations are descriptive statistics of flue gas attributes and confirmation/negation of linear relations between the pairs of attributes.

The obtained results of the research confirm that this approach is effective and fully applicable. At the same time, the results point to the possibility of mathematical-statistical methods used in the field of verification of fuel interchangeability in heat production.

Author Contributions: Conceptualization, M.H.; methodology, V.M. and G.F.; investigation, B.S. and G.B.; data analysis, B.S. and G.B.; writing—original draft preparation, V.M. and G.F.; writing—review and editing, M.H., G.F. and V.M. All authors have read and agreed to the published version of the manuscript.

Funding: This work was supported by the projects of the Scientific Grant Agency of the Ministry of Education, Science, Research and Sport of the Slovak Republic and the Slovak Academy of Sciences project No. VEGA 1/0403/18, No. VEGA 1/0600/20, No. VEGA 1/0264/21 and was supported by the projects of the Cultural and Educational Grant Agency of the Ministry of Education, Science, Research and Sport of the Slovak Republic and the Slovak Academy of Sciences project No. KEGA 012TUKE-4/2019, KEGA 013TUKE-4/2019 and APVV SK-SRB-18-0053.

Conflicts of Interest: The authors declare no conflict of interest.

References

1. Gadiraju, S.; Park, S.; Singh, P.; Pandit, J.; Ekkad, S.V.; Liberatore, F.; Srinivasan, R.; Ho, Y. Fuel Interchangeability Effects on the Lean Blowout for a Lean Premixed Swirl Stabilized Fuel Nozzle. In Proceedings of the ASME Turbo Expo: Turbomachinery Technical Conference and Exposition, Oslo, Norway, 11–15 June 2018; The American Society of Mechanical Engineers: New York, NY, USA, 2018; Volume 4b.
2. Vlček, J.; Velička, M.; Pyszko, R. Interchangeability of Mixed Gas Components for Heating Industrial Furnaces. In Proceedings of the 38th Meeting of Departments of Fluid Mechanics and Thermodynamics, Liptovsky Mikulas, Slovakia, 19–21 June 2019; pp. 1–5.
3. Ferguson, D.; Richard, G.A.; Straub, D. Fuel Interchangeability for Lean Premixed Combustion in Gas Turbine Engines. In *Combustion, Fuels and Emissions, Parts A and B, Proceeding of the ASME 2018 Turbo Expo, Oslo, Norway, 11–15 June 2018*; ASME: New York, NY, USA, 2018; Volume 3, pp. 973–981.
4. Park, S.; Choi, G.; Tanahashi, M. Combustion Characteristics of Syngas on Scaled Gas Turbine Combustor in Pressurized Condition: Pressure, H-2/CO Ratio, and N-2 Dilution of Fuel. *Fuel Process. Technol.* **2018**, *175*, 104–112. [CrossRef]
5. Maznoy, A.; Pichugin, N.; Yakovlev, I.; Fursenko, R.; Petrov, D.; Shy, S. (Steven). Fuel Interchangeability for Lean Premixed Combustion in Cylindrical Radiant Burner Operated in the Internal Combustion Mode. *Appl. Therm. Eng.* **2021**, *186*. [CrossRef]
6. Lopez-Anton, M.A.; Diaz-Somoano, M.; Ochoa-Gonzalez, R.; Martinez-Tarazona, M.R. Distribution of Trace Elements from a Coal Burned in Two Different Spanish Power Stations. *Ind. Eng. Chem. Res.* **2011**, *50*, 12208–12216. [CrossRef]
7. Klein, D.H.; Andren, A.W.; Bolton, N.E. Trace-Element Discharges from Coal Combustion for Power Production. *Water Air Soil Pollut.* **1975**, *5*, 71–77. [CrossRef]
8. Bool, L.E.; Helble, J.J. A Laboratory Study of The Partitioning of Trace-elements during Pulverized Coal Combustion. *Energy Fuels* **1995**, *9*, 880–887. [CrossRef]
9. Querol, X.; Juan, R.; Lopez-Soler, A.; Fernandez-Turiel, J.L.; Ruiz, C.R. Mobility of Trace Elements from Coal and Combustion Wastes. *Fuel* **1996**, *75*, 821–838. [CrossRef]
10. Helble, J.J. Trace-Element Behavior during Coal Combustion-Results of a Laboratory Study. *Fuel Process. Technol.* **1994**, *39*, 159–172. [CrossRef]
11. Davison, R.L.; Natusch, D.F.S.; Wallace, J.R.; Evans, C.A. Trace Elements in Fly Ash. Dependence of Concentration on Particle Size. *Environ. Sci. Technol.* **1974**, *8*, 1107–1113. [CrossRef]
12. Keramiotis, C.; Founti, M.A. An Experimental Investigation of Stability and Operation of a Biogas Fueled Porous Burner. *Fuel* **2013**, *103*, 278–284. [CrossRef]
13. Xiang, L.; Theotokatos, G.; Ding, Y. Investigation on Gaseous Fuels Interchangeability with an Extended Zero-dimensional Engine Model. *Energy Convers. Manag.* **2019**, *183*, 500–514. [CrossRef]
14. Pujihatma, P.; Hadi, S.P.; Rohmat, T.A. Combined Heat and Power-Multi-objective Optimization with an Associated Petroleum and Wet Gas Utilization Constraint. *J. Nat. Gas Sci. Eng.* **2018**, *54*, 25–36. [CrossRef]
15. Martinez, A.S.; Jani, A.; Dabdub, D. Emission Factor Estimation in Regional Air Quality Studies of Residential Natural Gas Fuel Interchangeability. *Fuel* **2014**, *119*, 129–140. [CrossRef]
16. Montgomery, D.C. *Design and Analysis of Experiments*, 8th ed.; Wiley: New York, NY, USA, 2012; ISBN 978-1-118-14692-7.
17. Hanke, J.E.; Reitsch, A.G. *Understanding Business Statistics*; Irwin: Homewood, IL, USA, 1991; ISBN 0-256-06627-2.

energies

MDPI

Article

Prosumers' Needs Satisfied Due to Cooperation with Offerors in the Context of Attitudes toward Such Cooperation

Agnieszka Izabela Baruk

Department of Management Systems and Innovation, Faculty of Management and Production Engineering, Lodz University of Technology, 90-924 Lodz, Poland; agnieszka.baruk@poczta.onet.pl

Abstract: The purpose of this article is to determine final purchasers' needs satisfied due to cooperation with offerors and the dependencies between these needs and previous behaviors and attitudes toward this cooperation. The results of the world literature analysis indicate a cognitive and research gap regarding the aspects mentioned. In order to reduce the gap, empirical studies were conducted, in which an online questionnaire was used to gather primary data. The research was implemented in the second half of 2020 among 1150 respondents representing Polish adult final purchasers. The data were subjected to quantitative analysis using statistical analysis and statistical testing, including exploratory factor analysis, cluster analysis, Pearson chi-square independence test, V-Cramer contingency coefficient analysis, and Kruskal–Wallis test. The results of the statistical analysis made it possible to verify six research hypotheses. Dependencies were found between needs satisfied due to cooperation with offerors and the following aspects: (1) purchasers' previous participation in cooperation with offerors, (2) purchasers' willingness to cooperate with offerors, and (3) the assessment of contemporary purchasers' readiness to cooperate with offerors. Willingness to cooperate with offerors differentiated all eleven needs satisfied due to cooperation with offerors analyzed in this study. Two other variables differentiated only a few of the needs analyzed. The results obtained from the research have a cognitive and applicability value. They contribute to theory of marketing and market behavior. They can also facilitate establishing and strengthening cooperation between offerors and final purchasers as important partners cooperating in the process of creating a marketing offer. This effect is very important in the case of shaping the cooperation between final purchasers and offerors of different products including energy ones. The originality of the approach proposed is evidenced by the fact it is the first time final purchasers' needs that can be satisfied due to cooperation with offerors have been analyzed in the context of attitudes and behavior reflecting purchasers' (1) previous participation in this cooperation, (2) willingness to cooperate with offerors, and (3) the assessment of contemporary final purchasers' readiness to cooperate with offerors.

Keywords: final purchaser; prosumer; offeror; needs; cooperation; attitudes; behaviors; co-creation

Citation: Baruk, A.I. Prosumers' Needs Satisfied Due to Cooperation with Offerors in the Context of Attitudes toward Such Cooperation. *Energies* **2021**, *14*, 7821. https://doi.org/10.3390/en14227821

Academic Editor: Dimitrios A. Georgakellos

Received: 26 October 2021
Accepted: 19 November 2021
Published: 22 November 2021

1. Introduction

In terms of the engagement and level of activity of final purchasers and offerors, the contemporary consumer market is characterized by a much greater symmetry compared with the past. Offerors used to be a more active party, whereas purchasers limited their activity mainly to buying products prepared by offerors without the participation of purchasers. This situation was reflected in a clear division in the scope of activities undertaken by purchasers and offerors and, thus, a clear division of market roles fulfilled by both parties.

The growing level of market awareness of final purchasers [1], resulting, among other things, from changes in society, including the development of information technologies, has changed this situation. Performing the role of passive recipients in line with the traditional paradigm of market functioning [2] ceased to be sufficient for purchasers, who began to show an open attitude toward participating in the creation of marketing offers together with offerors and even more often taking specific actions consisting of active participation in this process. Purchasers noticed that co-creating marketing offers allowed them not only to create products that better meet their growing requirements but also meet many other needs, especially those of a higher order.

Thus, it can be said that, today, meeting the assumptions of marketing orientation by offerors not only requires them to offer products in line with the expectations of purchasers but also presents them with much greater challenges including the necessity to protect the natural environment [3], implementing innovative social media marketing solutions [4], etc. What serves to meet the expectations of contemporary purchasers is the creation of conditions for them to play the role of active purchasers and even co-creators [5] of products and non-product elements of marketing offers. Frequently, the participation in these activities itself is valuable enough for purchasers to make them want to get involved [6]. Therefore, it is important to identify the needs met through the participation of purchasers in jointly creating offers with offerors and to analyze these needs in different contexts.

However, the results of the cognitive–critical analysis of world literature on the subject presented in the next part of this article indicate that, until now, these needs have not been considered in the context of purchasers' attitudes and behaviors, reflecting their previous experience as active participants in marketing activities and readiness for such activity. This conclusion also applies to publications on energy market. Therefore, in this study, an attempt was made to solve the following research problem: what needs of final purchasers are met through cooperation with offerors, and what are the dependencies between these needs and purchasers' previous attitudes and behaviors related to such cooperation?

The aim of this article is to identify final purchasers' needs satisfied through cooperation with offerors, the dependencies between these needs and previous behaviors in this area, and attitudes toward such activity. The article is structured to achieve this aim. In the second, theoretical part of this article, the results of the analysis of world literature on the subject are presented, which allow research hypotheses to be formed. The third and fourth parts present the methodology of primary research and the results, respectively. Next, an academic discussion is conducted, comparing the results obtained with the results of other authors' research, and the most important conclusions, theoretical and practical implications, and limitations of the research conducted, as well as directions for future research, are presented.

2. Literature Review

In the literature on the subject, apart from the notion of 'cooperation with offerors', many other terms are used, which some authors consider to be synonyms of cooperation, whereas other researchers claim that they cannot be used interchangeably [5]. These include concepts such as co-design, co-creation (inter alia Hansen [7]), co-production (inter alia Chatterjee, Rana, and Dwivedi [8]), co-working with consumers (inter alia France, Grace, Merrilees, and Miller [9]), etc. Each of them has one common feature, which is the active involvement [10] of the final purchasers in creating marketing values.

In this article, cooperation between final purchasers and offerors is defined as the joint creation of various elements of a marketing offer, mainly products, although obviously it may also apply to its non-product elements. Therefore, it fits into the approach presented by Prahalad and Ramaswamy [11], in whose opinion 'value co-creation' is a process of joint creation of values by various entities traditionally representing suppliers and customers. The final purchaser is defined in this article as a person purchasing a product. This term is intentionally used instead of the term 'consumer'. It is true that the literature usually uses this term in situations regarding the joint creation of values, and possibly the synonym 'customer' is used. However, the consumer is a person using a product, and the customer has a much broader meaning than the consumer or purchaser. This article uses it in terms of the cooperation of people buying products with offerors and having their needs met through this interaction. In turn, the offeror is a term referring to entities offering products on the consumer market, including producers, retailers, and service providers.

The cooperation between final purchasers and offerors reflects the growing level of activity of final purchasers who are looking for new ways to meet their growing expectations [12], not only toward products available on the consumer market but also toward the market role they play. The role of passive participants in this market, limiting their activity to purchasing the products offered, ceases to be enough for an increasing number of purchasers. Many purchasers would like to have a much greater impact on other market participants and the functioning of the market, including features of a marketing offer [13]. Therefore, they are increasingly playing the role of active market actors [14], involved in marketing activities previously undertaken by the offerors themselves or at least showing readiness for such activity.

Thus, the previously clear boundaries between the areas of offerors' activity and the area of purchasers' activity are blurred. Of course, purchasers cannot entirely take over the tasks performed by producers or traders, which result from independent technological, technical, and organizational limitations, etc. Purchasers cannot, therefore, fully replace offerors, but they can increasingly participate in marketing activities related to their preparation, which results, among other things, from dynamic development of communication technologies [15], including social media [16,17]. Purchasers as active participants in the modern consumer market, possessing valuable marketing potential [18], become prosumers, i.e., co-creators of products and other elements of a marketing offer through joining in creating their concept [19], design [20], modification, etc. However, it is difficult to agree with an approach presented by some researchers (among others, Tian, Shen, and Chen [21]) who believe that being a prosumer is about making products for one's own needs, since, in this case, no interaction takes place with any offeror whatsoever.

The cooperation between final purchasers and offerors brings various benefits to both parties [15,22]. The benefits can have a material and non-material dimension, and they are practically always much greater than benefits obtained in the case of the traditional division of market roles, when no cooperation takes place. The benefits that offerors achieve can be divided into economic, social, and image-related categories [23]. The benefits that active purchasers achieve through cooperation with offerors include obtaining products that better meet the expectations of recipients, which is often noted in the literature (inter alia Cheung and To [24]; Yi and Gong [25]). Better fulfillment of needs due to jointly prepared products is not the only positive effect of mutual cooperation. It even seems to not be the most important one, and thus it does not fulfill the role of the main motivator encouraging purchasers to engage in joint activities. Joining in the activities allows many other needs to be met, especially of a social nature [26,27] and psychological nature [28,29] and those related to purchasers' self-fulfillment through expanding and deepening their existing relational [30,31], social, emotional, and intellectual potential, and even their hedonistic needs [14], which are still mentioned mainly in relation to typical shopping behavior [32]. Of course, an active purchaser can achieve the possibility of costs reduction, which is emphasized in the case of energy market (for example by Koltsaklis, Panapakidis, Pozo, and Christoforidis [33]; Faia, Pinto, Vale, and Corchado [34]), but this material benefit is not considered to be the major stimulus of prosumeric activity in the comparison to the non-material benefits.

Thus, final purchasers become disposers of greater marketing potential, becoming at the same time more valuable partners for offerors [35] by sharing their experience [36], knowledge [19,37,38], etc. The resources made available by purchasers cannot be used without their integration with the offeror's resources [39]. The effect of the integration of the purchaser's resources with the offeror's corresponding resources is the creation of new experiences, new knowledge, etc., which leads to increasing the offeror's marketing potential. Therefore, cooperation becomes increasingly valuable in itself, both for active purchasers [5] and offerors. The effects of such cooperation, which include the creation of a community centered around a given offeror caring for the community's interests [14], are also increasingly valuable for both parties, as they are identified with common interests.

However, for cooperation between offerors and final purchasers to take place, there must be adequate conditions for its initiation. Both parties must be ready for it, showing an open attitude to cooperation [18] and thus to a re-definition of market roles traditionally assigned to each of them. In the case of an offeror, an open attitude must have both organizational and individual dimensions so as not to evoke professional stress among employees [40], which is mentioned in the literature as the main negative consequence of incorporating purchasers into cooperation. In the case of purchasers, openness to engaging in joint activities is certainly fostered by their previous market activity, especially positive experience acquired in relations with offerors, for example, on the occasion of typical shopping or communication activity, which is the first important step toward creative activity, forming prosumer activity [41,42]. One of the key incentives encouraging such activity is, of course, the assumption that it will meet many needs that could not be met if the scope of activity had been limited to shopping behavior only.

It should not be forgotten that the appearance of any dysfunctions in the interaction between purchasers and offerors can have negative effects. The consequence of this situation is not only failure to meet purchasers' needs but even worsening their feelings about their market role [43], which, in turn, can be a difficult barrier to overcome, preventing future cooperation [44]. Unfortunately, there is still relatively little research in the literature on the negative effects of cooperation. However, the possibility that negative effects will occur proves that the creation by offerors of the right conditions for the occurrence and strengthening of cooperation is a key factor determining the scope and nature of the effects of cooperation with final purchasers.

Until now, the world literature on the subject has considered cooperation between final purchasers and offerors in a variety of contexts, for example, by analyzing its scope [41], environment (especially virtual [30,45] and offline [41]), purchasers' competence to be prosumers [46], etc. Research has also been undertaken into the motives for purchasers to engage in joint activities with other entities, including other purchasers [47–49] and offerors [19,50]. Focus has been placed on the need for effective management of such cooperation, including purchasers' expectations of the cooperation [13]. Although playing the role of prosumers by final purchasers on energy market has been studied in recent years, researchers are not concerned with the issues of prosumer's needs satisfied due to cooperation with offerors in the context proposed in this article. For example, prosumer as an active participant of energy market is analyzed in the terms of technological solutions implemented and economic assessment [51]; legal framework of prosumption in countries of the UE [52]; democratizing access to the energy markets [53]; etc. This is a completely different approach than that proposed in this article by the author.

However, needs satisfied due to cooperation in the context of the previous creative behavior of final purchasers and their willingness to cooperate with offerors have not been studied. Therefore, one can speak of a cognitive and research gap in this respect, and filling the gap is important from the point of view of the theory and practice of marketing and market behavior. In order to fill the identified gap, an attempt is made in this article to achieve the goal of identifying final purchasers' needs satisfied through cooperation with offerors, dependencies between those needs and final purchasers' previous behavior in this respect, and their attitudes toward such activity.

In order to meet this goal, the following research hypotheses were checked:

Hypothesis 1 (H1). *There is dependence between respondents' previous active participation in cooperation with offerors and final purchasers' needs satisfied due to such cooperation.*

Hypothesis 2 (H2). *There is dependence between respondents' willingness to actively participate in cooperation with offerors and final purchasers' needs satisfied due to such cooperation.*

Hypothesis 3 (H3). *There is dependence between respondents' assessment of contemporary purchasers' willingness to actively cooperate with offerors and final purchasers' needs satisfied due to such cooperation.*

Hypothesis 4 (H4). *Respondents' previous active participation in cooperation with offerors is a feature differentiating final purchasers' needs satisfied due to such cooperation.*

Hypothesis 5 (H5). *Respondents' willingness to actively participate in cooperation with offerors is a feature differentiating final purchasers' needs satisfied due to such cooperation.*

Hypothesis 6 (H6). *Respondents' assessment of contemporary final purchasers' willingness to actively cooperate with offerors is a feature differentiating purchasers' needs satisfied due to such cooperation.*

3. Research Methodology

To achieve the research goal and verify the hypotheses, empirical research was conducted by means of the online survey method to collect primary data using the CAWI (Computer Assisted Web Interview) technique. The research was implemented in the second half of 2020 among 1150 representative of Polish adult final purchasers. The research had a nationwide geographical coverage and was of panel format. The sample was a representative quota of all Poles in terms of gender. Other sociodemographic features (age, education, and region) were maintained in a dispersion proportional to the distribution of a given feature in the general population, with a deviation of no more than ten respondents against the proportion for the distribution of the entire Polish population (based on Statistics Poland (in Polish—GUS) data and CAPI (Computer Assigned Personal Interview) population studies).

The subject scope of this article includes four variables: (1) final purchasers' needs satisfied due to cooperation with offerors, (2) final purchasers' previous participation in cooperation with offerors, (3) final purchasers' willingness to cooperate with offerors, and (4) the assessment of final purchasers' readiness to cooperate with offerors.

During the research, respondents were presented with a set of eleven needs that can be satisfied due to cooperation with offerors. The needs were separated on the basis of the results of cognitive–critical analysis of the literature. The list of needs that can be satisfied by final purchasers due to cooperation with offerors was based on a classification of values generated through 'value co-creation' proposed by Kuo and Feng [54]. In this article, the values were divided into the following four groups: cognitive (knowledge, information, etc.), social (relationships with other entities, etc.), self-assessment (status, reputation improvement, etc.), and hedonistic (satisfaction with interaction with other people, etc.). This list was supplemented with needs pre-identified based on the results of unstructured interviews preceding the survey.

Each of the eleven needs had to be ranked by respondents on a 5-grade Likert scale, which belongs to the most fundamental and most commonly used psychometric tools in the social sciences [55]. In this article, a five-level version was applied, where 5 indicated definitely yes, 4—rather yes, 3—neither yes nor no, 2—rather not, and 1—definitely not. Applying the Likert scale is a prerequisite for the application of average grade analysis and exploratory factor analysis.

The primary data collected were subjected to quantitative analysis, which included the following methods: average grade analysis, comparative analysis, exploratory factor analysis, cluster analysis, Pearson chi-square independence test, V-Cramer contingency coefficient analysis, and Kruskal–Wallis test.

Exploratory factor analysis is used to reduce the number of variables constituting primary data obtained from the survey and to detect structures in relationships between those variables to classify them [56]. In this article, this analysis was used to reduce the number of variables influencing the category of 'needs satisfied due to cooperation with offerors' and to detect internal dependencies between those variables. To extract the factors, a principal component method was used, and it was essential to determine the number of the components. To determine the number of common factors (the main components), the Kaiser Criterion was applied to leave only those factors with eigenvalues greater than 1. Each factor explains a certain level of overall variability of the system under consideration, determined by a percentage of variance that can be interpreted as a measure of explanation of the problem. The factors were rotated using the oblimin method. Within individual factors, the variables with the highest factorial loadings against given factors were distinguished (value ≥ 0.7 was assumed).

Factor analysis identifies hidden factors with features responsible for the perception of a problem included in a question. However, factor analysis does not facilitate the answer to whether the diversity in terms of separating individual groups (e.g., according to previous participation in cooperation with offerors, willingness to cooperate with offerors, and the assessment of final purchasers' readiness for cooperation with offerors) is statistically significant enough to state that the respondents' answers determined by the analyzed response are significantly different. The Kruskal–Wallis (KW) test, which is a non parametric equivalent of ANOVA, is used to answer this question. Therefore, the KW test was used in this study.

The data do not have to meet many statistical requirements for the KW test. In order to carry out the KW test, the following factors should be met [57]:

- The dependent variable should be measured at least on an ordinal scale (it can also be measured on a quantitative scale);
- Observations in the analyzed groups should be independent of each other, which means that a person in one group should not be included in another group (this requirement is met by dichotomous questions, which enable respondents to be divided into two separate groups, and single-choice questions).

The KW test, as a non-parametric counterpart of a one-factor ANOVA, is, therefore, used when the data does not meet the requirements for similar parametric tests and the data can be ordered according to specific criteria. The test checks whether the number of independent results from a group come from the same population or from a population with the same median. Individual samples do not have to be of the same number. The input data is an n-element statistical sample divided into 'k' disjointed groups with numbers ranging from n1 to nk.

The test is interpreted by comparing the value of '*p*' with the assumed level of significance (usually 0.05) or by analyzing the test's statistical value in case it is necessary to assess the 'power/intensity' differences between groups. High values of test statistics indicate differentiation in particular groups (i.e., against the equality hypothesis in particular groups), and the higher the values are, the greater the diversity.

In turn, cluster analysis is a multi-dimensional exploratory method. It consists of grouping objects into clusters in such a way that objects within one cluster have more common features with each other than in relation to objects from other clusters. Cluster analysis can be used for grouping people or other objects based on their value in a data set [58]. Cluster analysis can therefore be used for discovering data structures but without providing an explanation or interpretation. In other words, cluster analysis is used for detecting data structures without explaining why they exist.

The chi-square test was used in this study to determine whether there are statistically significant dependencies between the analyzed variables, whereas the V-Cramer coefficient determined the strength of dependencies between the analyzed variables. It is used when at least one variable has more than two values [59], i.e., when the contingency table has dimensions of at least 2×3.

Statistical analysis of the primary data collected was performed using the IBM SPSS Statistics Ver. 25.

4. Research Results

According to the research conducted, only two out of eleven analyzed needs that can be satisfied due to cooperation with offerors were definitely indicated by less than every fifth respondent (Table 1). These were the needs connected with a group of psychological needs relating to being recognized by other people. At the same time, the largest percentage of respondents indicated these needs as those that are not satisfied due to cooperation with offerors. Thus, these needs obtained average ratings of relatively the lowest values, taking the last two positions in the identified hierarchy of needs. It should be emphasized that, for each of the needs analyzed, the value of the standard deviation did not exceed one-third of the average value, which means that the average values accurately reflect the results obtained [60].

Table 1. Needs satisfied due to cooperation with offerors according to respondents.

Needs Satisfied	Symbol	5	4	5 + 4	3	2	1	2 + 1	Average Rating	Standard Deviation
I feel that I have real influence on an offer and/or company	P10.1	206	49.7	70.3	23.1	5.1	1.4	6.5	3.83	0.863
I feel needed	P10.2	29.4	47.4	76.8	20.0	1.7	1.4	3.1	4.02	0.833
I can check the practical usefulness of my ideas	P10.3	30.3	46.3	76.6	19.7	2.6	1.1	3.7	4.02	0.841
I get an offer that better meets my expectations	P10.4	27.4	49.4	76.8	18.0	3.1	2.0	5.1	3.97	0.873
I get an offer that better meets the expectations of my relatives and friends	P10.5	20.9	50.0	70.9	24.3	3.1	1.7	4.8	3.85	0.84
I acquire new experience	P10.6	38.3	45.1	83.4	13.1	1.7	1.7	3.4	4.17	0.844
I acquire new knowledge	P10.7	38.6	45.7	84.3	12.9	1.4	1.4	2.8	4.19	0.817
I acquire new skills	P10.8	37.1	45.7	82.8	13.1	2.0	2.0	4.0	4.14	0.863
I establish relationships with other people	P10.9	28.9	46.3	75.2	18.3	4.3	2.3	6.6	3.95	0.921
I can boast about my activity	P10.10	17.4	39.7	57.1	32.0	8.6	2.3	10.9	3.61	0.947
I get the recognition of other people	P10.11	17.7	47.7	65.4	28.0	5.7	0.9	6.6	3.76	0.840

where: 5—definitely yes; 4—rather yes; 3—neither yes nor no; 2—rather no; 1—definitely no.

Almost 40% of respondents strongly agreed that three other needs are satisfied due to cooperation with offerors. These needs reflected the acquisition of new elements of marketing potential by respondents in the form of (1) new knowledge, (2) new experience, and (3) new skills. These were the only needs to obtain the total percentage of positive responses amounting to over 80%, taking the first three positions in the hierarchy with average values exceeding 4.10.

In addition to the above three needs, two other needs analyzed obtained average ratings above 4.00. They referred to: (1) the feeling of being needed and (2) the opportunity to check the usefulness of respondents' ideas. It is worth noting that the need to receive an offer that better meets respondents' expectations took a relatively distant (sixth) position, with an average rating of 3.97.

In order to identify the internal hierarchy of final purchasers' needs that can be satisfied due to cooperation with offerors, an exploratory factor analysis was carried out for all respondents. Adequacy measure of the Kaiser-Meyer-Olkin (KMO) draw is 0,937, i.e., greater than 0.5 [61]; Bartlett's sphericity test is significant (variables are statistically significantly related); chi^2 is 2748,629; and p = 0.000. Based on the Kaiser criterion, four factors with eigenvalues exceeding 1 were identified, which in total explain almost 81% of the total variability of the analyzed system (Table 2). The first factor, with an eigenvalue of 5.819, explains over 62% of the total variability of the aspects studied. It includes three variables (Table 3) with factor loading values of at least 0.7.

Table 2. Hierarchy of factors according to their eigenvalues based on the Kaiser criterion (for the total of respondents).

Factor	Eigenvalue	Cumulative Eigenvalue	% of Total Eigenvalues (Variation)	Cumulative % of Eigenvalues
1	5.819	5.819	62.376	62.376
2	3.673	9.492	7.538	69.914
3	2.967	12.459	6.305	76.219
4	2.057	14.516	4.494	80.713

Table 3. Factor analysis of needs satisfied due to cooperation with offerors, according to respondents (for the total of respondents).

Analyzed Variable	Factor			
	1	2	3	4
I acquire new experience	**0.969**	−0.020	−0.053	−0.020
I acquire new skills	**0.956**	0.047	0.020	0.168
I acquire new knowledge	**0.712**	−0.010	0.092	−0.248
I can check the practical usefulness of my ideas	0.523	0.116	0.347	0.047
I establish relationships with other people	0.443	0.282	0.070	−0.286
I can boast about my activity	0.029	**0.877**	0.156	0.179
I get the recognition of other people	0.107	**0.717**	−0.080	−0.380
I feel that I have real influence on an offer and/or company	−0.058	0.075	**0.935**	0.064
I get an offer that better meets my expectations	0.226	−0.176	0.676	−0.296
I get an offer that better meets the expectations of my relatives and friends	0.188	0.197	0.616	0.031
I feel needed	0.075	0.201	0.406	−0.533

The variables that make up the first factor reflect the needs of acquiring new experience, skills, and knowledge, and thus expanding the existing marketing potential. It is worth recalling that the three variables listed took the first three positions among all needs analyzed, taking into account the values of their average scores.

The second of the identified factors explains a definitely smaller (over eight times) part of the total variability of the studied aspects. It includes two variables relating to the need of being recognized. These variables took the last two positions in the hierarchy identified on the basis of average rating values. However, the third factor is formed with only one variable, referring to the feeling of exerting a real influence on an offer and its offeror. This factor is of the relatively least significance, as evidenced by the lowest eigenvalue and a much smaller part of the system analyzed, which is explained by this factor. The variable forming the third factor took only the ninth place among all the needs analyzed. Within the fourth factor, no variable with a factor loading of 0.7 or more was identified.

It is worth noting that none of the identified factors included variables that reflected the possibility of getting an offer that better meets expectations and relate to relational aspects.

The results of cluster analysis for the total of respondents confirm the results of the exploratory factor analysis. The cluster with the least distance was identified for the variables corresponding to acquiring new experiences, acquiring new skills, and acquiring new knowledge (Figure 1), which forms the first, by far the most important, factor identified during factor analysis.

In the next stage of the analysis, three variables were included illustrating respondents' previous activity relating to cooperation with offerors, their willingness to undertake such cooperation, and their assessment on contemporary purchasers' readiness for such activity. As seen in Table 4, the vast majority of respondents had not previously participated in the process of jointly preparing marketing offers with offerors. Only 12.3% of all respondents showed activity in this area. However, more than half of the respondents declared an open attitude to such cooperation, and as many as 78.0% of respondents believed that purchasers as participants of the contemporary market are ready to co-create offers with offerors. The identified openness of respondents to joint creation of offers, alongside the fact that, for each of the eleven needs analyzed, in total, more than half of respondents believed that mutual cooperation with offerors allowed the needs to be met, is a positive premise conducive to cooperation between purchasers and offerors in the future.

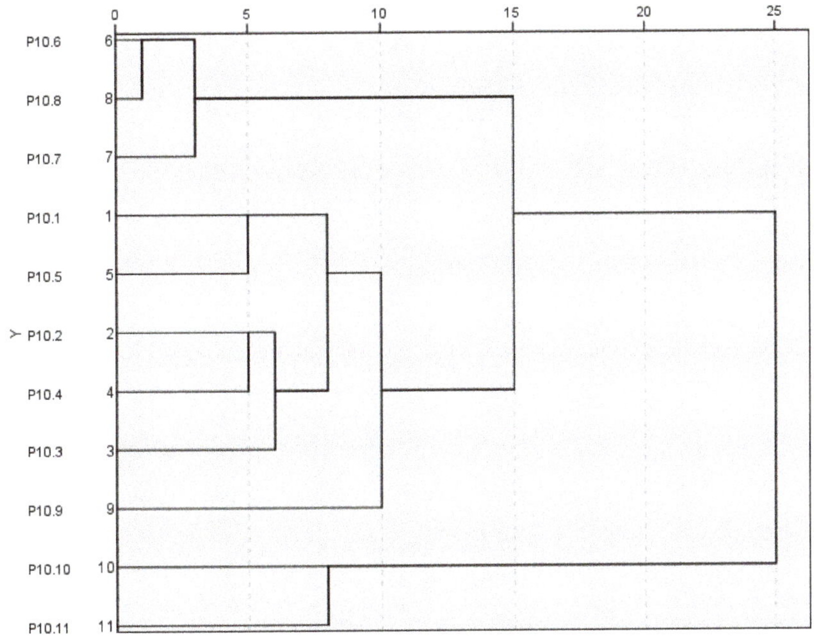

Figure 1. Dendrogram using the Ward link showing the structure of needs, which according to respondents, are satisfied due to cooperation with offerors. Where: symbols as in Table 1.

Table 4. Respondents' previous participation in cooperation with offerors, respondents' openness to such cooperation, and the assessment of contemporary final purchasers' readiness for cooperation (%).

Respondents' Opinions	Indications (%)	
	Yes	No
Respondents' previous active participation in cooperation with offerors while preparing marketing offers	12.3	87.7
Respondents' willingness for active participation in cooperation with offerors while preparing marketing offers	53.4	46.6
Respondents' assessment of contemporary purchasers' readiness for active cooperation with offerors while preparing marketing offers	78.0	22.0

In order to check whether there are statistically significant dependencies between the analyzed variables, statistical testing was performed. It revealed that, between each of the three analyzed aspects reflecting respondents' behavior and attitudes toward cooperation with offerors, and the needs satisfied due to this cooperation, there is a statistically significant dependence (Table 5). Thus, in relation to respondents, the research hypotheses H1, H2, and H3, respectively, are valid. However, these are not strong dependencies. The value of the V-Cramer coefficient is, in each, case less than 0.3, i.e., the limit value between the weak and medium strength of the dependence between the analyzed variables [59]. The relatively strongest dependence occurs in the case of willingness for cooperation, and the weakest is in the case of respondents' assessment of contemporary purchasers' readiness for joint creation of offers with offerors.

Table 5. Analysis of dependencies between respondents' attitudes toward co-creation of marketing offers with offerors and the needs satisfied due to such cooperation.

Analyzed Dependence	Pearson chi^2 Test Value	V-Cramer Coefficient Value	Level of Significance 'p'
Previous participation in cooperation with offerors vs. the needs satisfied due to such cooperation	89.483	0.152	**0.000**
Willingness for cooperation vs. needs satisfied due to such cooperation	223.221	0.241	**0.000**
Respondents' assessment of purchasers' readiness for cooperation with offerors vs. needs satisfied due to such cooperation	78.238	0.143	**0.000**

In the next stage of the research process, the Kruskal–Wallis test was conducted in order to check possible differentiation between respondents' opinions on the needs satisfied due to the co-creation of a marketing offer by purchasers with offerors according to the following aspects: (1) undertaking such an activity before (Table 6), (2) willingness to undertake such cooperation (Table 7), and (3) assessment of contemporary purchasers' readiness for such cooperation (Table 8). Statistically significant diversity occurred for seven needs in the first case, all needs in the second case, and six needs in the third case. For respondents, research hypotheses H4, H5, and H6 were, therefore, valid only for these variables. It is worth adding that previous activity and the assessment of purchasers' readiness to cooperate with offerors did not differentiate both variables reflecting the needs of recognition and the possibility of being applauded, i.e., the needs that took the last two positions in the hierarchy identified during the research.

Table 6. Analysis of significance of differences between respondents' opinions relating to the needs satisfied due to cooperation with offerors, according to the criterion of respondents' previous participation in such cooperation.

Analyzed Variable	Participation in Cooperation	Ranks	KW Test Value	Level of Significance 'p'
I feel that I have real influence on an offer and/or company	Yes	201.14	3.678	0.055
	No	171.91		
I feel needed	Yes	225.08	13.691	**0.000**
	No	168.56		
I can check the practical usefulness of my ideas	Yes	208.94	6.188	**0.013**
	No	170.82		
I get an offer that better meets my expectations	Yes	211.95	7.462	**0.006**
	No	170.39		
I get an offer that better meets the expectations of my relatives and friends	Yes	199.56	3.255	0.071
	No	172.13		
I acquire new experience	Yes	207.73	5.856	**0.016**
	No	170.99		
I acquire new knowledge	Yes	208.93	6.334	**0.012**
	No	170.82		
I acquire new skills	Yes	209.51	6.510	**0.011**
	No	170.74		
I establish relationships with other people	Yes	218.40	10.120	**0.001**
	No	169.49		
I can boast about my activity	Yes	189.53	1.050	0.306
	No	173.53		
I get the recognition of other people	Yes	203.47	4.336	0.037
	No	171.58		

Table 7. Analysis of significance of differences between respondents' opinions relating to the needs satisfied due to cooperation with offerors, according to the criterion of respondents' willingness to undertake such cooperation.

Analyzed Variable	Willingness to Cooperate	Ranks	KW Test Value	Level of Significance 'p'
I feel that I have real influence on an offer and/or company	Yes No	201.46 145.72	30.876	**0.000**
I feel needed	Yes No	201.18 146.04	30.074	**0.000**
I can check the practical usefulness of my ideas	Yes No	203.56 143.31	35.676	**0.000**
I get an offer that better meets my expectations	Yes No	200.15 147.22	27.941	**0.000**
I get an offer that better meets the expectations of my relatives and friends	Yes No	191.75 156.86	12.160	**0.000**
I acquire new experience	Yes No	199.36 148.13	26.284	**0.000**
I acquire new knowledge	Yes No	197.43 150.35	22.314	**0.000**
I acquire new skills	Yes No	197.02 150.81	21.343	**0.000**
I establish relationships with other people	Yes No	192.60 155.88	13.172	**0.000**
I can boast about my activity	Yes No	189.77 159.13	8.891	**0.003**
I get the recognition of other people	Yes No	190.04 158.82	9.604	**0.002**

Table 8. Analysis of significance of differences between respondents' opinions relating to the needs satisfied due to cooperation with offerors, according to the criterion of the assessment of purchasers' readiness for such cooperation.

Analyzed Variable	Assessment of Purchasers' Readiness for Cooperation	Ranks	KW Test Value	Level of Significance 'p'
I feel that I have real influence on an offer and/or company	Yes No	183.37 147.61	8.763	**0.003**
I feel needed	Yes No	181.46 154.38	5.001	**0.025**
I can check the practical usefulness of my ideas	Yes No	180.75 156.87	3.867	**0.049**
I get an offer that better meets my expectations	Yes No	185.22 141.05	13.424	**0.000**
I get an offer that better meets the expectations of my relatives and friends	Yes No	184.56 143.38	11.680	**0.001**
I acquire new experience	Yes No	179.30 162.02	2.063	0.151
I acquire new knowledge	Yes No	179.36 161.83	2.132	0.144
I acquire new skills	Yes No	180.33 158.38	3.320	0.068
I establish relationships with other people	Yes No	181.02 155.94	4.237	**0.040**
I can boast about my activity	Yes No	179.82 160.19	2.516	0.113
I get the recognition of other people	Yes No	179.56 161.12	2.310	0.129

5. Discussion

As was shown in the theoretical part of this article, active final purchasers' needs satisfied due to cooperation with offerors have not been studied in the context of purchasers' behaviors and attitudes reflecting their previous experience as active participants of marketing activities and reflecting purchasers' readiness for such activity. The results of the research presented in this article indicate that there is statistically significant dependence between respondents' previous active participation in cooperation with offerors and the needs satisfied due to such cooperation. Statistically significant dependence was also identified between respondents' readiness to cooperate with offerors and the needs satisfied as a result of such cooperation, as well as between opinions on purchasers' readiness to cooperate with offerors and the needs satisfied as a result of this cooperation.

It should be emphasized that other authors have so far focused on analyzing specific reasons for purchasers' willingness to engage in value co-creation. These reasons included altruism [62], self-complacency and social respect [63], social position [64], material rewards [65], costs reduction in the case of energy co-creation [66], satisfaction with performing creative tasks, and even the opportunity to find friends [28], although the last two reasons relate to cooperation with other purchasers, not with offerors. Thus, the subject of these studies was different. The reason for joining in the cooperation is not, however, tantamount to the needs satisfied. It can merely be equated with the needs that purchasers expect to be satisfied due to becoming engaged in cooperation. In addition, the studies of the authors mentioned above did not analyze dependencies between these motivators and the variables proposed in this article.

It is worth adding that the research conducted by Kolomiiets, Krzyżanowska, and Mazurek [28] identified that satisfaction with performing creative tasks was accompanied by the reason for the possibility of obtaining an exceptional and unique product. However, the results of factor analysis conducted as part of the research underlying this article show that the issue of obtaining a product that better meets purchasers' expectations was not found in any of the identified factors. Therefore, these aspects were not related to other needs of the respondents, which can be satisfied through cooperation with offerors.

In turn, Nadeem, Juntunen, Shirazi, and Hajli [30] found that trust does not affect the intention (readiness) to jointly create value. However, in their research, they focused only on 'value co-creation' in the context of the sharing economy and not on cooperation with offerors in relation to the needs satisfied due to this cooperation. In addition, the scope of this research was different, as only representatives of Generation Y were included. These authors did not examine dependencies between needs satisfied due to cooperation with offerors and the variables included in this article, either. It is true that trust can be considered as a result of experiences resulting from previous activity of purchasers as value co-creators. However, results of the research conducted by the author of this article indicate that there is dependence between previous involvement in cooperation with offerors and the needs satisfied as a result.

The results of the research conducted by the author of this study, and in turn, those conducted by other researchers (among others Pan and Holland [67]; Mitręga and Małecka [68]), in relation to the satisfaction achieved by final purchasers due to their involvement in cooperation with offerors, showed that the purchasers' involvement in cooperation is conducive to the increase in customer satisfaction. Such needs as a sense of authentic influence on an offer and/or company and acquiring new knowledge is part of achieving satisfaction. It is worth recalling that the latter of these variables took the first position among the total needs analyzed, creating with the other two variables the most important factor identified during the factor analysis.

In turn, Ranjan and Read [69] and Oertzen, Odekerken-Schröder, Brax, and Mager [70] found that purchasers' pro-social orientation positively affects their involvement in joint value creation. Therefore, they examined the issues of cooperation in a different context than the context of needs satisfied through this cooperation proposed in this article. They focused on the motivators, not on effects. It is worth adding that cooperation may also take place between offerors in the form of coopetition. However, coopetition has a completely different nature and scope than prosumeric cooperation. Moreover, it does not include final purchasers. That's why the considerations (conducted inter alia by Czakon, Mucha-Kuś, and Sołtysik [71]) on the abovementioned concept do not focus on meeting their needs.

Some researchers (among others, Leclercq, Poncin, and Hammedi [14]; Kuo and Feng [54]) admittedly examined the effects of purchasers' cooperation with other entities. However, these studies had a different subject scope (they related to cooperation with other purchasers within purchaser community) and a different object scope (they related to online activity), and they related to values obtained due to this cooperation and not the needs satisfied as a result. It is true that the value generated allows a specific need or needs to be satisfied; however, it is not synonymous with these needs, being only a means to satisfy them.

6. Conclusions

In conclusion, it was revealed in this article that it was of key importance for respondents to satisfy the needs of acquiring new knowledge, experience, and skills, i.e., of improving their marketing potential. The three listed needs took the first three positions in the hierarchy of needs that can be satisfied as a result of cooperation with the offeror, creating the first and most important factor identified during the factor analysis. This means that there is a group of respondents who attach importance to the possibility of satisfying only those needs through cooperation with offerors. Obtaining an offer better suited to the expectations of a given person and the expectations of their family members owing to cooperation between purchasers and offerors came in relatively later positions. In addition, neither of these needs entered any of the four factors identified during factor analysis.

It is true that almost 88% of respondents had not previously been involved in any activity of cooperating with offerors, but the majority of respondents were willing to participate in such activities and believed that contemporary purchasers are ready to get involved in this type of activity. It is worth noting that a clearly larger proportion of respondents positively referred to readiness to cooperate with other purchasers, compared to the percentage of people expressing such an opinion in relation to themselves.

This article also revealed that there are statistically significant dependencies between the needs satisfied due to cooperation with offerors and each of the three analyzed variables, i.e., previous participation in cooperation with offerors, willingness to participate in such cooperation, and assessment of contemporary final purchasers' readiness to undertake this cooperation. Thus, for respondents, the research hypotheses H1, H2, and H3 proved to be valid.

For all the needs analyzed, statistically significant differences between respondents' answers regarding the needs satisfied due to cooperation with offerors were identified only in the case of respondents' willingness to undertake such cooperation. Regarding the other two variables reflecting, respectively, previous participation in cooperation with offerors and the assessment of final purchasers' readiness to undertake such cooperation, statistically significant diversity was identified only for some of the needs. In the case of respondents, research hypotheses H4, H5, and H6 proved to be valid only for these variables. It is worth noting that the results of the analysis of the diversity of responses regarding the needs satisfied due to cooperation with offerors due to the criterion of the assessment of purchasers' readiness to begin cooperation with offerors indicate no diversity in the case of 'acquiring new knowledge', 'acquiring new experience', and 'acquiring new skills'.

7. Implications, Limitations, and the Directions of Future Research

The results of the research conducted and conclusions drawn from it constitute an important contribution to the theory of marketing and theory of market behavior. They fill the knowledge gap identified during the cognitive–critical analysis of world literature on the subject. They show the hierarchy of needs that are satisfied due to cooperation with offerors and indicate the existence of homogeneous groups of people who perceive a possibility of meeting the same needs in this cooperation. Identifying dependencies between needs satisfied due to cooperation with offerors and the three analyzed variables also enriches the existing knowledge of marketing and market behavior.

The results of the research also have important practical implications, which constitute valuable tips for managers. They indicate, among other things, that, within respondents, there are homogeneous purchaser groups whose representatives see the opportunity to meet the same needs in cooperation with offerors. Representatives of the first of the identified factors, who paid attention to the possibility of increasing their marketing potential by acquiring new knowledge, experience, and skills, may be particularly valuable partners for offerors including offering energy products, especially in the situation of increasing problems in this market. Knowledge of the identified distribution of respondents, especially of the aforementioned group, can definitely help managers effectively use the marketing potential of final purchasers in the process of initiating mutually beneficial cooperation when creating a marketing offer including, for example, activities in the scope of promoting renewable energy sources. Their use is part of the concept of responsible consumption. It should be clearly emphasized that, in order to be able to talk about responsible consumption, final purchasers must have a sense of also meeting their different needs through active participation in creating the marketing offer together with the offerors. It is especially important for managers on the energy market because proposed approach was not used before in the case of these products.

Obviously, this research has some limitations. These include, first of all, its scope, the fact that only adults were considered and that the needs satisfied due to the cooperation between final purchasers and offerors were considered in the context of attitudes toward such cooperation. Recognizing these limitations will guide future research that will analyze minors and will attempt to analyze needs satisfied due to cooperation with offerors in other contexts.

Funding: This research received no external funding.

Data Availability Statement: Not applicable.

Conflicts of Interest: The author declares no conflict of interest.

References

1. Grubor, A.; Marić, D. Contemporary consumer in the global environment. In *CBU International Conference Proceedings*; CBU Research Institute: Prague, Czech Republic, 2015; Volume 3, pp. 28–36. [CrossRef]
2. O'Hern, M.S.; Rindfleisch, A. Customer co-creation: A typology and research agenda. *Rev. Mark. Res.* **2010**, *6*, 84–106. [CrossRef]
3. Jacobs, S.B. The energy prosumer. *Ecol. Law Q.* **2016**, *43*, 519–579.
4. Szymański, G. Marketing activities of local food producers in e-commerce. *Sustainability* **2021**, *13*, 9406. [CrossRef]
5. Fernandes, T.; Remelhe, P. How to Engage Customers in Co-Creation: Customers' Motivations for Collaborative Innovation. *J. Strateg. Mark.* **2016**, *24*, 311. [CrossRef]
6. Chandler, J.; Chen, S. Prosumer motivations in service experiences. *J. Serv. Theory Pract.* **2015**, *25*, 220–239. [CrossRef]
7. Hansen, A.V. Value co-creation in service marketing: A critical (re)view. *Int. J. Innov. Stud.* **2019**, *3*, 73–83. [CrossRef]
8. Chatterjee, S.; Rana, N.P.; Dwivedi, Y.K. Assessing consumers' co-production and future participation on value co-creation and business benefit: An F-P-C-B model perspective. *Inf. Syst. Front.* **2021**. [CrossRef]
9. France, C.; Grace, D.; Merrilees, B.; Miller, D. Customer brand co-creation behavior: Conceptualization and empirical validation. *Mark. Intell. Plan.* **2018**, *36*, 334–348. [CrossRef]
10. Schallehn, H.; Seuring, S.; Strähle, J.; Freise, M. Defining the antecedents of experience co-creation as applied to alternative consumption models. *J. Serv. Manag.* **2019**, *30*, 209–251. [CrossRef]
11. Prahalad, C.K.; Ramaswamy, V. Co-creation experiences: The next practice in value creation. *J. Interact. Mark.* **2004**, *18*, 5–14. [CrossRef]
12. Rajagopal, R. *Transgenerational Marketing. Evolution, Expansion, and Experience*; Palgrave Macmillan: Cham, Switzerland, 2020. [CrossRef]
13. Fellesson, M.; Salomonson, N. The expected retail customer: Value co-creator, co-producer or disturbance? *J. Retail. Consum. Serv.* **2016**, *30*, 204–211. [CrossRef]
14. Leclercq, T.; Poncin, I.; Hammedi, W. The engagement process during value co-creation: Gamification in new product-development platforms. *Int. J. Electron. Commer.* **2017**, *21*, 454–488. [CrossRef]

15. García Haro, M.A.; Martínez Ruiz, M.P.; Martínez Cañas, R. The effects of the value co-creation process on the consumer and the company. *Expert J. Mark.* **2014**, *2*, 68–81.
16. Sharma, R.R.; Kaur, B. E-mail viral marketing: Modeling the determinants of creation of 'viral infection'. *Manag. Decis.* **2020**, *58*, 112–128. [CrossRef]
17. Rashid, Y.; Waseem, A.; Akbar, A.A.; Azam, F. Value co-creation and social media: A systematic literature review using citation and thematic analysis. *Eur. Bus. Rev.* **2019**, *31*, 761–784. [CrossRef]
18. Ranjan, K.R.; Read, S. Value co-creation: Concept and measurement. *J. Acad. Mark. Sci.* **2016**, *44*, 290–315. [CrossRef]
19. Ziemba, E.; Eisenbardt, M. Prosumers' eagerness for knowledge sharing with enterprises—A Polish study. *Online J. Appl. Knowl. Manag.* **2014**, *2*, 40–58.
20. Atakan, S.S.; Bagozzi, R.P.; Yoon, C. Consumer participation in the design and realization stages of production: How self-production shapes consumer evaluations and relationships to products. *Int. J. Res. Mark.* **2014**, *31*, 395–408. [CrossRef]
21. Tian, J.; Shen, L.; Chen, Y. A study on customer prosumption concept and its impact on enterprise value co-creation. *Theor. Econ. Lett.* **2017**, *7*, 2040–2053. [CrossRef]
22. Ida, E. The role of customers' involvement in value co-creation behaviour is value co-creation the source of competitive advantage? *J. Compet.* **2017**, *9*, 51–66. [CrossRef]
23. Piligrimiene, Z.; Dovaliene, A.; Virvilaite, R. Consumer engagement in value co-creation: What kind of value it creates for company? *Eng. Econ.* **2015**, *26*, 452–460. [CrossRef]
24. Cheung, M.F.; To, W.M. Customer involvement and perceptions: The moderating role of customer co-production. *J. Retail. Consum. Serv.* **2011**, *18*, 271–277. [CrossRef]
25. Yi, Y.; Gong, T. Customer value co-creation behavior: Scale development and validation. *J. Bus. Res.* **2013**, *66*, 1279–1284. [CrossRef]
26. Dellaert, B.G.C. The consumer production journey: Marketing to consumers as co-producers in the sharing economy. *J. Acad. Mark. Sci.* **2019**, *47*, 238–254. [CrossRef]
27. Gierszewska, G.; Seretny, M. Sustainable behavior—The need of change in consumer and business attitudes and behavior. *Found. Manag.* **2019**, *11*, 197–208. [CrossRef]
28. Kolomiiets, O.; Krzyżanowska, M.; Mazurek, G. Customer disposition to value co-creation activities. *The case of the clothing industry. J. Manag. Bus. Adm. Cent. Eur.* **2018**, *26*, 30–48. [CrossRef]
29. Hsieh, S.H.; Chang, A. The psychological mechanism of brand co-creation engagement. *J. Interact. Mark.* **2016**, *33*, 13–26. [CrossRef]
30. Nadeem, W.; Juntunen, M.; Shirazi, F.; Hajli, N. Consumers' value co-creation in sharing economy: The role of social support, consumers' ethical perceptions and relationship quality. *Technol. Forecast. Soc. Chang.* **2020**, *151*, 119786. [CrossRef]
31. Cao, H.; Zheng, Q.; Wei, F. Classification of customer participation behavior based on multi perspective. *J. Mod. Manag. Sci.* **2013**, *3*, 41–44.
32. Adomaviciute, K. Relationship between utilitarian and hedonic consumer behavior and socially responsible consumption. *Contemp. Mark.* **2013**, *18*, 754–760. [CrossRef]
33. Koltsaklis, N.; Panapakidis, I.P.; Pozo, D.; Christoforidis, G.C. A prosumer model based on smart home energy management and forecasting techniques. *Energies* **2021**, *14*, 1724. [CrossRef]
34. Faia, R.; Pinto, T.; Vale, Z.; Corchado, J.M. Prosumer community portfolio optimization via aggregator: The case of the iberian electricity market and portuguese retail market. *Energies* **2021**, *14*, 3747. [CrossRef]
35. Grönroos, C.; Ravald, A. Service as business logic: Implications for value creation and marketing. *J. Serv. Manag.* **2011**, *22*, 5–22. [CrossRef]
36. Agrawal, A.K.; Rahman, Z. Roles and resource contributions of customers in value co-creation. *Int. Strateg. Manag. Rev.* **2015**, *3*, 144–160. [CrossRef]
37. Dong, B.; Sivakumar, K.; Evans, K.R.; Zou, S. Effect of customer participation on service outcomes: The moderating role of participation readiness. *J. Serv. Res.* **2015**, *18*, 160–176. [CrossRef]
38. Koniorczyk, G. Customer knowledge in (co)creation of product. A case study of IKEA. *J. Econ. Manag.* **2015**, *22*, 107–120.
39. Xie, K.; Wu, Y.; Xiao, J.; Hu, Q. Value co-creation between firms and customers: The role of big data-based cooperative assets. *Inf. Manag.* **2016**, *53*, 1034–1048. [CrossRef]
40. Johlke, M.C.; Iyer, R. A model of retail job characteristics, employee role ambiguity, external customer mind-set, and sales performance. *J. Retail. Consum. Serv.* **2013**, *20*, 58–67. [CrossRef]
41. Baruk, A. The effect of consumers' ethnocentric attitudes on their willingness for prosumption. *Heliyon* **2019**, *5*, e02015. [CrossRef]
42. Lam, S. S-K. Multi-layered identities by social media and prosumption practices in digital and participatory communication. *J. Digit. Media Interact.* **2019**, *2*, 7–22. [CrossRef]
43. Blut, M.; Heirati, N.; Schoefer, K. The dark side of customer participation: When customer participation in service co-development leads to role stress. *J. Serv. Res.* **2020**, *23*, 156–173. [CrossRef]
44. Sugathan, P.; Ranjan, K.R.; Mulky, A.G. Atypical shifts post-failure: Influence of co-creation on attribution and future motivation to co-create. *J. Interact. Mark.* **2017**, *38*, 64–81. [CrossRef]
45. Tu, J.; Zhang, M. Empirical study of the effect of interaction on co-created value in virtual community. *J. Hunan Univ. (Nat. Sci.)* **2013**, *11*, 114–119.

46. Cova, B.; Cova, V. On the road to prosumption: Marketing discourse and the development of consumer competencies. *Consum. Mark. Cult.* **2012**, *15*, 149–168. [CrossRef]
47. Martineau, E.; Arsel, Z. Managing communities of co-creation around consumer engagement styles. *J. Assoc. Consum. Res.* **2017**, *2*, 179–195. [CrossRef]
48. Stokburger-Sauer, N.E.; Wiertz, C. Online consumption communities: An introduction. *Psychol. Mark.* **2015**, *32*, 235–239. [CrossRef]
49. Chu, S.; Kim, Y. Determinants of consumer engagement in electronic word of mouth (eWOM) in social networking sites. *Int. J. Advert.* **2011**, *30*, 47–75. [CrossRef]
50. Schüler, M.; Maier, M.F.; Liljedal, K.T. Motives and barriers affecting consumers' co-creation in the physical store. *Int. Rev. Retail Distrib. Consum. Res.* **2020**, *30*, 289–310. [CrossRef]
51. Milčiuviene, S.; Kiršiene, J.; Doheijo, E.; Urbonas, R.; Milčius, D. The role of renewable energy prosumers in implementing energy justice theory. *Sustainability* **2019**, *11*, 5286. [CrossRef]
52. Inês, C.; Guilherme, P.L.; Esther, M.-G.; Swantje, G.; Stephen, H.; Lars, H. Regulatory challenges and opportunities for collective renewable energy prosumers in the EU. *Energy Policy* **2020**, *138*, 111212. [CrossRef]
53. Moura, R.; Brito, M.C. Prosumer aggregation policies, country experience and business models. *Energy Policy* **2019**, *132*, 820–830. [CrossRef]
54. Kuo, Y.-F.; Feng, L.-H. Relationships among community interaction characteristics, perceived benefits, community commitment, and oppositional brand loyalty in online brand communities. *Int. J. Inf. Manag.* **2013**, *33*, 948–962. [CrossRef]
55. Joshi, A.; Kale, S.; Chandel, S.; Pal, D.K. Likert scale: Explored and explained. *Br. J. Appl. Sci. Technol.* **2015**, *7*, 396–403. [CrossRef]
56. Abdi, H.; Williams, L.J. Principal component analysis. *Wiley Interdiscip. Rev. Comput. Stat.* **2010**, *2*, 433–459. [CrossRef]
57. Ostertagova, E.; Ostertag, O.; Kovač, J. Methodology and application of the Kruskal-Wallis test. *Appl. Mech. Mater.* **2014**, *611*, 115–120. [CrossRef]
58. Rezenkova, H. Cluster analysis of economic data. *Statistica* **2014**, *94*, 73.
59. King, B.M.; Rosopa, P.J.; Minium, E.W. *Statistical Reasoning in the Behavioral Sciences*; Wiley: Hoboken, NJ, USA, 2018.
60. Variance and Standard Deviation. Available online: https://www150.statcan.gc.ca/n1/edu/power-pouvoir/ch12/5214891-eng.htm (accessed on 25 May 2021).
61. Hadi, N.U.; Naziruddin, A.; Ilham, S. An easy approach to exploratory factor analysis: Marketing perspective. *J. Educ. Soc. Res.* **2016**, *6*, 215–223. [CrossRef]
62. Füller, J.; Hutter, K.; Fries, M. Crowdsourcing for goodness sake: Impact of incentive preference on contribution behaviour for social innovation. *Adv. Int. Mark.* **2012**, *11*, 137–159. [CrossRef]
63. Nambisan, S.; Baron, R.A. Virtual customer environments: Testing a model of voluntary participation in value co-creation activities. *Prod. Innov. Manag.* **2009**, *26*, 388–406. [CrossRef]
64. Chen, L.; Marsden, J.R.; Zhang, Z. Theory and analysis of company-sponsored value co-creation. *J. Manag. Inf. Syst.* **2012**, *29*, 141–172. [CrossRef]
65. Aarikka-Stenroos, L.; Jaakkola, E. Value co-creation in knowledge intensive business services: A dyadic perspective on the joint problem-solving process. *Ind. Mark. Manag.* **2012**, *41*, 15–26. [CrossRef]
66. Annuk, A.; Yaïci, W.; Lehtonen, M.; Ilves, R.; Kabanen, T.; Miidla, P. Simulation of energy exchange between single prosumer residential building and utility grid. *Energies* **2021**, *14*, 1553. [CrossRef]
67. Pan, B.; Holland, R. A mass customised supply chain for the fashion system at the design-production interface. *J. Fash. Mark. Manag. Int. J.* **2006**, *10*, 345–359. [CrossRef]
68. Mitręga, M.; Małecka, A. Prosumption among Young Consumers. Some Research Insights from Poland. Available online: https://www.verbraucherforschung.nrw/sites/default/files/2017-10/DOI%2010.15501%20978-3-86336-918-7_22-mitrega-malecka.pdf (accessed on 19 May 2021).
69. Ranjan, K.R.; Read, S. Bringing the individual into the co-creation of value. *J. Serv. Mark.* **2019**, *33*, 904–920. [CrossRef]
70. Oertzen, A.-S.; Odekerken-Schröder, G.; Brax, S.A.; Mager, B. Co-creating services—Conceptual clarification, forms and outcomes. *J. Serv. Manag.* **2018**, *29*, 641–679. [CrossRef]
71. Czakon, W.; Mucha-Kuś, K.; Sołtysik, M. Coopetitive platform: Common benefits in electricity and gas distribution. *Energies* **2021**, *14*, 7113. [CrossRef]

energies

MDPI

Article

Culture-Based Green Workplace Practices as a Means of Conserving Energy and Other Natural Resources in the Manufacturing Sector

Helena Bulińska-Stangrecka [1,*] and Anna Bagieńska [2]

1 Faculty of Administration and Social Sciences, Warsaw University of Technology, Politechniki 1, 00-661 Warsaw, Poland

2 Faculty of Engineering Management, Bialystok University of Technology, 15-351 Bialystok, Poland; a.bagienska@pb.edu.pl

* Correspondence: helena.stangrecka@pw.edu.pl

Abstract: The purpose of this research is to analyze the role of organizational culture in fostering green practices in the workplace while investigating the mediating role of intrinsic motivation in the context of energy conservation. Based on a cross-sectional quantitative study with a sample of 203 employees from the manufacturing sector, the hypothesized relationships were verified. Based on the mediation analysis, statistical analyses revealed positive relationships between organizational culture and green workplace practices, as well as organizational culture and intrinsic motivation. Additionally, the study found that intrinsic motivation mediates the relationship between organizational culture and green workplace practices. This study supported the importance of organizational culture in enhancing green workplace practices aimed at conserving energy and natural resources. The underlying mechanism behind the significant positive effect of intrinsic motivation on proenvironmental behavior in the workplace was also identified. The research demonstrates the importance of an organizational culture to reinforce green practices in the workplace. Furthermore, based on the used models, this study illustrates the importance of each organizational culture dimension: leadership, sense of community, communication, collaboration, and structure in promoting green behavior. Additionally, the research suggests a key role of intrinsic motivation in this relationship. This study provides valuable guidance for the implementation of specific environmental measures in companies.

Keywords: green workplace practices; organizational culture; energy saving; conservation of natural resources; intrinsic motivation

Citation: Bulińska-Stangrecka, H.; Bagieńska, A. Culture-Based Green Workplace Practices as a Means of Conserving Energy and Other Natural Resources in the Manufacturing Sector. *Energies* **2021**, *14*, 6305. https://doi.org/10.3390/en14196305

Academic Editor: Dimitrios A. Georgakellos

Received: 3 August 2021
Accepted: 24 September 2021
Published: 2 October 2021

1. Introduction

Energy is critical to manufacturing sector organizations and often represents a significant cost to both the business and the environment. Taking steps to better manage energy consumption not only saves money for the organization, but also benefits society. One of the objectives of ISO 50001:2018 Energy Management System is, among other things, to develop policies for more efficient energy use and continuous improvement of energy management [1,2]. The measures taken for sustainable energy consumption result from the proenvironmental policy adopted by the company [3,4] or ecoefficiency of production [5] and corporate social responsibility [6]. Increased awareness of resource overexploitation, environmental degradation, and climate change is driving public and business interest in reducing adverse environmental impacts [7]. Ecofriendly behavior in organizations includes programs for recycling, reuse, waste minimization, reducing energy consumption, minimizing water consumption, reducing carbon dioxide generation, and measuring environmental impact [8]. Therefore, organizations implement a strategy of responsible use of not only energy, but also other natural resources. Successful implementation of a corporate strategy that involves the least possible negative impact on the environment and a commitment to conserve natural resources requires the involvement/involvement of all employees. Employees play a key role in putting the adopted strategy into practice because the result achieved is a function of the decisions, behavior, and performance of the members of the organization [9–11].

Green workplace practices as a way to conserve natural resources, including energy, is a trend that enables this strategy from the employees' perspective. Employees play a central role in their organization's green practices because they contribute to the reduction of energy and other resource consumption through the daily implementation of green behaviors in the workplace. Green employee behavior refers to environmentally friendly actions in the workplace such as rational use of resources, e.g., reducing energy consumption, water consumption, recycling, participation in environmental initiatives, etc. [12–14].

Green workplace practices can be defined as the implementation of individual, voluntary proenvironmental employee behavior in in organizations. Moreover, employees' green behaviors are entirely voluntary and are not included in formal job descriptions.

Proenvironmental behavior of the employee pertains to the prosocial dimension of work, because the employee, through his attitude, noticing the possibility of ecological behavior, engaging in various forms of resource conservation, consciously tries to minimize the negative impact of his actions on the environment [15]. Values and beliefs that guide employees in making green choices and assessing the benefits of environmentally friendly actions vary; they can range from hedonic values (which make the employee focus on what makes him feel good), to egoistic values (the employee focuses on actions that increase his resources), and altruistic values (the employee focuses on actions that benefit others) [16]. However, the main driver of proenvironmental behavior is the employee's conviction that what he or she does today has long-term consequences for society and future generations.

The undertaking of proenvironmental actions by an employee may result from his/her intrinsic motivation or extrinsic motivation [17]. The extrinsic motivational factors for proenvironmental activities are the possible personal benefits, rewards, or other consequences resulting from the achievement of, for example, certain standards [18]. An employee may also engage in proenvironmental activities in the workplace primarily for intrinsic satisfaction, e.g., printing double-sided or turning off the computer/light when leaving work, thereby promoting resource conservation. In this case, the source of such voluntary behavior is the employee's intrinsic motivation. Intrinsic motivation can stem from the belief that voluntary green actions are necessary, effective, and appropriate [19,20]. Intrinsic motivation can be a robust predictor of proenvironmental action [18]; therefore, this factor is considered in this study. Environment-friendly behavior is an employee choice, although it is embedded in their daily activities [10,21], so it is important to know the factors that support the implementation of green practices in the workplace.

Previous research on the factors influencing proenvironmental behavior in the workplace have included individual factors, such as employee's personal attitudes and norms [22,23], intraorganizational factors, such as implemented norms and their control [24], and environmental training of employees [25]. Furthermore, it has been emphasized that actions should be taken within the organization to help reinforce the results of the introduced resource-saving programs and initiatives through the developed green innovation strategy [26] sustainable innovation [27,28] and by changing daily habits in the workplace [29]. Employees' taking eco-logical actions at work can be supported or undermined by the organization's culture. Organizational culture is an important contextual factor that affects employees' attitudes in multifaceted ways. Previous research has highlighted that organizational culture and environmental management are closely related [30,31]. An organization can improve its environmental performance by creating a culture that supports environmental management [32,33].

However, this is an as-yet unexplored topic, and research on this topic is still in its nascent stage. As indicated by Norton et al. [20] and Banwo and Du [24], there is a lack of research linking intrinsic employee motivation, organizational culture, and proenvironmental practices in the workplace. The literature on energy conservation in manufacturing primarily focuses on manufacturing technologies [34,35] and energy efficiency processes [36,37], and less on the roles of employees in developing green workplace practices. To address this research gap, our study examines the roles of organizational culture in promoting green behavior. We also examine the relationship of intrinsic motivation to these workplace practices. The purpose of this study is to analyze a research problem concerning the role of organizational culture in facilitating green workplace practices mediated by intrinsic motivation.

The study used empirical data obtained from a questionnaire administered to 203 employees in the manufacturing sector. Comprehensive mediation analysis based on the linear regression method allowed us to verify the research problem.

The contribution of this research is related to the identification of the relationship between organizational culture and employee behavior directed towards energy and natural resource conservation in the workplace. These are, to the best of our knowledge, the first studies on this topic. Additionally, this study provides guidance to managers in designing desirable policies in building green workplace practices based on the various dimensions of organizational culture. Our research indicates which organizational culture dimensions should be focused on in order to promote resource-efficient behavior by employees in the workplace. They therefore help to plan the development of green workplace practices in contemporary organizations.

2. Background

2.1. Green Workplace Practices

Green workplace practices (GWP), also referred to as environmentally sustainable workplace practices, are those activities that result in the efficient use of resources and ensure that environmental degradation resulting from human activities and workplace operations is minimized [38]. GWPs are practices that lead to saving energy and water, minimizing the use of plastic bottles, reducing paper consumption, conducting waste management, and recycling [39,40]. Minimizing energy consumption in the production process refers, among other things, to using energy-efficient machinery and equipment, shortening the production process, and increasing production efficiency. Reducing energy consumption in the workplace, on the other hand, may involve, for example, the use of LED lighting, switching off unnecessary lights, disconnecting equipment before the employee leaves the workplace, etc.

Previous studies show that conserving energy and other resources or using them efficiently reduces the negative impact of energy use on the environment [41–44] and positively affects a company's financial performance. Therefore, paying attention to the efficient use of energy in the workplace can positively affect both the environment and the company.

Research on the determinants of an employee's energy-saving behavior in the workplace shows that it is different than in the household [45–47]. In a household, an individual's behavior to reduce energy consumption leads directly to financial savings. In a company, the importance of monetary savings becomes less important for individual employees because they do not perceive the direct impact as in their household [48,49]. For employees to want to adopt GWPs, organizations must focus on the employee-level determinants/phenomena that will lead to the implementation of sustainable work practices [50,51]. Previous research has indicated that environmental knowledge is a predictor of employee attitudes toward GWPs, but benefits to the organization and the environment will arise when there is a sustained change in attitudes and intentions toward the GWPs being implemented [52]. Therefore, it is important to examine what contributes to employees' post-environmental behavior [53] and green practices in the workplace.

2.2. Organizational Culture

The phenomenon of organizational culture refers to the specific and unique identity of an organization. It is assumed that the culture is a way of organizing the functioning of the company [54], a certain collective programming of employees [55] or meanings that have meaning for members of the company [56]. Organizational culture can be thought of as a pattern of shared assumptions that a group has developed based on its own experiences and that guide the actions of its members [57]. Thus, organizational culture is a unique defining characteristic of an organization and an indication of the distinct ways in which a particular company operates, based on established norms and values, manifested in its organizational processes.

In the scholarly debate, it has become accepted that because of the elusiveness of the concept of organizational culture, the manifestations of it are studied [58]. In particular, studies of culture are concerned with different perspectives of organizational culture analysis, such as verifying different models of culture according to organizational focus and organizational processes (clan, adhocracy, hierarchical, and market) [59], examining the interrelationship of organizational culture and national cultures [60], assigning organizational cultures to separate models based on the degree of risk of their functioning and the speed of information received from the market [54], or following the dynamics of cultural change in digital transformation processes [61]. Despite many existing studies, however, there is still no acceptance of a single model of culture [62]. Hence, there are attempts to develop a framework for analyzing organizational culture which will meet the needs of both scholars and practitioners. The measurement of culture is usually carried out by identifying values [63] or dimensions [64]. According to the functionalistic–systematic paradigm, the dimensions of organizational culture are manifestations of deeper-rooted elements such as values or beliefs that exhibit themselves in specific practices [65]. From this perspective, the analysis of organizational culture through dimensions focuses on the functions and structures of particular activities, which are manifestations of values and shared assumptions. In this approach, organizational culture is a specific organizational context that affects the management of the organization [66]. In this study, we examine the dimensions of organizational culture that have been identified based on previous empirical research [67]. The investigated dimensions were developed on the basis of studies of organizational cultures that were conducted, among others, in the manufacturing sector [68,69]. Based on this process approach to organizational culture, seven dimensions of organizational culture were identified, which allow to diagnose key areas of company functioning: sense of community, strategic orientation, leadership, team collaboration, communication, team structure, and informal relationships [64]. The each of these aforementioned dimensions is an important element that has been used in empirical analyses of organizational cultures. Sense of community refers to the positive atmosphere that characterizes an organization in which its members feel integrated and feel a sense of participation in the workplace community [70–72]. Strategic orientation has also been selected as a determinant of organizational culture dimensions in several published empirical studies [57,73–75]. Likewise, the leadership dimension has been used repeatedly as a measure of organizational culture [76–78]. Similarly, teamwork is a well-established indicator of organizational culture [79–82]. Communication as a dimension of culture allows for effective diagnosis of information dissemination practices within an organization [69,80,83–89]. Furthermore, many of the extant studies have confirmed that team structure is an important distinguishing feature of organizational cultures [71,90,91]. Moreover, interpersonal relationships among organizational members are a validated factor of organizational culture [88]. Thus, in conclusion, the above dimensions provide a basic framework for process analysis of an organization's culture.

Extant research confirms that an organization's culture is critical to achieving effectiveness [92,93]. Alignment between values and culture ensures greater performance because employees feel integrated into the established norms [94]. It has been demonstrated that the organizational culture also supports the development of a competitive advantage for the company [95]. Moreover, as Linnenluecke and Griffiths' [63] research has confirmed, organizational culture also affects the sustainability of organization. Indeed, a systematic analysis of the literature has shown that there are ten elements that link the sustainability of an organization to its culture [96]. In addition, a recent study of 137 companies found that organizational sustainability is linked to organizational culture [97]. Moreover, studies conducted in manufacturing companies confirm that organizational culture is related to environmental performance [98]. Therefore, in light of the above-described relationships, it seems reasonable to hypothesize a positive association between organizational culture and GWP. It can be hypothesized that an organizational culture focused on conservation of energy and natural resources will support employees' green behavior. Such a culture that is oriented towards environmental care and conservation of natural resources can be referred to as a green organizational culture. Supporting environmental goals through the organization's culture aligns employees' environmental practices. Thus, the accepted hypothesis is:

Hypothesis 1 (H1). *(Green) Organizational culture is positively related to GWP.*

2.3. Intrinsic Motivation

Ryan and Deci (2000) emphasize that motivation activates energy required to initiate action [99]. In organizational terms, motivation influences the intensity, duration, direction, and persistence of work behaviors [100]. There are different theoretical approaches to motivation. According to self-determination theory (SDT), an individual's motivation depends on whether it is caused by extrinsic (e.g., financial incentives) or intrinsic factors (e.g., dedication to work, passion) [101]. SDT stresses that if an individual enjoys a task and is interested in it, he/she will do it with more commitment and perseverance [102].

SDT explains how employees perform tasks because they are interesting and enjoyable to them [101]. Tasks driven by intrinsic motivation are perceived by employees as autonomous, whereas extrinsically motivated tasks tend to be perceived as controlled and imposed [103]. Therefore, employees who are intrinsically motivated to perform a given task will be more persistent, regardless of extrinsic motivation factors. Altogether, SDT emphasizes the role of intrinsic motivation, shaped by the performance of interesting tasks, and related to the commitment and persistence of employees in completing them. Through the lens of SDT, stimulating intrinsic motivation is an important challenge to not only increase the effectiveness of attitude formation but also to reinforce new habits. Moreover, from an SDT perspective, the workplace is a context that can effectively support intrinsic motivation [99]. Therefore, providing support from the environment, such as the organizational culture, to inspire interest in a given task can increase intrinsic motivation and, consequently, improve the performance of that task. Thus, the SDT perspective indicates the role of organizational culture in enhancing employees' intrinsic motivation to achieve specific goals.

Previous research indicates that intrinsic motivation is associated with higher levels of goal attainment and job satisfaction [101]. In summary, when an employee is intrinsically motivated, he or she self-regulates his or her performance, and his or her dedication to the task at hand is high. In the workplace, intrinsic motivation is understood as a specific type of autonomous motivation that is characterized by an employee's commitment to action with a full sense of desire, will, and choice [104]. Moreover, intrinsic motivation is characterized by a strong assessment of personal commitment [105].

It pertains to actions for which motivation is directed from one' s behavior. With respect to intrinsic motivation, it is the spontaneous experiences of enjoyment associated with a particular behavior themselves that provide the "rewards" [104]. Employees can be intrinsically motivated for at least some of their work, if not all of it. Intrinsically motivated individuals produce high-quality job performance. Intrinsic motivation also contributes to wellbeing [104,106].

Given that the culture of an organization provides an important context for employee actions, it can also affect employee motivation. Previous research has shown that organizational culture influences attitudes toward work [107]. Therefore, it seems that organizational culture could provide a structure to support the internalization of certain attitudes that lead to the performance of tasks with greater dedication and perseverance. For example, research in the education sector has demonstrated that organizational culture supports increased employee motivation [108]. Therefore, in the light of SDT theory, the culture of an organization is the context that shapes employees' attitudes and can enhance their intrinsic motivation. Building on SDT's concept of contextual influence on motivations for action, we propose that employees will engage more in green workplace practices if the organizational culture reinforces proenvironmental attitudes. This leads to the hypothesis:

Hypothesis 2 (H2). *(Green) Organizational culture is positively related to employees' intrinsic motivation.*

The next issue for consideration is the relationship of intrinsic motivation to green workplace practices. According to SDT, those employees who are more convinced of the need for natural resource conservation initiatives, and have an intrinsic interest in not doing so, will be more likely to practice green behaviors at work. Empirical research proves that there is a relationship between intrinsic motivation and environmental action [109]. Moreover, the literature demonstrates the verified role of intrinsic motivation in shaping green creativity [110] and proenvironmental behavior at work [111]. Interest and passion, or perceiving an issue as enjoyable, forms the basis of intrinsic motivation, which activates individuals to pursue given actions [104]. Thus, employees interested in environmental issues should be more inclined to undertake and sustain environmental activities. Based on the above discussion, employees who are intrinsically motivated will be more committed to implementing green practices in the workplace:

Hypothesis 3 (H3). *Employees' intrinsic motivation is positively related to GWP.*

Top management support and organizational culture are key variables that influence employees' behaviors regarding environmental issues in the workplace [49,112].

Empirical studies show that through the mediating mechanism of green intrinsic motivation, leadership influences green behavior [111,113], and through the mediating role of intrinsic motivation, green transformational leadership influences green creativity [110]. Venhoeven, Bolderdijk, and Steg's [114] empirical findings suggest that decision-makers can use people's intrinsic motivation to promote sustainable actions. Therefore, we hypothesize that intrinsic motivation is the underlying mechanism mediating the relationship between an organization's culture supporting resource conservation and GWP.

Hypothesis 4 (H4). *Intrinsic motivation mediates the relationship between organizational culture and GWP.*

3. Materials and Methods

3.1. Sampling and Research Context

The survey was conducted among employees in the manufacturing sector. Manufacturing is one of the four major types of industries detailed in the Annual Energy Outlook Reference 2021 [115]. Information on the energy consumption of U.S. manufacturing plants shows that manufacturing consumes 77% of all energy used by industry [116]. In the European Union, manufacturing consumes one-third of total energy consumption [117]. Although the final consumption of manufacturing industries in EU 28 has decreased from 323.4 Mtoe in 2000 to 271.5 Mtoe in 2018 [118], it still needs to be reduced.

In the manufacturing sector, energy consumption includes both the production process and energy used for the general and administrative needs of the business. Studies show that energy consumption in nonresidential buildings in Europe has increased by 74% over 20 years [119]. In contrast, the use of computers in the workplace has contributed to almost 30% of the increase in energy consumption in offices [120]. Reducing energy demand in workplaces is important in reducing total energy consumption [121,122].

It is very important to reduce energy consumption in this sector, mainly because of the impact of consumption on the environment and the efficiency of operations. The introduced standard ISO 50001:2018 Energy Management System indicates that energy management should be concerned with the efficient management of energy in all forms. In addition, ISO 50001:2018 indicates an important need to increase the role of top management and emphasizes the role of the organization's culture in implementing this standard into practice [1].

The questionnaire was conducted online. A self-administered survey was used for data collection. It was preceded by a pilot study conducted on 12 employees from a small business organization. The feedback from this pilot study helped to adjust the questionnaire to make sure that it was understandable.

The study was conducted online in July 2021, due to the prevailing COVID-19 pandemic and imposed restrictions. The target group for this questionnaire included employees in the manufacturing industry in Poland. A purposeful sampling method was used [123]. This method is based on obtaining and selecting respondents who are rich in information with respect to the issue being analyzed [124]. It involves identifying individuals who have sufficient knowledge and experience to provide reliable answers to questions [125]. The goal of this method is to obtain information that is specialized and contextualized. The survey was addressed to employees of manufacturing companies, currently employed and of legal age, who agreed to participate anonymously in the study. The authors sent an email to HR managers of manufacturing companies operating in Poland with background information and a link to an online questionnaire. After being approved by HR managers, the link to the questionnaire was sent to employees. A total of 203 employees of manufacturing companies participated in the study. The profile of respondents consists of 35.5 per cent females and 64.5 per cent males. Of those surveyed, 36% have a high school diploma, 23.2% have a bachelor's degree, 10.3% have a master's degree, 4.9% have an MBA, and 15.8% have completed an engineering degree. The sample included employees of different ages. A total of 29.2% of the respondents were up to 25 years old, 35.7% were 26–35 years old, 24.2% were 36–50 years old, and 10.9% were over 50 years old. The following positions were represented among the respondents: top management 6.4% (13), middle management 28.6% (58), assistant 21.7% (44), specialist 14.3% (29), and other 29.1% (59).

3.2. Measures

The questionnaire consisted of two parts. The first included questions about the demographic data of the participants such as gender, age, education, and job position. The second part of the questionnaire (Appendix A) contained constructs to examine the variables in the study.

The measurement of the independent variable (X) of organizational culture was prepared on the basis of an empirically verified instrument [64]. The dimensions used in this measure were previously examined in an empirical study of IT companies. The quantitative survey used to identify the organizational culture dimensions was conducted on a sample of 92 IT organizations, randomly selected from the 302 best IT organizations operating in Poland. The data obtained from the questionnaire were analyzed using exploratory factor analysis [67]. This method allows the identification of real descriptive factors reflecting the data. Based on exploratory factor analysis, seven dimensions were identified for analyzing organizational culture These were then confirmed using a qualitative method in the form of interviews. The qualitative research was conducted on a group of 20 respondents representing the surveyed IT organizations, employed in managerial and specialist positions. The analysis of the content of interviews has allowed to positively verify the model of organizational culture obtained as a consequence of quantitative research [67]. Such application of both quantitative and qualitative methods in diagnosing the dimensions of culture is a recognized research practice [126,127] that permits a comprehensive understanding of the ephemeral phenomenon of organizational culture [57]. The dimensions of this instrument, described in detail above, are used as indicators that diagnose organizational culture. The identified dimensions for the measurement of organizational culture align with the areas by which organizational cultures are analyzed: sense of community [70–72], strategic orientation [57,73–75], leadership [76–78], team collaboration [79–82], communication [69,80,83–89], team structure [71,91,92], and interpersonal relationships [88]. Therefore, this tool is suitable for studying the characteristics of organizational culture. Moreover, some of the dimensions have been previously applied to the study of the culture in the industrial sector [62,63], which indicates that their relevance to this study can be assumed.

The first step in examining the dimensions of organizational culture was initially to diagnose from the extant literature the areas that might accurately characterize organizational culture. Subsequently, questionnaire questions were constructed for evaluation by respondents. After carrying out the pilot study, the measure was used in the actual study. The questions were adapted to the context of the study concerning energy conservation and waste-free use of natural resources (Appendix A).

The independent variable (Y) of GWP was measured using an instrument adapted from an existing questionnaire [128]. It consisted of 10 questions illustrating behaviors regarding saving energy and taking actions that conserve natural resources. Individual work characteristics lead to the availability of different proenvironmental behaviors [122], hence this study selected the most frequently mentioned workplace practices for conservation of natural resources [129,130], as well as general formulations on the use of environmentally friendly practices. The precise formulation of the questions enabled the examination of GWPs and the assessment of the use of specific actions in practice compared, for example, to the approach presented by Bissing-Olson et al., which uses only a general measurement of employee engagement in the initiative to act in environmentally-friendly ways at work [131]. In addition, the precise formulation of questions makes it easier for respondents to give accurate answers and reduces the level of ambiguity, contributing to more reliable results [132].

The measurement of the mediator (M) of intrinsic motivation was based on a measure developed by Fagan, Neill, and Wooldridge [132]. Individual items in this measure are consistent with selected questions of the interest/enjoyment subscale, considered a self-report measure of intrinsic motivation, from the Intrinsic Motivation Inventory [133]. The use of self-reports of interest and enjoyment of activities, per se, is a common approach to measuring extrinsic motivation [134]. Intrinsic motivation was operationalized by reference to workplace research [135]. In addition, referring to the concept of intrinsic motivation, the selected questions are those that pertain to interest in a particular aspect of work, as interest and enjoyment are indicated as inherent components of intrinsic motivation [103]. It permits to emphasize the motivation originating from interest or enjoyment in the activity rather than from expectation of reward, recognition, or expected performance improvement. Although the instrument chosen is relatively short, it has been successfully used in empirical studies [135]. Moreover, the choice of a short questionnaire was dictated by the need to take into account practical aspects of the questionnaire, such as the time needed to complete them. Long questionnaires can lead to reduced data quality due to respondents' fatigue and decreased cognitive engagement [136]. The questions were developed using a five-point Likert scale (Appendix A).

3.3. Measurement Model

Considering the fact that the present research analyzes the perceptions of the investigated variables by employees in the manufacturing sector and the questionnaires were self-reported, this may have led to problems of common method variance [137]. According to Podsakoff, MacKenzie, and Podsakoff [138], performing certain statistics procedures may provide a remedy to detect whether this problem actually occurred in the study. In this study, anonymity and confidentiality of data were ensured for the purpose of this study, as anonymity can reduce the bias [139]. In addition, thorough instructions were ensured during the design of the questionnaire [132]. We also applied an ex-post remedy by using a Harman's single-factor test [140]. Analysis of all indicators loaded into a single factor showed that together they account for 41.35% of the variance. According to the recommended limit, it should be less than 60% [141]. Therefore, the results suggest that common method variance is not a problem in this study.

Model validation was conducted as recommended by MacKenzie, Podsakoff, and Podsakoff [142], and includes an assessment of the goodness-of-fit of the model, validity assessment, and construct reliability. Subsequently, a confirmatory factor analysis (CFA) was carried out according to standard procedures [143], the results of which are shown in Table 1. As recommended by Kline [144], the following goodness-of-fit measures of the model were used: chi-square and degrees of freedom, RMSEA, CFI, and SRMR. The results obtained indicated a good model fit.

Table 1. CFA results.

Factor	Indicator	Estimate	SE	*p*	Stand. Estimate
Organizational culture	oc1	1.130	0.0652	<0.001	0.931
	oc2	1.138	0.0634	<0.001	0.949
	oc3	0.881	0.0731	<0.001	0.741
	oc4	0.972	0.0745	<0.001	0.784
	oc5	0.990	0.0757	<0.001	0.785
	oc6	0.900	0.0810	<0.001	0.699
	oc7	0.959	0.0773	<0.001	0.753
Intrinsic motivation	im1	0.728	0.0541	<0.001	0.817
	im2	0.849	0.0550	<0.001	0.902
	im3	0.568	0.0523	<0.001	0.700
Green workplace practices	gwp1	0.523	0.0483	<0.001	0.687
	gwp2	0.594	0.0528	<0.001	0.708
	gwp3	0.613	0.0545	<0.001	0.708
	gwp4	0.505	0.0552	<0.001	0.603
	gwp5	0.597	0.0529	<0.001	0.710
	gwp6	0.683	0.0566	<0.001	0.745
	gwp7	0.622	0.0502	<0.001	0.760
	gwp8	0.719	0.0488	<0.001	0.851
	gwp9	0.567	0.0539	<0.001	0.673
	gwp10	0.554	0.0628	<0.001	0.585
CFA Goodness-of-Fit Statistics (Overall model fit indices)					
χ^2 = 488 (p = <0.001); df = 167; χ^2/df = 2.92; CFI = 0.923; RMSEA = 0.077; SMRS = 0.621					

Notes: SE: standard error; *p*: significance; χ^2: chi-square; df: degrees of freedom; CFI: comparative fit index (CFI); SMRS: standardized root mean residual.

To test the model, an analysis of average variance extracted (AVE) of constructs and convergent validity and composite reliability was conducted, using questions as suggested by Hair et al. [143]. The reliability of the questions was measured by investigating the loadings for each. It was found to be adequate because all loadings are equal to or above 0.5 [143]. The results of composite reliability and AVE are presented in Table 2.

Table 2. Evaluation of the measurement model.

	Factor Loadings	Cronbach's Alpha	AVE	CR
Organizational culture	0.931 0.949 0.741 0.784 0.785 0.699 0.753	0.935	0.657562	0.92997
Intrinsic motivation	0.817 0.902 0.7	0.843	0.657031	0.85046
Green workplace practices	0.687 0.708 0.708 0.603 0.71 0.745 0.76 0.851 0.673 0.585	0.876	0.347483	0.872928

Notes: The construct items are used to explain the construct. AVE: average variance extracted; CR: composite reliability.

Although in one variable (green workplace practices) the average variance extracted is below 0.5, it is acceptable because the convergent validity value is above 0.6 [145]. Therefore, the model analysis is acceptable and allows further operations on the data.

Discriminant validity was assessed by analyzing the correlations between constructs and the variances and covariances of the constructs. Each square root of the AVE of every variable should be higher than the correlation between each construct [145]. The analysis performed confirmed discriminant validity for all the constructs (Table 3).

Table 3. Square root of the average variance extracted (AVE) and correlations matrix.

	GreenWP		IntrinsicMotiv		Culture
GreenWP	(0.811)				
IntrinsicMotiv	0.623	***	(0.809)		
Culture	0.543	***	0.476	***	(0.589)

Note: *** $p < 0.001$. In the parenthesis are the square root of AVE.

The verification of the hypotheses adopted in the study was based on the Hayes [146] PROCESS procedure. The PROCESS macro in the R studio [147] program was used. The use of macros enabled to conduct advanced analysis of mediation using bootstrapping and estimation of confidence intervals.

The study used a bootstrapping approach of 5000 bootstrapped samples. According to the method, the effect of organizational culture as an independent variable on green workplace practices dependent variable was verified by considering intrinsic motivation as a mediator of this relationship. Based on the method developed by Hayes [146], the total, direct, and indirect effects of this relationship were analyzed. A 95% confidence interval (CI) analysis was used as an indicator of the statistical significance of the calculation. When the difference between the lower (LLCI) and upper (ULCI) confidence interval does not contain zero, such mediation test is assumed to be statistically significant. In this study, a partial mediation assumption was made. Thus, it was assumed that the indirect effect βyx.m does not fall below zero, and the mediation analysis is statistically significant (p level).

4. Results

4.1. Hypothesis Testing

The first step concerning the analysis was the verification of hypothesis 1—(Green) organizational culture is positively related to GWP. It was conducted based on linear regression analysis. This analysis showed that the organizational culture has a significant impact on the green workplace practices ($\beta = 0.358$; ($F(1,201) = 83.928$; $p < 0.001$). Furthermore, it demonstrated that organizational culture explains 29 percent of variance ($R^2 = 0.294$) in green workplace practices. Thus, hypothesis H1 was positively verified.

Furthermore, a one-way ANOVA was conducted to assess the relationship of various organizational culture dimensions with green workplace practices. The analyses revealed statistically significant differences between the following dimensions of organizational culture and the degree of declared application of green workplace practices: sense of community ($F(4,198) = 13.8$, $p < 0.001$), strategic orientation ($F(4,198) = 15.7$, $p < 0.001$), leadership ($F(4,198) = 13.2$, $p < 0.001$), team collaboration ($F(4,198) = 16.2$, $p < 0.001$), communication ($F(4,198) = 14.8$, $p < 0.001$), structure of teams ($F(4,198) = 9.28$, $p < 0.001$), informal relationships ($F(4,198) = 17.5$, $p < 0.001$).

Post hoc comparisons using the Turkey test found that mean score for sense of community differs significantly at the lowest level of green workplace practices (MD = 3.44, SD = 0.871) and high level of green workplace practices (MD = 4.54, SD = 0.492). Thus, declarations of a low sense of community are statistically associated with low practice of green workplace practices, while high sense of community is associated with the highest use of green practices.

Additionally, post hoc tests showed that there was significant differentiation in the level of strategic orientation to a of green workplace practices. While high levels of green practices (MD = 3.45, SD = 0.851) are associated with high levels of strategic orientation, low levels of strategic orientation are associated with low levels of green practices (MD = 4.61, SD = 0.523).

Leadership also differentiates the level of green workplace practices. The tests show a significant difference between low levels of green practices (MD = 3.37, SD = 0.857) and the highest (MD = 4.49, SD = 0.482) and the levels of leadership.

In turn, team collaboration also differentiates the mean level of green workplace practices. Statistically, a differentiation is observed between team collaboration and the lowest level (MD = 3.33, SD = 0.892) of green practices and the highest level (MD = 4.49, SD = 0.471).

Furthermore, the communication also differentiates between the lowest (MD = 3.51, SD = 0.793) and highest (MD = 4.53, SD = 0.564) levels of green practices in the workplace.

The average level of the team structure also indicates a statistically significant differentiation in the level of green practices. The differences occur especially between the lowest level (MD = 3.53, SD = 0.809) of green practices and the highest (MD = 4.23, SD = 0.688) in relation to the mean value of the team structure.

Finally, team relationships also differentiate green practices. Post hoc tests demonstrate that high levels (MD = 3.51, SD = 0.861) of green practices are appropriate for high levels of relationships, while low levels of green practices (MD = 4.48, SD = 0.505) are for low levels.

The analysis of the obtained data confirms that each dimension of organizational culture (sense of community, strategic orientation, leadership, communication, teamwork, team structure and relationships) significantly differentiate the use of green workplace practices. A high level of each of the organizational culture dimensions discussed above is indicative of high implementation of green practices.

Further regression analysis results also confirmed the relationship between organizational culture and intrinsic motivation (H2: *(Green) Organizational culture is positively related to employees' intrinsic motivation*) ($\beta = 0.353$; ($F(1,201) = 58.806$; $p < 0.001$; $R^2 = 0.226$). The hypothesis assuming the relationship between intrinsic motivation and GWP was also positively verified (H3: *Employees' intrinsic motivation is positively related to GWP*) ($\beta = 0.556$; ($F(1,201) = 128$; $p < 0.001$; $R^2 = 0.389$).

4.2. Mediation Analysis

Hypothesis 4 (intrinsic motivation mediates the relationship between organizational culture and GWP) was verified according to the Hayes procedure described above [146]. Detailed results of the mediation analysis are presented in Table 4. The results of the mediation analysis indicate that the total effect of the study is statistically significant (βyxm = 0.359; LLCI = 0.282; ULCI = 0.436; $p < 0.001$). Moreover, the direct effect was also statistically significant, but its value was reduced in comparison to the total effect (βyx = 0.210; LLCI = 0113; ULCI = 0.286); $p < 0.001$). The analyzed model explains almost half of the variance in green workplace practices ($R^2 = 0.467$). In addition, the static significance is confirmed by the ratio of the indirect to total effect of X on Y: β = 0.413; LLCI = 0.088; ULCI = 0.258.

Table 4. Total, direct, and indirect links between organizational culture and green workplace practices through intrinsic motivation.

Effect [β]	SE	t	p	Bootstrap 95% Confidence Interval (CI) LLCI	ULCI
Total effect (βyx): organizational culture (X) on green workplace practices (Y)					
0.359 Fp = 87.590 *** $R^2 = 0.467$	0.039	9.161	<0.001	0.282	0.436
Direct effect: organizational culture (X) on green workplace practices (Y)					
0.210	0.038	5.420	<0.001	0.134	0.286
Indirect effect (βyx.m) organizational culture (X) on green workplace practices (Y) through intrinsic motivation (M)					
Intrinsic motivation 0.148	0.030			0.133	0.286

Notes: *** $p < 0.001$; lower-level confidence interval (LLCI); upper-level confidence interval (ULCI); number of bootstrap samples for bias corrected bootstrap confidence intervals: 5000; level of confidence for all confidence intervals in output: 95%. N = 203.

In summary, the results of the mediation analysis carried out indicate a mediation analysis with 5000 bootstrapped samples on statistically significant total (βxm = 0.148; LLCI = 0.133; ULCI = 0.286; $p < 0.001$), direct (βyx = 0.210. LLCI = 0113; ULCI = 0.286), and indirect effect (direct (βyx.m = 0.148; LLCI = 0133; ULCI = 0.286). The interrelationship is illustrated in Figure 1. Additionally, the kappa-squared value for the indirect effect is 0.236, which suggests a large mediating effect [148].

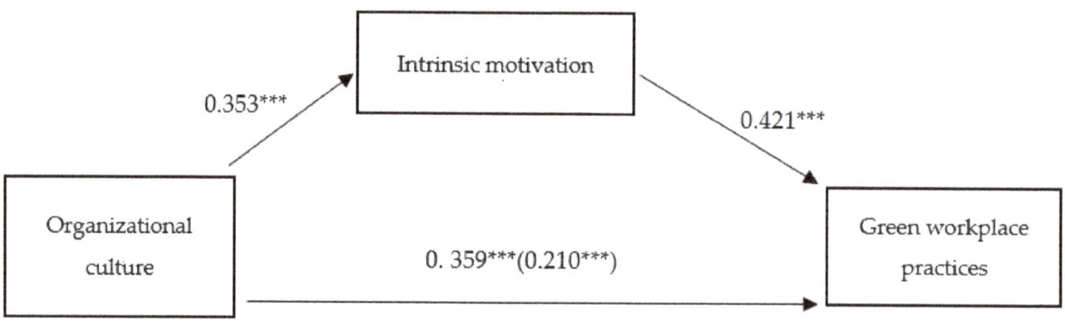

Figure 1. Parallel mediation model (N = 203). Indirect effects of organizational culture on green workplace practices through intrinsic motivation. Standardized effects estimates are presented. The effects on the direct path from organizational culture to green workplace practices depict the direct effect and the total effect (in brackets). *** $p = 0.210$, $p = 0.359$, $p < 0.001$.

Taken together, the results of the mediation analysis discussed above demonstrate that organizational culture is statically significant in supporting workplace green practices.

Intrinsic motivation plays an important role in mediating this relationship and explains almost half of the (46.7%) variance in GWP.

The obtained results provided positive verification of hypothesis 1— *(Green) Organizational culture is positively related to GWP* and hypothesis 4—*Intrinsic motivation mediates the relationship between organizational culture and GWP*. More importantly, the findings provide support for the assumption that implementing green practices in the workplace is based on the organization's culture, through intrinsic motivation.

5. Discussion

This study has several important findings for both the theory and practice of organizational management. Firstly. this study found that organizational culture has a significant positive impact on GWP in a manufacturing sector organization. This is consistent with the findings of Young et al. [112], which indicate that organizational culture is a key driver of employee behavior regarding environmental issues. In a similar vein to the results of Bakhsh et al. [98], this study demonstrates that the effective formation of environmental behavior among employees in the manufacturing sector is based on the organizational culture of a particular company. Moreover, the obtained results allow us to assume that organizational culture can also be a major barrier to the implementation of energy and natural resource conservation policies in the company. Thus, this supports the findings of Trianni et al. [149], suggesting that the lack of alignment of organizational culture with green actions is a source of resistance to the introduction of environmentally friendly practices. However, this study is the first to point to specific culture dimensions that support GWP implementation. As the findings indicate, both a sense of community with respect to green initiatives and strategic orientation towards green goals, as well as leadership support for green actions, cooperation oriented towards environmentally friendly initiatives, communication of the need for green actions within the organization, and structural support for environmental solutions, including informal relationships, significantly support GWP in companies. Moreover, the study is the first to identify the core features of an organizational culture for successful implementation of GWP from a process perspective.

Secondly, this study also confirms the important role of intrinsic motivation, which mediates the relationship between organizational culture and GWP. In this aspect, this study corresponds to Venhoeven, Bolderdijk, and Steg's [114] research, which confirms the role of intrinsic motivation in promoting sustainable initiatives. People who are environmentally conscious, including in their work, are more likely to help save the natural resources, conserve energy, and engage in environmentally friendly activities. Employees undertake these efforts not because of external rewards, but because it gives them satisfaction. They feel what is referred to as a "helper high" [150]. Additionally, if employees believe in the need for environmentally friendly actions, they are more committed to them, and the change in their attitude is more permanent [151]. Therefore, these results suggest that it is through the implementation of appropriate policies in the specific dimensions of an organizational culture (sense of community, strategic orientation, leadership, team collaboration, communication, structure, and informal relationships) underpinned by undertaking measures that support intrinsic motivations for environmentally friendly behavior, that the successful implementation of GWP is possible.

Thirdly, addressing only the day-to-day activities related to saving energy and other resources in their workplaces is important but not sufficient. It is much more important to understand the reasons that influence the long-term implementation of environmentally friendly practices in companies. This research provides clear support for what measures are needed to effectively implement GWP among employees. In line with the model of Lulfs and Hahn (2013), which suggests that organizational context and individual circumstances, perceived behavioral control, personal norms, attitudes, intentions, and habits are key predictors of proenvironmental behavior in the workplace [152], these results demonstrate that the role of employee attitudes should be considered in the successful implementation of green policies. Our research shows which specific management interventions for shaping organizational culture processes can support the employees' environmental behavior and how to reinforce the desired behavioral patterns of employees concerning the conservation of energy and natural resources.

Finally, in manufacturing companies, energy efficiency is characterized by a rather interdisciplinary set of organizational problems [153]. Abdelaziz et al. [154] discussed three pathways to improve energy efficiency: regulation, technology, and management. Energy and environmental management play an important role in contemporary challenges for manufacturing companies [155–158]. This study addresses this issue and illustrates how to support GWP.

5.1. Theoretical Contribution

Our study extends the body of research on organizational culture and broadens understanding of the role of culture in shaping employees' attitudes. This study contributes to the emerging debate on how to influence employees' intrinsic motivation through the formation of organizational culture. Building on SDT theory in the context of organizational behavior, the findings raise an important point, supporting the notion that organizational culture significantly influences employees to undertake green practices. Thus, they point to contextual sources that foster the adoption of individual ecological actions. They also emphasize the mechanism of cultural drivers affecting green behavior through intrinsic motivation. From a theoretical perspective, the present investigation is consistent with the findings of Faraz et al. [111] and Li et al. [110], whose research indicates the role of intrinsic motivation in workplace environmental initiatives. Moreover, the results of the study make a significant contribution to the literature by demonstrating that particular dimensions of organizational culture (sense of community, strategic orientation, leadership, team collaboration, communication, team structure, and informal relationships) help promote proenvironmental behavior. Thus, in order to foster green attitudes among employees, actions must be taken in each of these seven dimensions of organizational culture, potentially enhancing not only GWP but also intrinsic motivation to conserve natural resources.

5.2. Practical Contribution

With the growing importance of green employee behavior in organizations today, supporting green practices in the workplace is an important aspect of organizational management [10,11]. The organizational culture can be instrumental in increasing the green work behavior of employees. This study has significant practical implications by showing that aligning the various dimensions of organizational culture leads to successful implementation of the GWP. In addition, it emphasizes the role of employees' intrinsic motivation in undertaking green actions in the workplace. Saving energy and conserving natural resources is a major challenge for modern organizations. By highlighting concrete practices, this study provides valuable guidance for the implementation of specific environmental measures in companies. In particular, this can be accomplished by creating a green atmosphere and fostering a sense of awareness of the importance of environmental issues among employees. Companies should also cultivate a strong attitude towards conserving natural resources. Moreover, by running internal campaigns to promote a green mindset, organizations can foster internal motivation among employees to undertake green initiatives and sustain GWP.

5.3. Limitations and Future Research

This study provides an interesting and useful theoretical framework for understanding the mechanisms that facilitate proenvironmental behavior. However, this study has some limitations. Although it establishes an important framework for promoting GWP based on organizational culture mediated by employee intrinsic motivation, there may be other variables that can also influence this relationship. Future studies may address other potential factors affecting GWP. Another limitation is that the data source was a survey, which means that caution must be taken when generalizing the results. Still another limitation of this study is the use of a measure that was constructed by the authors. This may provide some constraints in generalizing the results, and therefore this measure needs further empirical verification. Moreover, in order to strengthen the validity of the results, it is recommended that more longitudinal studies and qualitative research be carried out to investigate in-depth employee attitudes. It will be particularly interesting to undertake future qualitative research to capture respondents' individual perspectives on the relationship between organizational culture and GWP. The use of qualitative interviews allows one to understand the perceptions of individual respondents [122] and provides in-depth insight into the analyzed phenomenon. The different employee approaches to GWP can also be analyzed in terms of intrinsic and extrinsic motivation. Examining the relationship between organizational culture and GWP in other sectors may also be an important direction for future research.

6. Conclusions

This study set out to examine the contextual factors that promote GWP in the industrial sector. Moreover, it focused on the critical factors that facilitate influence on employees' attitudes and behaviors. The study demonstrated that organizational culture has a significant positive impact on GWP. In addition, our findings highlight the role of intrinsic motivation, which strengthens GWP. In conclusion, this study empirically examined some unexplored aspects of energy policy implementation in manufacturing organizations. It established the relationship between organizational culture and GWP in manufacturing companies. Additionally, it pointed out the intrinsic motivation mechanism that mediates the link between organizational culture and GWP. In essence, this study helps managers in the manufacturing industry who are interested in promoting GWP by identifying specific areas of organizational culture that enhance employees' green behavior.

Our paper and its proposed theoretical framework bridge the gap in the literature regarding the relationship between organizational culture, intrinsic motivation, and GWP [20,24]. Moreover, it brings the analysis of employees' proenvironmental behavior to the manufacturing context. This research has resulted in the development of a new instrument to study the green behavior of employees.

The findings are particularly relevant to the measures taken by companies. This applies especially to the introduction of appropriate regulations, which can reinforce the desired environmental behavior of employees [14,159]. This framework can be effectively used by both scholars and practitioners to increase GWP. The growing importance of supporting green activities in organizations further emphasizes the need to understand the mechanisms that promote GWP in the manufacturing context.

Author Contributions: Conceptualization: H.B.-S. and A.B.; methodology: H.B.-S.; software: H.B.-S.; validation: H.B.-S. and A.B.; formal analysis: H.B.-S.; investigation: H.B.-S.; resources: H.B.-S. and A.B.; data curation: H.B.-S. and A.B.; writing—original draft preparation: H.B.-S. and A.B.; writing—review and editing: H.B.-S. and A.B.; visualization: A.B.; funding acquisition: H.B.-S. and A.B. All authors have read and agreed to the published version of the manuscript.

Funding: This research is supported by: Warsaw University of Technology—Open Science Program IDUB and Bialystok University of Technology—grant number WZ/WIZ-INZ/1/2020.

Institutional Review Board Statement: Not applicable.

Informed Consent Statement: Not applicable.

Data Availability Statement: Data are available on request from the corresponding author.

Conflicts of Interest: The authors declare no conflict of interest.

Appendix A

Instrument for the construct of (green) organizational culture. Intrinsic motivation and green workplace practices

Organizational culture

1. In my organization, everyone as a community feels a responsibility to conserve natural resources.
2. In my organization, there is a strategy that supports the conservation of natural resources.
3. In my organization, leaders support conservation of natural resources.
4. In my organization, team collaboration emphasizes resource conservation (e.g., reducing paperwork, reducing commuting).
5. In my organization, we communicate information about natural resource conservation initiatives.
6. In my organization, the structure of teams includes resource conservation (e.g., through the ability to work remotely).
7. In my organization, in informal relationships with co-workers we encourage each other to conserve natural resources.

Intrinsic motivation

1. Saving resources at work makes me feel good about myself.
2. Saving resources at work makes me happy.
3. I am interested in how I can conserve natural resources.

Green workplace practices

1. At work, I take steps to conserve natural resources by saving paper.
2. At work, I take steps to conserve natural resources by saving water.
3. At work, I take steps to conserve natural resources by saving electricity.
4. At work, I take steps to conserve natural resources by recycling.
5. At work, I do apply environmentally friendly practices.
6. At work, I avoid wasting electricity.
7. At work, I avoid wasting water.
8. At work, I avoid wasting natural resources.
9. At work, I take steps to conserve natural resources through other initiatives.
10. At work, I take steps to u conserve natural resources through my commuting choices.

References

1. ISO 50001:2018. Energy Management Systems—Requirements with Guidance for Use. Available online: https://www.iso.org/standard/69426.html (accessed on 10 March 2021).
2. Poveda-Orjuela, P.P.; García-Díaz, J.C.; Pulido-Rojano, A.; Cañón-Zabala, G. ISO 50001: 2018 and Its Application in a Comprehensive Management System with an Energy-Performance Focus. *Energies* **2019**, *12*, 4700. [CrossRef]
3. Chen, Y.; Tang, G.; Jin, J.; Li, J.; Paillé, P. Linking market orientation and environmental performance: The influence of environmental strategy, employee's environmental involvement, and environmental product quality. *J. Bus. Ethics* **2014**, *127*, 479–500. [CrossRef]
4. Awan, U.; Imran, N.; Munir, G. Sustainable Development through Energy Management: Issues and Priorities in Energy Savings. *Res. J. Appl. Sci. Eng. Technol.* **2014**, *7*, 424–429. [CrossRef]
5. Moraes, S.d.S.; Chiappetta Jabbour, C.J.; Battistelle, R.A.G.; Rodrigues, J.M.; Renwick, D.S.W.; Foropon, C.; Roubaud, D. When knowledge management matters: Interplay between green human resources and eco-efficiency in the financial service industry. *J. Knowl. Manag.* **2019**, *23*, 1691–1707. [CrossRef]
6. Hori, S.; Shinozaki, M.; Nogata, D.; Fujita, T. The role of CSR in promoting companies' energy-saving actions in two Asian cities. *Energy Policy* **2014**, *69*, 116–121. [CrossRef]
7. Kim, A.; Kim, Y.; Han, K.; Jackson, S.E.; Ployhart, R.E. Multilevel influences on voluntary workplace green behavior: Individual differences, leader behavior, and coworker advocacy. *J. Manag.* **2017**, *43*, 1335–1358. [CrossRef]
8. Safari, A.; Salehzadeh, R.; Panahi, R.; Abolghasemian, S. Multiple pathways linking environmental knowledge and awareness to employees' green behavior. *Corp. Gov.* **2020**, *18*, 81–103. [CrossRef]
9. Dilchert, S.; Ones, D.S. Environmental sustainability in and of organizations. *Ind. Organ. Psychol.* **2012**, *5*, 503–511. [CrossRef]
10. Alzaidi, S.M.; Iyanna, S. Developing a conceptual model for voluntary pro-environmental behavior of employees. *Soc. Responsib. J.* Available online: https://doi.org/10.1108/SRJ-11-2020-0477 (accessed on 4 February 2021).
11. Dobrovič, J.; Čabinová, V.; Gallo, P.; Partlová, P.; Váchal, J.; Balogová, B.; Orgonáš, J. Application of the DEA Model in Tourism SMEs: An Empirical Study from Slovakia in the Context of Business Sustainability. *Sustainability* **2021**, *13*, 7422. [CrossRef]

12. De Roeck, K.; Farooq, O. Corporate social responsibility and ethical leadership: Investigating their interactive effect on employees' socially responsible behaviors. *J. Bus. Ethics* **2017**, *151*, 923–939. [CrossRef]
13. Rayner, J.; Morgan, D. An empirical study of 'green' workplace behaviours: Ability. Motivation and opportunity. *Asia Pac. J. Hum. Resour.* **2018**, *56*, 56–78. [CrossRef]
14. Ahmed, M.; Zehou, S.; Raza, S.A.; Qureshi, M.A.; Yousufi, S.Q. Impact of CSR and environmental triggers on employee green behavior: The mediating effect of employee well-being. *Corp. Soc. Responsib. Environ. Manag.* **2020**, *27*, 2225–2239. [CrossRef]
15. Afsar, B.; Badir, Y.; Kiani, U.S. Linking spiritual leadership and employee pro-environmental behavior: The influence of workplace spirituality, intrinsic motivation, and environmental passion. *J. Environ. Psychol.* **2016**, *45*, 79–88. [CrossRef]
16. Steg, L. Values, norms, and intrinsic motivation to act proenvironmentally. *Annu. Rev. Environ. Resour.* **2016**, *41*, 277–292. [CrossRef]
17. Kasser, T.; Ryan, R.M. Further examining the American dream: Differential correlates of intrinsic and extrinsic goals. *Personal. Soc. Psychol. Bull.* **1996**, *22*, 80–87. [CrossRef]
18. Tabernero, C.; Hernández, B. Self-efficacy and intrinsic motivation guiding environmental behavior. *Environ. Behav.* **2011**, *43*, 658–675. [CrossRef]
19. Ture, R.S.; Ganesh, M.P. Pro-environmental behaviours at workplace: An empirical study in Indian manufacturing organizations. *Benchmarking Int. J.* **2018**, *25*, 3743–3766. [CrossRef]
20. Norton, T.A.; Parker, S.L.; Zacher, H.; Ashkanasy, N.M. Employee green behavior a theoretical framework, multilevel review, and future research agenda. *Organ. Environ.* **2015**, *28*, 103–125. [CrossRef]
21. Yuriev, A.; Boiral, O.; Francoeur, V.; Paille, P. Overcoming the barriers to pro-environmental behaviors in the workplace: A systematic review. *J. Clean. Prod.* **2018**, *182*, 379–394. [CrossRef]
22. Ruepert, A.; Keizer, K.; Steg, L.; Maricchiolo, F.; Carrus, G.; Dumitru, A.; Mira, R.G.; Stancu, A.; Moza, D. Environmental considerations in the organizational context: A pathway to pro-environmental behaviour at work. *Energy Res. Soc. Sci.* **2016**, *17*, 59–70. [CrossRef]
23. Steg, L.; De Groot, J.I.; Dreijerink, L.; Abrahamse, W.; Siero, F. General antecedents of personal norms, policy acceptability, and intentions: The role of values, worldviews, and environmental concern. *Soc. Nat. Resour.* **2011**, *24*, 349–367. [CrossRef]
24. Banwo, A.O.; Du, J. Workplace pro-environmental behaviors in small and medium-sized enterprises: An employee level analysis. *J. Glob. Entrep. Res.* **2019**, *9*, 34. [CrossRef]
25. Zibarras, L.D.; Coan, P. HRM practices used to promote proenvironmental behavior: A UK survey. *Int. J. Hum. Resour. Manag.* **2015**, *26*, 2121–2142. [CrossRef]
26. Sun, Y.; Sun, H. Green Innovation Strategy and Ambidextrous Green Innovation: The Mediating Effects of Green Supply Chain Integration. *Sustainability* **2021**, *13*, 4876. [CrossRef]
27. Awan, U. Steering for Sustainable Development Goals: A Typology of Sustainable Innovation. In *Industry, Innovation and Infrastructure. Encyclopedia of the UN Sustainable Development Goals*; Leal Filho, W., Azul, A., Brandli, L., Lange Salvia, A., Wall, T., Eds.; Springer: Cham, Switzerland, 2020. [CrossRef]
28. Ullah, S.; Ahmad, N.; Khan, F.; Badulescu, A.; Badulescu, D. Mapping Interactions among Green Innovations Barriers in Manufacturing Industry Using Hybrid Methodology: Insights from a Developing Country. *Int. J. Environ. Res. Public Health* **2021**, *18*, 7885. [CrossRef]
29. Michalek, G.; Thronicker, I.; Yildiz, Ö.; Schwarze, R. Habitually green: Integrating the concept of habit into the design of pro-environmental interventions at the workplace. In *Nachhaltigkeits Management Forum*; Springer: Berlin/Heidelberg, Germany, 2019; Volume 27, pp. 113–124. [CrossRef]
30. Jabbour, C.J.C. How green are HRM practices, organizational culture, learning and teamwork? A Brazilian study. *Ind. Commer. Train.* **2011**, *43*, 98–105. [CrossRef]
31. Fernández, E.; Junquera, B.; Ordiz, M. Organizational culture and human resources in the environmental issue: A review of the literature. *Int. J. Hum. Resour. Manag.* **2003**, *14*, 634–656. [CrossRef]
32. Ojo, A.O.; Tan, C.N.-L.; Alias, M. Linking green HRM practices to environmental performance through pro-environment behaviour in the information technology sector. *Soc. Responsib. J.* **2020**. [CrossRef]
33. Renwick, D.W.S.; Redman, T.; Maguire, S. Green Human Resource Management: A Review and Research Agenda. *Int. J. Manag. Rev.* **2013**, *15*, 1–14. [CrossRef]
34. Wang, J.; Fei, Z.; Chang, Q.; Li, S. Energy Saving Operation of Manufacturing System Based on Dynamic Adaptive Fuzzy Reasoning Petri Net. *Energies* **2019**, *12*, 2216. [CrossRef]
35. Urban, W.; Łukaszewicz, K.; Krawczyk-Dembicka, E. Application of Industry 4.0 to the Product Development Process in Project-Type Production. *Energies* **2020**, *13*, 5553. [CrossRef]
36. Khan, A.M.; Jamil, M.; Salonitis, K.; Sarfraz, S.; Zhao, W.; He, N.; Mia, M.; Zhao, G. Multi-Objective Optimization of Energy Consumption and Surface Quality in Nanofluid SQCL Assisted Face Milling. *Energies* **2019**, *12*, 710. [CrossRef]
37. Nota, G.; Nota, F.; Peluso, D.; Lazo, A.T. Energy Efficiency in Industry 4.0: The Case of Batch Production Processes. *Sustainability* **2020**, *12*, 6631. [CrossRef]
38. Epstein, M.J. *Making Sustainability Work: Best Practices in Managing and Measuring Corporate Social. Environmental and Economic Impacts*; Routledge: London, UK, 2018.

39. Filimonau, V.; De Coteau, D.A. Food waste management in hospitality operations: A critical review. *Tour. Manag.* **2019**, *71*, 234–245. [CrossRef]
40. Norton, T.A.; Zacher, H.; Parker, S.L.; Ashkanasy, N.M. Bridging the gap between green behavioral intentions and employee green behavior: The role of green psychological climate. *J. Organ. Behav.* **2017**, *38*, 996–1015. [CrossRef]
41. Dincer, I.; Rosen, M.A. Energy, environment and sustainable development. *Appl. Energy* **1999**, *64*, 427–440. [CrossRef]
42. Scheraga, J.D. Energy and the environment Something new under the sun? *Energy Policy* **1994**, *22*, 798–803. [CrossRef]
43. Omer, A.M. Energy use and environmental impacts: A general review. *J. Renew. Sustain. Energy* **2009**, *1*, 53101. [CrossRef]
44. Hoyo-Montaño, J.A.; Valencia-Palomo, G.; Galaz-Bustamante, R.A.; García-Barrientos, A.; Espejel-Blanco, D.F. Environmental Impacts of Energy Saving Actions in an Academic Building. *Sustainability* **2019**, *11*, 989. [CrossRef]
45. Pothitou, M.; Varga, L.; Kolios, A.J.; Gu, S. Linking energy behaviour, attitude and habits with environmental predisposition and knowledge. *Int. J. Sustain. Energy* **2017**, *36*, 398–414. [CrossRef]
46. Steg, L.; Vlek, C. Encouraging pro-environmental behaviour: An integrative review and research agenda. *J. Environ. Psychol.* **2009**, *29*, 309–317. [CrossRef]
47. Littleford, C.; Ryley, T.J.; Firth, S.K. Context, control and the spillover of energy use behaviours between office and home settings. *J. Environ. Psychol.* **2014**, *40*, 157–166. [CrossRef]
48. Griskevicius, V.; Cantú, S.M.; van Vugt, M. The Evolutionary Bases for Sustainable Behavior: Implications for Marketing, Policy, and Social Entrepreneurship. *J. Public Policy Mark.* **2012**, *31*, 115–128. [CrossRef]
49. Russell, S.V.; Evans, A.; Fielding, K.S.; Hill, C. Turn It Off: An Action Research Study of Top Management Influence on Energy Conservation in the Workplace. *Front. Psychol.* **2016**, *7*, 389. [CrossRef]
50. Chan, E.S.; Hon, A.H.; Chan, W.; Okumus, F. What drives employees' intentions to implement green practices in hotels? The role of knowledge, awareness, concern and ecological behaviour. *Int. J. Hosp. Manag.* **2014**, *40*, 20–28. [CrossRef]
51. Alshaabani, A.; Naz, F.; Magda, R.; Rudnák, I. Impact of Perceived Organizational Support on OCB in the Time of COVID-19 Pandemic in Hungary: Employee Engagement and Affective Commitment as Mediators. *Sustainability* **2021**, *13*, 7800. [CrossRef]
52. Haider, S.; Fatima, F.; Bakhsh, K.; Ahmed, M. Effect of intervention on employees' intentions to use environmentally sustainable work practices: A field experiment. *J. Environ. Manag.* **2019**, *248*, 109334. [CrossRef] [PubMed]
53. Unsworth, K.L.; Dmitrieva, A.; Adriasola, E. Changing behaviour: Increasing the effectiveness of workplace interventions in creating pro-environmental behaviour change. *J. Organ. Behav.* **2013**, *34*, 211–229. [CrossRef]
54. Deal, T.E.; Kennedy, A.A. *Corporate Cultures: The Rites and Rituals of Corporate Life*; Addison-Wesley Publishing Company: Boston, MA, USA, 1982.
55. Hofstede, G. *Culture's Consequences: Comparing Values. Behaviours. Institutions and Organizations across Nations*, 2nd ed.; Sage Publications: Thousand Oaks, CA, USA, 2001.
56. Turner, B.A. *Exploring the Industrial Subculture*; Macmillan: London, UK, 1971.
57. Schein, E.H. *Organizational Culture and Leadership*; Jossey-Bass: San Francisco, CA, USA, 2004.
58. Buchanan, D.A.; Huczynski, A. *Organizational Behaviour*, 9th ed.; Person: London, UK, 2017.
59. Deshpande', R.; Farley, J.U.; Webster, F.E. Corporate culture, customer organization, and innovativeness in Japanese firms: A quadrad analysis. *J. Mark.* **1993**, *57*, 23–37. [CrossRef]
60. Hofstede, G.; Hofstede, G.J.; Minkov, M. *Cultures and Organizations: Software of the Mind: Intercultural Cooperation and Its Importance for Survival*, 3rd ed.; McGraw-Hill Professional: New York, NY, USA, 2010.
61. Kanter, M.R. *Evolve! Succeding in the Digital Culture of Tomorrow*; Harvard Business School Press: Boston, MA, USA, 2001.
62. Sánchez-Cañizares, S.M.; Muñoz, M.; Ángel, A.; López-Guzmán, T. Organizational culture and intellectual capital: A new model. *J. Intellect. Cap.* **2007**, *8*, 409–430. [CrossRef]
63. Linnenluecke, M.K.; Griffiths, A. Corporate sustainability and organizational culture. *J. World Bus.* **2010**, *45*, 357–366. [CrossRef]
64. Bulińska-Stangrecka, H. Kultura organizacji wirtualnych. Charakterystyka e-kultury. *Acta Univ. Lodz. Folia Oeconomica* **2016**, *3*, 55–69. [CrossRef]
65. Sułkowski, Ł. Elements of Organizational Culture—Theoretical and methodological problems. *Management* **2012**, *16*, 63–71. [CrossRef]
66. Sulkowski, Ł. The functionalist understanding of culture in management. *Organ. Manag.* **2014**, *1B*, 25–26.
67. Bulińska-Stangrecka, H. *E-kultura. Model i Analiza Kultury Organizacji Wirtualnych*; Oficyna Wydawnicza Politechniki Warszawskiej: Warszawa, Poland, 2018.
68. Brockmann, C.; Birkholz, A. Industry culture in construction and manufacturing. In Proceedings of the Joint International Conference on Construction Culture, Innovation and Management (CCIM), Dubai, United Arab Emirates, 26–29 November 2006; Available online: https://gpc.stanford.edu/publications/industrial-culture-construction-and-manufacturing (accessed on 23 August 2021).
69. Nahm, A.Y.; Vonderembse, M.A.; Koufteros, X.A. The Impact of Organizational Culture on Time-Based Manufacturing and Performance. *Decis. Sci.* **2004**, *35*, 579–607. [CrossRef]
70. Chalofsky, N.; Griffin, M.G. Work-life Programs and Organizational Culture: The Essence of Workplace Community. Available online: https://eric.ed.gov/?id=ED492293 (accessed on 21 August 2021).
71. Garmendia, J.A. The Impact of Corporate Culture on Company Performance. *Curr. Sociol.* **2004**, *52*, 1021–1038. [CrossRef]

72. Park, W.-W. The Corporate Culture Change Campaigns in Korea: Lessons from Their Failures. *Asia Pac. Bus. Rev.* **2001**, *7*, 89–110. [CrossRef]
73. Marcoulides, G.A.; Heck, R.H. Organizational Culture and Performance: Proposing and Testing a Model. *Organ. Sci.* **1993**, *4*, 209–225. [CrossRef]
74. Denison, D.R.; Mishra, A.K. Toward a Theory of Organizational Culture and Effectiveness. *Organ. Sci.* **1995**, *6*, 204–223. [CrossRef]
75. Gordon, G.G.; Ditomaso, N. Predicting corporate performance from organizational culture. *J. Manag. Stud.* **1992**, *29*, 783–798. [CrossRef]
76. Cameron, K.S.; Quinn, R.E. *Diagnosing and Changing Organizational Culture: Based on the Competing Values Frame-Work*, 3rd ed.; Jossey-Bass: San Francisco, CA, USA, 2011.
77. Smit, I. Assessment of Cultures: A Way to Problem Solving or a Way to Problematic Solutions. In *The International Handbook of Organizational Culture and Climate*; John Willey & Sons, Ltd.: Chichester, UK; New York, NY, USA; Weinheim, Germany; Brisbane, Australia; Singapore; Toronto, ON, Canada, 2001; pp. 165–180.
78. Shearer, C.S.; Hames, D.S.; Runge, J.B. How CEOs influence organizational culture following acquisitions. *Leadersh. Organ. Dev. J.* **2001**, *22*, 105–113. [CrossRef]
79. Etchegaray, J.M.; Thomas, E.J.; Profit, J. Preoccupation with failure and adherence to shared baselines: Measuring high-reliability organizational culture. *J. Patient Saf. Risk Manag.* **2019**, *24*, 147–152. [CrossRef] [PubMed]
80. Veiga, J.; Lubatkin, M.; Calori, R.; Very, P. Research Note Measuring Organizational Culture Clashes: A Two-Nation Post-Hoc Analysis of a Cultural Compatibility Index. *Hum. Relat.* **2000**, *53*, 539–557. [CrossRef]
81. Sarros, J.C.; Gray, J.; Densten, I.L.; Cooper, B. The Organizational Culture Profile Revisited and Revised: An Australian Perspective. *Aust. J. Manag.* **2005**, *30*, 159–182. [CrossRef]
82. Lee, S.K.J.; Yu, K. Corporate culture and organizational performance. *J. Manag. Psychol.* **2004**, *19*, 340–359. [CrossRef]
83. Van Der Post, W.Z.; De Coning, T.J.; Smit, E.V. An instrument to measure organizational culture. *S. Afr. J. Bus. Manag.* **1997**, *28*, 147–168. [CrossRef]
84. Horwitz, F.M.; Andersen, K.; Bezuidenhout, A.; Cohen, S.; Kirsten, F.; Mosoeunyane, K.; Smith, N.; Thole, K.; van Heerden, A. Due diligence neglected: Managing human resources and organizational culture in mergers and acquisitions. *S. Afr. J. Bus. Manag.* **2002**, *33*, 1–10. [CrossRef]
85. Sørensen, J.B. The Strength of Corporate Culture and the Reliability of Firm Performance. *Adm. Sci. Q.* **2002**, *47*, 70. [CrossRef]
86. Carmeli, A. The Link Between Organizational Elements, Perceived External Prestige and Performance. *Corp. Reput. Rev.* **2004**, *6*, 314–331. [CrossRef]
87. Carleton, J.R.; Lineberry, C.S. *Achieving Post-Merger Success: A Stakeholder's Guide to Cultural Due Diligence, Assessment, and Integration*; John Wiley & Sons, Inc.: New York, NY, USA, 2004; 213p.
88. Lemon, M.; Sahota, P. Organizational culture as a knowledge repository for increased innovative capacity. *Technovation* **2004**, *24*, 483–498. [CrossRef]
89. Sparrow, P.R. Developing Diagnostics for High Performance Organization Cultures. In *The International Handbook of Organizational Culture and Climate*; John Willey & Sons, Ltd.: Chichester, UK; New York, NY, USA; Weinheim, Germany; Brisbane, Australia; Singapore; Toronto, ON, Canada, 2001; pp. 85–103.
90. Cacciattolo, K. Understanding Organisational Cultures. *Eur. Sci. J.* **2014**, *2*, 1–7.
91. Child, J. What Determines Organization Performance? The universals vs. the it-all-depends. *Organ. Dyn.* **1974**, *3*, 2–18. [CrossRef]
92. Heskett, J. *The Culture Cycle: How to Shape the Unseen Force That Transforms Performance*; FT Press: Upper Saddle River, NJ, USA, 2012.
93. Moczydłowska, J.; Sadkowska, J. Project Culture as a Key Project Success Factor: The Perspective of Polish Project Managers. *Wseas Trans. Bus. Econ.* **2021**, *18*, 822–837. [CrossRef]
94. Gochhayat, J.; Giri, V.N.; Suar, D. Influence of Organizational Culture on Organizational Effectiveness: The Mediating Role of Organizational Communication. *Glob. Bus. Rev.* **2017**, *18*, 691–702. [CrossRef]
95. Jardon, C.M.; Martínez-Cobas, X. Leadership and Organizational Culture in the Sustainability of Subsistence Small Businesses: An Intellectual Capital Based View. *Sustainability* **2019**, *11*, 3491. [CrossRef]
96. Isensee, C.; Teuteberg, F.; Griese, K.-M.; Topi, C. The relationship between organizational culture, sustainability, and digitalization in SMEs: A systematic review. *J. Clean. Prod.* **2020**, *275*, 122944. [CrossRef]
97. Dyck, B.; Walker, K.; Caza, A. Antecedents of sustainable organizing: A look at the relationship between organizational culture and the triple bottom line. *J. Clean. Prod.* **2019**, *231*, 1235–1247. [CrossRef]
98. Magsi, H.B.; Ong, T.S.; Ho, J.A.; Hassan, A.F.S. Organizational Culture and Environmental Performance. *Sustainability* **2018**, *10*, 2690. [CrossRef]
99. Ryan, R.M.; Deci, E.L. Self-determination theory and the facilitation of intrinsic motivation, social development, and well-being. *Am. Psychol.* **2000**, *55*, 68–78. [CrossRef]
100. Pinder, C. *Work Motivation in Organizational Behavior*, 2nd ed.; Psychology Press: New York, NY, USA, 2008.
101. Jex, S.M.; Britt, T.W. *Organizational Psychology: A Scientist-Practitioner Approach*; John Wiley & Sons: New York, NY, USA, 2008.
102. Deci, E.L. *Intrinsic Motivation*; Plenum: New York, NY, USA, 1975.
103. Rigby, C.S.; Ryan, R.M. Self-Determination Theory in Human Resource Development: New Directions and Practical Considerations. *Adv. Dev. Hum. Resour.* **2018**, *20*, 133–147. [CrossRef]

104. Deci, E.L.; Olafsen, A.H.; Ryan, R.M. Self-Determination Theory in Work Organizations: The State of a Science. *Annu. Rev. Organ. Psychol. Organ. Behav.* **2017**, *4*, 19–43. [CrossRef]
105. Fischer, C.; Malycha, C.P.; Schafmann, E. The Influence of Intrinsic Motivation and Synergistic Extrinsic Motivators on Creativity and Innovation. *Front. Psychol.* **2019**, *10*, 137. [CrossRef]
106. Burton, K.D.; Lydon, J.E.; D'Alessandro, D.U.; Koestner, R. The differential effects of intrinsic and identified motivation on well-being and performance: Prospective, experimental, and implicit approaches to self-determination theory. *J. Pers. Soc. Psychol.* **2006**, *91*, 750–762. [CrossRef]
107. Janus, K.; Browning, S.L. The Effect of Professional Culture on Intrinsic Motivation Among Physicians in an Academic Medical Center. *J. Health Manag.* **2014**, *59*, 287–304. [CrossRef]
108. Al-Sada, M.; Al-Esmael, B.; Faisal, M.N. Influence of organizational culture and leadership style on employee satisfaction, commitment and motivation in the educational sector in Qatar. *EuroMed J. Bus.* **2017**, *12*, 163–188. [CrossRef]
109. Graves, L.M.; Sarkis, J.; Gold, N. Employee proenvironmental behavior in Russia: The roles of top management commitment, managerial leadership, and employee motives. *Resour. Conserv. Recycl.* **2019**, *140*, 54–64. [CrossRef]
110. Li, W.; Bhutto, T.A.; Xuhui, W.; Maitlo, Q.; Zafar, A.U.; Bhutto, N.A. Unlocking employees' green creativity: The effects of green transformational leadership, green intrinsic, and extrinsic motivation. *J. Clean. Prod.* **2020**, *255*, 120229. [CrossRef]
111. Faraz, N.A.; Ahmed, M.; Ying, M.; Mehmood, S.A. The interplay of green servant leadership, self-efficacy, and intrinsic motivation in predicting employees' pro-environmental behavior. *Corp. Soc. Responsib. Environ. Manag.* **2021**, *28*, 1171–1184. [CrossRef]
112. Young, W.; Davis, M.; McNeill, I.M.; Malhotra, B.; Russell, S.; Unsworth, K.; Clegg, C.W. Changing Behaviour: Successful Environmental Programmes in the Workplace. *Bus. Strat. Environ.* **2015**, *24*, 689–703. [CrossRef]
113. Luu, T.T. Building employees' organizational citizenship behavior for the environment: The role of environmentally-specific servant leadership and a moderated mediation mechanism. *Int. J. Contemp. Hosp. Manag.* **2019**, *31*, 406–426. [CrossRef]
114. Venhoeven, L.A.; Bolderdijk, J.W.; Steg, L. Why going green feels good. *J. Environ. Psychol.* **2020**, *71*, 101492. [CrossRef]
115. Annual Energy Outlook. 2021. Available online: https://www.eia.gov/outlooks/aeo/ (accessed on 10 May 2021).
116. Use of Energy Explained Energy Use in Industry. Available online: https://www.eia.gov/energyexplained/use-of-energy/industry.php (accessed on 15 May 2021).
117. What Kind of Energy Do We Consume in the EU? Available online: https://ec.europa.eu/eurostat/cache/infographs/energy/bloc-3a.html?lang=en (accessed on 15 May 2021).
118. Key Indicators. Energy Intensity of Manufacturing Industry. Available online: https://www.indicators.odyssee-mure.eu/online-indicators.html (accessed on 10 May 2021).
119. Europe's Buildings under the Microscope, BPIE. 2011. Available online: https://www.bpie.eu/publication/europes-buildings-under-the-microscope/ (accessed on 20 August 2021).
120. Murtagh, N.; Nati, M.; Headley, W.R.; Gatersleben, B.; Gluhak, A.; Imran, M.A.; Uzzell, D. Individual energy use and feedback in an office setting: A field trial. *Energy Policy* **2013**, *62*, 717–728. [CrossRef]
121. Bedwell, B.; Leygue, C.; Goulden, M.; McAuley, D.; Colley, J.; Ferguson, E.; Banks, N.; Spence, A. Apportioning energy consumption in the workplace: A review of issues in using metering data to motivate staff to save energy. *Technol. Anal. Strat. Manag.* **2014**, *26*, 1196–1211. [CrossRef]
122. Kotsopoulos, D.; Bardaki, C.; Lounis, S.; Papaioannou, T.; Pramatari, K.; Pucihar, A.; Borštnar, M.K.; Kittl, C.; Ravesteijn, P.; Clarke, R.; et al. Designing an IoT-enabled Gamification Application for Energy Conservation at the Workplace: Exploring Personal and Contextual Characteristics. In Proceedings of the Bled eConference, Bled, Slovenia, 18–21 June 2017. [CrossRef]
123. Palinkas, L.A.; Horwitz, S.M.; Green, C.A.; Wisdom, J.P.; Duan, N.; Hoagwood, K. Purposeful Sampling for Qualitative Data Collection and Analysis in Mixed Method Implementation Research. *Adm. Policy Ment. Health* **2015**, *42*, 533–544. [CrossRef] [PubMed]
124. Patton, M.Q. *Qualitative Research and Evaluation Methods*, 3rd ed.; Sage Publications: Thousand Oaks, CA, USA, 2002.
125. Cresswell, J.W.; Plano Clark, V.L. *Designing and Conducting Mixed Method Research*, 2nd ed.; Sage: Thousand Oaks, CA, USA, 2011.
126. Fey, C.F.; Denison, D.R. Organizational Culture and Effectiveness: Can American Theory Be Applied in Russia? *Organ. Sci.* **2003**, *14*, 686–706. [CrossRef]
127. Janićijević, N. Methodological approaches in the research of organizational culture. *Èkon. Anal.* **2011**, *56*, 69–99. [CrossRef]
128. Morgan, D.; Rayner, J. Development of a scale measure for green employee workplace practices. *JNBIT* **2019**, *17*, 1–25.
129. Inoue, Y.; Alfaro-Barrantes, P. Pro-environmental Behavior in the Workplace: A Review of Empirical Studies and Directions for Future Research. *Bus. Soc. Rev.* **2015**, *120*, 137–160. [CrossRef]
130. Bin Saeed, B.; Afsar, B.; Hafeez, S.; Khan, I.; Tahir, M.; Afridi, M.A. Promoting employee's proenvironmental behavior through green human resource management practices. *Corp. Soc. Responsib. Environ. Manag.* **2019**, *26*, 424–438. [CrossRef]
131. Bissing-Olson, M.J.; Iyer, A.; Fielding, K.; Zacher, H. Relationships between daily affect and pro-environmental behavior at work: The moderating role of pro-environmental attitude. *J. Organ. Behav.* **2012**, *34*, 156–175. [CrossRef]
132. Chang, S.-J.; Van Witteloostuijn, A.; Eden, L. Common method variance in international business research. *J. Int. Bus. Stud.* **2010**, *41*, 178–184. [CrossRef]
133. Intrinsic Motivation Inventory. Available online: https://selfdeterminationtheory.org/intrinsic-motivation-inventory/ (accessed on 10 January 2021).

134. Ryan, R.M.; Deci, E. Intrinsic and Extrinsic Motivations: Classic Definitions and New Directions. *Contemp. Educ. Psychol.* **2000**, *25*, 54–67. [CrossRef] [PubMed]
135. Fagan, M.H.; Neill, S.; Wooldridge, B.R. Exploring the Intention to Use Computers: An Empirical Investigation of the Role of Intrinsic Motivation. Extrinsic Motivation. and Perceived Ease of Use. *J. Comput. Inf. Syst.* **2008**, *48*, 31–37. [CrossRef]
136. Revilla, M.; Ochoa, C. Ideal and Maximum Length for a Web Survey. *Int. J. Mark. Res.* **2017**, *59*, 557–565. [CrossRef]
137. Lindell, M.K.; Whitney, D.J. Accounting for common method variance in cross-sectional research designs. *J. Appl. Psychol.* **2001**, *86*, 114–121. [CrossRef]
138. Podsakoff, P.M.; MacKenzie, S.B.; Podsakoff, N.P. Sources of Method Bias in Social Science Research and Recommendations on How to Control It. *Annu. Rev. Psychol.* **2012**, *63*, 539–569. [CrossRef]
139. Konrad, A.M.; Linnehan, F. Formalized Hrm Structures: Coordinating Equal Employment Opportunity Or Concealing Organizational Practices? *Acad. Manag. J.* **1995**, *38*, 787–820. [CrossRef]
140. Podsakoff, P.M.; MacKenzie, S.B.; Lee, J.Y.; Podsakoff, N.P. Common method biases in behavioral research: A critical review of the literature and recommended remedies. *J. Appl. Psychol.* **2003**, *88*, 879–903. [CrossRef]
141. Tehseen, S.; Ramayah, T.; Sajilan, S. Testing and Controlling for Common Method Variance: A Review of Available Methods. *J. Manag. Sci.* **2017**, *4*, 142–168. [CrossRef]
142. MacKenzie, S.B.; Podsakoff, P.M.; Podsakoff, N.P. Construct Measurement and Validation Procedures in MIS and Behavioral Research: Integrating New and Existing Techniques. *MIS Q.* **2011**, *35*, 293. [CrossRef]
143. Hair, J.F.; Hult, G.T.M.; Ringle, C.; Sarstedt, M. *A Primer on Partial Least Squares Structural Equation Modelling (PLS-SEM)*; Sage: Thousand Oaks, CA, USA, 2016.
144. Kline, R.B. *Principles and Practice of Structural Equation Modeling*, 2nd ed.; Guilford Publications: New York, NY, USA, 2005.
145. Fornell, C.; Larcker, D.F. Evaluating Structural Equation Models with Unobservable Variables and Measurement Error. *J. Mark. Res.* **1981**, *18*, 39–50. [CrossRef]
146. Hayes, A.F. *Introduction to Mediation. Moderation. And Conditional Process Analysis: A Regression-Based Approach*; Guilford Press: New York, NY, USA; London, UK, 2018.
147. R Core Team. *R: A Language and Environment for Statistical Computing*; R Foundation for Statistical Computing: Vienna, Austria, 2018; Available online: https://www.R-project.org (accessed on 10 January 2021).
148. Preacher, K.J.; Kelley, K. Effect size measures for mediation models: Quantitative strategies for communicating indirect effects. *Psychol. Methods* **2011**, *16*, 93–115. [CrossRef] [PubMed]
149. Trianni, A.; Cagno, E.; Farné, S. Barriers, drivers and decision-making process for industrial energy efficiency: A broad study among manufacturing small and medium-sized enterprises. *Appl. Energy* **2016**, *162*, 1537–1551. [CrossRef]
150. van der Linden, S. Intrinsic motivation and pro-environmental behaviour. *Nat. Clim. Chang.* **2015**, *5*, 612–613. [CrossRef]
151. Palmer, C.; Souza, G.I.; Laray, E.; Viana, V.; Hall, A. Participatory policies and intrinsic motivation to conserve forest commons. *Nat. Sustain.* **2020**, *3*, 620–627. [CrossRef]
152. Lulfs, R.; Hahn, R. Corporate greening beyond formal programs. initiatives. and systems: A conceptual model for voluntary pro-environmental behavior of employees. *Eur. Manag. Rev.* **2013**, *10*, 83–98. [CrossRef]
153. Rotzek, J.N.; Scope, C.; Günther, E. What energy management practice can learn from research on energy culture? *Sustain. Account. Manag. Policy J.* **2018**, *9*, 515–551. [CrossRef]
154. Abdelaziz, E.A.; Saidur, R.; Mekhilef, S. A review on energy saving strategies in industrial sector. *Renew. Sustain. Energy Rev.* **2011**, *15*, 150–168. [CrossRef]
155. Amundsen, A. Joint management of energy and environment. *J. Clean. Prod.* **2000**, *8*, 483–494. [CrossRef]
156. Antunes, P.; Carreira, P.; da Silva, M.M. Towards an energy management maturity model. *Energy Policy* **2014**, *73*, 803–814. [CrossRef]
157. Tachmitzaki, E.V.; Didaskalou, E.A.; Georgakellos, D.A. Energy Management Practices' Determinants in Greek Enterprises. *Sustainability* **2019**, *12*, 133. [CrossRef]
158. Korauš, A.; Gombár, M., Vagaská, A.; Šišulák, S.; Černák, F. Secondary Energy Sources and Their Optimization in the Context of the Tax Gap on Petrol and Diesel. *Energies* **2021**, *14*, 4121. [CrossRef]
159. Michalek, G.; Meran, G.; Schwarze, R.; Yildiz, Ö. Nudging as a New "Soft" Policy Tool: An Assessment of the Definitional Scope of Nudges, Practical Implementation Possibilities and Their Effectiveness. *Economics*. Available online: http://www.economics-ejournal.org/economics/discussionpapers/2016-18/ (accessed on 15 September 2021).

energies

MDPI

Article

Food Choices and Their Impact on Health and Environment

Katarzyna Szalonka [1,*], Elżbieta Stańczyk [2], Anna Gardocka-Jałowiec [3], Paweł Waniowski [4], Agata Niemczyk [5] and Zofia Gródek-Szostak [6]

1 Faculty of Law, Administration and Economics Institute of Economic Sciences, University of Wroclaw, ul. Uniwersytecka 22/26, 50-145 Wrocław, Poland
2 Dolnoslaskie Centre for Regional Surveys Statiscal Office in Wroclaw, Oławska 31, 50-950 Wrocław, Poland; e.stanczyk@stat.gov.pl
3 Faculty of Economics and Finance, University of Bialystok, Warszawska 63, 15-062 Bialystok, Poland; a.gardocka@uwb.edu.pl
4 Department of Marketing Research, Wroclaw University of Economics and Business, ul. Komandorska 118/120, 53-345 Wrocław, Poland; pawel.waniowski@ue.wroc.pl
5 Department of Tourism, Institute of Management at the Cracow University of Economics, 27 Rakowicka Street, 31-510 Cracow, Poland; agata.niemczyk@uek.krakow.pl
6 Department of Economics and Enterprise Organization, Cracow University of Economics, ul. Rakowicka 27, 31-510 Cracow, Poland; grodekz@uek.krakow.pl
* Correspondence: katarzyna.szalonka@uwr.edu.pl

Citation: Szalonka, K.; Stańczyk, E.; Gardocka-Jałowiec, A.; Waniowski, P.; Niemczyk, A.; Gródek-Szostak, Z. Food Choices and Their Impact on Health and Environment. *Energies* **2021**, *14*, 5460. https://doi.org/10.3390/en14175460

Academic Editor: Dimitrios A. Georgakellos

Received: 30 July 2021
Accepted: 30 August 2021
Published: 2 September 2021

Abstract: Food choices are complex and highly variable, even over short periods, as they are influenced by numerous psychological, social, and cultural factors, in addition to biological and economic ones. Consumer choices are increasingly complex because of the growing quantity and variety of available food products, which also affects individuals' environments. This paper is part of a larger study on health-related food choices, and it discusses how food choices affect the environment. To achieve the research goal, classes of respondents that are homogeneous in their food choices were identified. The authors used an algorithm to build classification trees and found that health status is determined by respondents' age and food consumption habits. The paper demonstrates that understanding individual nutritional choices is a prerequisite for changing consumption habits and shaping healthy behavior, which is in line with the principle of sustainable development through sustainable consumption. The findings are relevant to public health researchers and practitioners who wish to understand the relationship between nutritional practices and health in line with sustainable development.

Keywords: environment; sustainable consumption; food choices; health; decision trees; classification trees

1. Introduction

Our goal is to examine the impact of food choices on health and the environment. Food consumption accounts for almost one-third of households' total environmental impact. These environmental impacts include climate change, water pollution, water scarcity, etc. Understanding consumers' preferences for food products is essential for bettering food policy [1,2] whose primary goal should be to improve human health [3] and promote sustainable development [4–6]. Contemporary European Union legislation on food information tries to take into account the right of consumers to reliable information but also stresses the freedom of choice of every human being [7]. Hence, it is important to have more comprehensive and accurate information, not just to provide information on labels [8]. Providing more accurate health-related information can increase people's propensity to eat more healthily, especially in the case of obese people [9]. Not only is the availability of food information important, so are the costs of searching for that information. Reducing these costs may increase the likelihood of consumers choosing healthier products [10]. More and more consumers pay attention to the way goods and services are produced, as well as the degree of their environmental friendliness, in accordance with the principles of sustainable development on the effects of producers' actions for the environment and eco-innovation [11–13].

Developed countries are seeing more and more problems resulting from an unhealthy diet, including obesity, cardiovascular diseases, and cancer [14]. Therefore, many, including state authorities, are pressuring people to change their eating habits [1,15]. State authorities have an opportunity to conduct information campaigns, set different tax rates for different products, and regulate the food market. According to research conducted in Great Britain, only information campaigns are effective [16].

Tastes and preferences influence food choices throughout a person's life. A love for sweetness and an aversion to bitter and sour tastes are present in humans from an early age since they are innate [17]. However, research clearly shows that parents (especially mothers) pay attention to children's food choices, as they are an important part of parenting and feel obliged, in part because of external pressure, to provide their children with healthy food [18].

Food preferences and aversions develop from experience, attitudes, beliefs, and expectations [19]. Hence, one may notice changes in one's food consumption preferences. After a period of fascination in the 1970s and 1980s with highly processed foods, consumers began to take an interest in natural products and in the origins of food ingredients in line with the principles of sustainable development [20]. Good food choices affect many areas of people's lives around the world. Paradoxically, in developed countries, excessive weight and obesity kill more people than deficient weight does [21,22]. In 2016, 39 percent of adults eighteen years or older were overweight and 13 percent were obese [21].

Due to the increasing wealth of society and progressive urbanization, people are consuming more animal protein and processed food (rich in fats and sugars) and less whole grain and other sources of fiber [23]. Understanding the individual motives that drive certain food choices is key to being able to change consumption habits, shape healthy behavior, and become more sustainable [24]. Each modification of the diet can lead to both benefits (improvement in health) and losses (less enjoyment of food), and a permanent change can only be expected when, in the consumers' view, the benefits are higher than the losses [25]. This can foster a positive attitude toward ecological foods, which, according to CAWI research conducted in Poland in 2014, are perceived as healthier, more environmentally friendly, tastier, and of better quality, as they are subject to more stringent controls and is produced in a more traditional way [26].

In modern times, overeating, or emotional eating, is common, usually as a response to everyday stress [27], intensified even further by the COVID-19 pandemic. However, studies conducted in the United States on 13 February 2019 and 31 March 2020—before and in the beginning of the pandemic—did not find significant changes in food consumption between the two dates, except for a 14 percent increase in sugar consumption among adults, which could have been caused not so much by a change in eating habits as by an increase in purchases of processed foods containing more sugar [28]. The relationship between food expenditure and income seems to be important. People with higher incomes are more likely to change their preferences and are therefore less stable in their choices. This is because the cost of error decreases with income [29]. People who change preferences pay less attention to price and more often consider novelty an important attribute of food products [29–31].

Recently, an increase in food consumption that might not be justified by human needs has been associated with harm to the environment. Among other things, it is responsible for 20–30 percent of greenhouse gas emissions [32,33]. One of the reasons people change their eating habits is by reading information on product labels about products' contribution to environmental damage. Thus far, this information has mainly concerned the carbon footprint of producing a given food item, but the degree to which it changes food choices remains unknown and requires further research [34].

Contemporary shopping and consumption habits reflect not only the need to maintain one's current standard of living, but also snobbery or imitation or buying in stock [35].

The latter applies especially to food that has a strict use-by date. This applies not only to end consumers, but to the entire supply chain. In the United States, 31 percent of food, corresponding to USD 161 billion, is wasted at the retail level [36].

2. Materials and Methods

2.1. Research Material

We conducted our research in March and April 2020 with a questionnaire taken by 428 respondents—Polish residents aged 15–65, and aged 65 and over. The sample population was controlled in terms of the place of residence (province), age group, and gender. The questionnaire consisted of thirty-eight questions grouped thematically into physical activity, nutrition prophylaxis, food preferences, and general health.

To identify what food choices affect health and to identify classes of respondents that are relatively homogeneous in their food choices, we used an algorithm for building classification trees. Classification trees (also known as decision trees) are one example of data mining. Breiman et al. [37] introduced its use in regression analysis.

2.2. Methodology Research

The tree method, a data-analysis tool, has been widely applied to many research fields, not only statistics and econometrics. More and more often it is used in medicine—for example, for determining survival probabilities [38,39]. The use of classification trees in strategies of segmentation of individuals—for example, recipients of health services—in relation to appropriate segmentation into homogeneous subgroups may constitute the basis for targeted interventions by the health service [40,41], or consumers of food products, in particular with regard to the assessment of the symptoms of food addiction (overeating, mindless eating, etc.) for a healthy life [42].

In the present study, the method of classification trees allows us to divide the surveyed respondents into classes and thus to determine whether they belong to the classes of the qualitative dependent variable (health status) on the basis of measurements of explanatory variables (food choices). The classification tree algorithm analyzes the relationship of each explanatory variable (a food choice) with the dependent variable (health status). The variables that were selected for the model and that describe the division in the appropriate nodes of the tree and profiling the relevant subsets of the community under study are also the variables that determine the assessment of health.

The classification tree is a graphical presentation of the recursive group division method. In each node, the relation of division into successive subgroups is checked. The left branch shows the subgroup for which the relation is true, and the right branch shows the other respondents (for whom the relation is not true). The variable and the relations of the division were selected in such a way as to optimize the homogeneity of the division with regard to the dependent variable (that is, health status assessment).

At the lowest level, which illustrates the final division of the population in the study, there are end nodes—lists containing information about the number of respondents assigned to individual classes, the expected value of the dependent variable (health status assessment), and the histograms of the frequency distribution of the dependent variable (see Figure 1). In order to arrive at a relatively simple tree, we stopped the procedure of recursive group division before the segments and classes became fully homogeneous. For this purpose, we applied the fast algorithm for classification trees (FACT) direct-stop rule for 4 percent of the population. The explained (dependent) variable, Y, is a subjective assessment of health whose values were assigned based on the respondents' answers: 1 (good or very good health) or 0 (very bad health, bad health, or neither bad nor good health (so-so)). The explanatory (independent) variables, or food choices (X1, ... , X18), are:

- X1: eating three meals a day (that is, breakfast, lunch, dinner) (1 = no; 2 = rather not; 3 = depends on the situation at work and at home; 4 = yes; 5 = definitely yes; 6 = more than three meals a day);

- X2: fruit consumption (0 = I do not eat fruit; 1 = I eat it occasionally; 2 = I eat it every few months; 3 = I eat it every few days; 4 = I eat it once a day; 5 = I eat it several times a day);
- X3: vegetable consumption (0 = I do not eat vegetables; 1 = I eat them occasionally; 2 = I eat them two times a day; 3 = I eat them three times a day; 4 = I eat them four times a day; 5 = I eat them five times a day; 6 = I eat them more than five times a day);
- X4: fish consumption (0 = I do not eat fish; 1 = I eat it sporadically; 2 = I eat it every few months; 3 = I eat it every few days; 4 = I eat it once a day; 5 = I eat it several times a day);
- X5: meat consumption (0 = I do not eat meat; 1 = I eat it sporadically; 2 = I eat it every few months; 3 = I eat it every few days; 4 = I eat it once a day; 5 = I eat it several times a day);
- X6: dairy consumption (0 = I do not eat dairy; 1 = I eat it occasionally; 2 = I eat it every few months; 3 = I eat it every few days; 4 = I eat it once a day; 5 = I eat it several times a day);
- X7: consumption of gluten-containing bread (that is, bread made of wheat or rye flour) (0 = I do not eat it; 1 = I eat it sporadically; 2 = I eat it every few months; 3 = I eat it every few days; 4 = I eat it once a day; 5 = I eat it several times a day);
- X8: consumption of sweets and cakes (0 = I do not eat sweets and cakes; 1 = I eat them sporadically; 2 = I eat them every few months; 3 = I eat them every few days; 4 = I eat them once a day; 5 = I eat them several times a day);
- X9: snacking between meals (0 = I do not snack between meals; 1 = I do so sporadically; 2 = I do so every few months; 3 = I do so once a day; 4 = I do so twice a day; 5 = I do so several times a day);
- X10: gluten-free daily diet (0 = no; 1 = yes);
- X11: a lactose-free daily diet (0 = no; 1 = yes);
- X12: an egg-free daily diet (0 = no; 1—yes);
- X13: low-fat daily diet (0 = no; 1 = yes);
- X14: low-energy daily diet (0 = no; 1 = yes);
- X15: low-carbohydrate (high protein and fat) daily diet (0 = no; 1 = yes);
- X16: high-protein daily diet (0 = no; 1 = yes);
- X17: slimming daily diet (0 = no; 1 = yes);
- X18: reading the information on food labels carefully before buying food products (1 = I do not; 2 = I'd rather not; 3 = very occasionally; 4 = yes; 5 = definitely yes).

The control variables (X19, ... X20) are:

- X19: gender (F = female; M = male);
- X20: age (up to eighteen years old; nineteen to twenty-four; twenty-five to thirty-nine; forty to fifty-four; fifty-five to sixty-four; sixty-five or older);
- X21: education (1 = basic completed, no education; 2 = lower secondary; 3 = basic vocational; 4 = postsecondary; secondary vocational and general education; 5 = higher);
- X22: professional activity (1 = a school student, a university student, or a person in training or an unpaid internship; 2 = unemployed person; 3 = retiree on a pension; 4 = professionally inactive for other reasons; 5 = a person running a farm home, a housekeeper, or a head of a family; 6 = a working person who helps with family activities);
- X23: the level of household income that allows one to meet their basic food supply needs (0 = hard to say; 1 = definitely not; 2 = rather not; 4 = yes; 5 = definitely yes).

The calculations were performed in the software Statistica, in the case of classification trees, using the classification and regression trees (C&RT) method for exhaustive search for one-dimensional divisions.

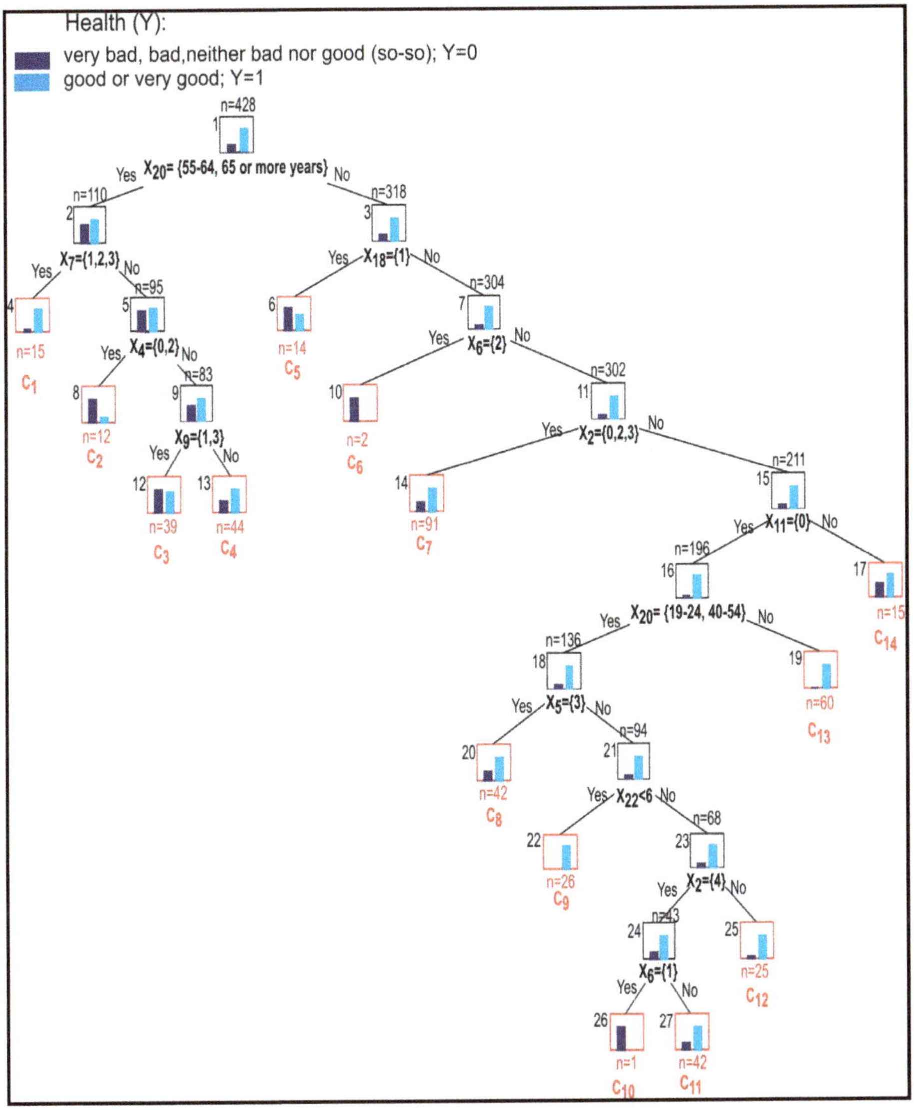

Figure 1. Classification tree for health assessment.

3. Results

All participants, 428, were controlled in terms of the place of residence (province), age group, and gender. Seventy percent of participants were women and 30 percent men; 75.4 percent of respondents lived in cities with more than twenty thousand residents (20 percent in cities with between twenty thousand and one hundred thousand, and 55 percent in cities with over one hundred thousand), and 25 percent were residents of towns with up to twenty thousand residents. Seventy-one percent of the respondents were working at the time of the questionnaire, 20 percent were students, and 7 percent were on a pension. The age distribution of the respondents is presented in Table 1.

Table 1. Age distribution of the respondents.

Age Range	Number	%
Up to 18	11	2.6
19–24	119	27.8
25–39	89	20.8
40–54	99	23.1
55–64	73	17.1
65 years old or more	37	8.6

Among the respondents, 19.2 (Table 2) percent assessed their health as very good and 55.1 percent as good (74.3 percent in total). People with very bad or bad health constituted a small share of the remaining respondents.

Table 2. Self-assessment of health status.

Age (X_{20})	How do You Assess Your Health? (Y)				
	Very Bad	Bad	Neither Bad nor Good (so-so) In %	Good	Very Good
Up to 24	1.5	3.8	16.9	46.9	30.8
25–39	0.0	0.0	12.4	57.3	30.3
40–54	0.0	3.0	20.2	65.7	11.1
55–64	0.0	4.1	37.0	56.2	2.7
65 years old or more	0.0	8.1	37.8	48.6	5.4
Total	0.5	3.3	22.0	55.1	19.2

Respondents' self-assessment of health in relation to selected food choices is presented in Table 3. People who assessed their health as good or very good most often declared that they ate fruit sporadically, vegetables three to four times a day, fish daily, meat once every few months, and dairy products occasionally. For comparison, people assessing their health condition as at best so-so (neither good nor bad) most often declared that they consumed no fruit, vegetables sporadically or one to two times a day, deficient amounts of fish, meat daily, and dairy products once every few months. These observations clearly demonstrate the different self-assessments of people who follow different patterns of nutrition.

Table 3. Self-assessment of health status and selected food choices.

Specification	(Y) Self-Assessment of Health Status	
	Neither Bad nor Good (so-so)	Good or Very Good
	(X_2) Fruit consumption	
I do not eat fruit	75.0	25.0
I eat it sporadically	14.8	85.2
I eat it once every few months	33.3	66.7
I eat it once every few days	32.7	67.3
I eat it once a day	24.2	75.8
I eat it a few times a day	20.7	79.3
	(X_3) Vegetable consumption	
I do not eat vegetables	20.0	80.0
I eat them sporadically	26.7	73.3

Table 3. *Cont.*

Specification	(Y) Self-Assessment of Health Status	
	Neither Bad nor Good (so-so)	**Good or Very Good**
	(X_3) Vegetable consumption	
I eat them once a day	26.8	73.2
I eat them 2 times a day	30.7	69.3
I eat them 3 times a day	17.1	82.9
I eat them 4 times a day	18.8	81.3
I eat them 5 times a day	16.7	83.3
I eat them more than 5 times a day	20.0	80.0
	(X_4) Fish consumption	
I do not eat fish	26.7	73.3
I eat it sporadically	23.3	76.7
I eat it once every few months	33.3	66.7
I eat it once every few days	25.3	74.7
I eat it once a day	—	100.0
I eat it a few times a day	—	100.0
	(X_5) Meat consumption	
I do not eat meat	12.5	87.5
I eat it sporadically	10.3	89.7
I eat it once every few months	—	100.0
I eat it once every few days	24.6	75.4
I eat it once a day	32.1	67.9
I eat it a few times a day	25.0	75.0
	(X_6) Dairy consumption	
I do not eat dairy	55.6	44.4
I eat it sporadically	20.0	80.0
I eat it once every few months	100.0	—
I eat it once every few days	27.3	72.7
I eat it once a day	25.3	74.7
I eat it a few times a day	22.8	77.2
	(X_1) Eating three meals a day (breakfast, lunch, dinner)	
Definitely not	16.7	83.3
Rather no	33.3	66.7
Rather yes	22.1	77.9
Definitely yes	25.9	74.1
Depends on the situation at work and at home	27.7	72.3

After constructing the classification tree, we ranked the importance of variables based on one-dimensional divisions (where 0 means low importance and 100 means high importance), as shown in Table 4. The table reveals the regularities described above for people who declare themselves to be in good or very good health.

Table 4. Ranking of importance of variables for good or very good health.

Variable	Importance
X_{20}: age	100
X_5: meat consumption	95
X_6: dairy consumption	90
X_2: fruit consumption	86
X_1: eating three meals a day (breakfast, lunch, dinner)	83
X_{18}: reading food label information carefully before purchasing products	70
X_3: vegetable consumption	65
X_9: snacking between meals	65
X_{11}: lactose-free daily diet	63
X_{22}: professional activity	63
X_7: consumption of bread with gluten	49
X_{21}: education	47
X_{13}: low-fat daily diet	45
X_{23}: the level of household income that allows one to meet their basic food needs X_4: fish consumption	40 35
X_{19}: gender	32
X_{12}: egg-free daily diet	30
X_8: consumption of sweets and cakes	29
X_{15}: low-carbohydrate (high protein and fat) daily diet	26
X_{16}: high-protein daily diet	18
X_{10}: gluten-free daily diet	17
X_{17}: slimming daily diet	6
X_{14}: low-energy daily diet	4
X_{17}: slimming daily diet X_{14}: low-energy daily diet	6 4

According to the model we constructed (Figure 1) and the ranking of explanatory variables (Table 4), apart from the control variables, the variables that most influenced whether someone rated themselves as being in good or very good health were age and frequency of fruit, dairy, and meat consumption. In contrast, low-energy daily diet and slimming daily diet had the least influence.

Finally, after applying the stop rule, the structure of the classification tree included ten significant variables that had the greatest classification power in the model development (which is equivalent to saying they were crucial in dividing the entire population into classes of respondents that differed significantly in their health status). Significant variables that in the classification tree model were the basis for the first divisions of the respondents turned out to be:

- X20: age;
- X7: consumption of bread with gluten (that is, bread made from wheat or rye flour);
- X18: reading the information on food labels carefully before buying products;
- X4: fish consumption;
- X6: dairy consumption;
- X9: snacking between meals;
- X2: fruit consumption;
- X11: a lactose-free daily diet;
- X22: professional activity;

- X5: meat consumption.

Self-assessment of health status (the dependent variable) was therefore most strongly determined by age (age of sixty-five years or more). In the first subset of respondents (aged fifty-five or over, numbering 110 respondents), 80.2 percent rated themselves as in good or very good health; in the second (with 318 respondents aged less than fifty-five years), 57.2 percent did.

These subgroups were further divided, with the sets of significant variables selected for both submodels being significantly different. In the group of people aged 55 and over (left side of the figure), the consumption of gluten-containing bread, i.e., traditional wheat or rye flour, had the greatest impact on good or very good health status, as well as consumption of fish or consumption of dairy products. In the group of people under 55 (right side of the figure), the variable of carefully reading the information on food labels before buying them, as well as fruit and meat consumption or professional activity, had an additional impact on good or very good self-rating of health.

As a result of the procedure, with the use of appropriate quality measures for the division of the group of respondents, 14 classes of respondents were distinguished (C1, … C14)—see end nodes (lists) in Figure 1 and a set of classes in Table 5.

Table 5. Respondents' classes.

Class	Characteristics	Share of People Declaring They Have Good or Very Good Health in a Given Class
C_1	Number = 15 units (3.5% of the total number of respondents) Age 55 or older Consumption of bread with gluten at most once every few days	80.0%
C_2	Number = 12 units (2.8% of the total number of respondents) Age 55 or older For the vast majority, the consumption of gluten-free bread more often than once every few days Consumption of fish every few months	18.2%
C_3	Number = 39 units (8.9% of the total number of respondents) Age 55 or older For the vast majority, the consumption of gluten-free bread more often than once every few days Consuming fish more than once every few months Snacks between meals occasionally or once every few days	46.2%
C_4	Number = 44 units (10.3% of the total number of respondents) Age 55 or older For the vast majority, the consumption of gluten-free bread is more frequent than once every few days Consuming fish more than once every few months The vast majority snacked between meals at least once a day	68.2%
C_5	Number = 14 units (3.3% of the total number of respondents) Age under 55 Failure to carefully read the information on food labels before purchasing food products	12.9%
C_6	Number = 2 units (0.5% of the total number of respondents) Age under 55 Respondents read the information on food labels carefully before buying them Consumption of dairy products every few months	0%
C_7	Number = 91 units (21.3% of the total number of respondents) Age under 55 Respondents read the information on food labels carefully before buying products Consumption of dairy products more than once every few months Fruit consumption every few months or every few days	74.1%

Table 5. *Cont.*

Class	Characteristics	Share of People Declaring They Have Good or Very Good Health in a Given Class
C_8	Number = 42 units (9.8% of the total number of respondents) Age 19–24 or 40–54 Respondents read the information on food labels carefully before buying them Consumption of dairy products more than once every few months Fruit consumption more than once every few months or once every few days Not a lactose-free daily diet Consumption of meat every few days	71.4%
C_9	Number = 26 units (6.1% of the total number of respondents) Age 19–24 or 40–54 Respondents read the information on food labels carefully before buying products Consumption of dairy products more than once every few months Fruit consumption more than once every few months or once every few days Not a lactose-free daily diet For the vast majority, meat consumption more often every few days Not people who work or help in self-employed family activities	100%
C_{10}	Number = 1 unit (0.2% of the total number of respondents) Age 19–24 years Respondents read the information on food labels carefully before buying products Occasional consumption of dairy products Consumption of fruit every few days Not a lactose-free daily diet Meat consumption more than once every few days Persons who work or help in self-employed family activities	0%
C_{11}	Number = 42 units (9.8% of the total number of respondents) Age 19–24 or 40–54 Respondents read the information on food labels carefully before buying them Consumption of dairy products more than once every few months Fruit consumption more than once every few days Not a lactose-free daily diet Meat consumption more often than every few days People who work or help in self-employed family activities	81.1%
C_{12}	Number = 25 units (5.8% of the total number of respondents) Age 19–24 or 40–54 Respondents read the information on food labels carefully before buying products Consumption of dairy products more than once every few months Consumption of fruit more than once every few days Not a lactose-free daily diet Meat consumption more often than every few days People who work or help in self-employed family activities	96.0%
C_{13}	Number = 60 units (14.0% of the total number of respondents) Age under 19 or 25–39 Respondents read the information on food labels carefully before buying products Consumption of dairy products more than once every few months Consumption of fruit every few days Not a lactose-free daily diet	96.7%

Table 5. *Cont.*

Class	Characteristics	Share of People Declaring They Have Good or Very Good Health in a Given Class
C_{14}	Number = 15 units (3.5% of the total number of respondents) Age under 55 Respondents read the information on food labels carefully before buying products Consumption of dairy products more than once every few months Consumption of fruit every few days Not a lactose-free daily diet	60.0%

Our research on the effects of food choices on health leads to three main conclusions. First, considering the respondents aged fifty-five years or over, the largest share of people declaring themselves to be in good or very good health (80.0 percent) was respondents consuming bread with gluten only once every few days (class C1). In contrast, the smallest share of people who assessed their health condition as good or very good (18.2 percent) was recorded in class C2, which included people who consumed bread with gluten more than once every few days and fish only once every few months.

Second, among the classes distinguished in the classification tree model, the C9 class, grouping people with good and very good health (100 percent), had as common features that they were aged nineteen to twenty-four or forty to fifty-four, they carefully read the information on food labels before purchasing food, they consumed dairy products more than once every few months, they consumed fruit more than once every few months or every few days, and their daily diet was not lactose free. The vast majority consumed meat more often than once every few days. These are people who do not work. Third, one of the classes dominated by people who did not assess their health condition as good or very good was the class of respondents who were under fifty-five years old and declared that they did not carefully read the information on the labels of food products before buying them (C5).

4. Discussion

In light of our results, consuming healthy food products has a clear role in the healthy functioning of an individual, as suggested by Hippocrates's principle "Let your food be your medicine" [43]. This thesis is confirmed by low consumption of meat or gluten-containing products and high consumption of vegetables, fruit, and fish. Each of these types of food has advantages and disadvantages [44]. Moreover, an important role in the pro-health trend is played not so much by information on product labels [45,46] as by reading them carefully before buying [47,48]. Nevertheless, as stated above, self-assessment of one's health status is most strongly determined by age. Since food choices are influenced by established consumption habits rather than by example [49], good consumption habits must be established within respective age groups.

Much older generations reached good physical condition, for example, thanks to unprocessed food (food products consumed in childhood by older respondents were quite different in nutritional value than modern meat, dairy products, vegetables, and fruit). Young people are more interested in the healthfulness of their diet. Modern technologies in healthcare are useful: "In the health-care setting, technological change has seen for example, the introduction of electronic health records, mobile health apps, mobile computing" [50]. McCamley et al. [51] found that implementing an electronic medical record system increased the quality, availability, and accessibility of data for nutrition assessments and increased time efficiency. The internet and social media have significantly changed the way consumers access health information. Additionally, social media allows users to interact and gives them an opportunity to acquire and share information. Evidence indicates that younger age is significantly associated with using the internet as the first source of health information, blogging about health, and using social media for health reasons [52].

Therefore, one can contest the thesis of Pliner and Mann [49], quoted earlier, that mainly because of technology, food choices in the twenty-first century are shaped by the behavior and examples of other people. We hope that the examples in question will rationally affect quality of life by promoting patterns of sustainable consumption.

Ensuring sustainable food consumption can be seen as a generic goal that can be supported by most Sustainable Development Goals [53]. Additionally, hope in this regard especially concerns the young generation, who are susceptible to digital influences and treated as future consumers whose habits relating to sustainable food consumption can be associated with large-scale global concerns related to sustainability [54].

5. Conclusions

The research shows that the health of every human being is significantly influenced by the consumption of appropriate products and paying attention to the information contained in food labels and making one's own choices on this basis. A lack of knowledge may affect consumption choices, but is unlikely to be a dominant factor in nutritional differences, especially in making decisions about home or out-of-home nutrition choices [55]. However, as our research shows, food choices are not always significantly influenced by consumption habits established and examples of other people. This means that the possibilities of active and effective shaping of food choices are significant as they depend to a greater extent on shaping factors (e.g., on providing rational information) than on the socio-cultural standards established.

The problem is worth being examined more seriously as promoting healthy eating is of great importance for sustainable development since the costs of obesity treatment place a huge burden on the healthcare system and are mostly financed by public entities [56].

The analysis conducted confirmed the usefulness of classification trees in the segmentation of respondents due to the assessment of health condition and in distinguishing (classes) of enterprise profiles in terms of key nutritional choices. The advantage of this method was an ability to present data graphically and the ease of interpretation of the model obtained. This study naturally has some limitations. The authors are aware of the weaknesses of the model applied which are revealed in the instability of the classification tree model, as in extreme cases even slight changes in the empirical data set in subsequent research may lead to different divisions of the respondent population. The fact that this paper investigates the Poles, which means that its generalization to other counties is limited, is also considered a limitation of the research. Therefore, comparative studies in other counties are potential areas of future research.

Author Contributions: Conceptualization, K.S., E.S., A.G.-J., P.W., A.N., Z.G.-S.; methodology, K.S., E.S., A.G.-J., P.W., A.N., Z.G.-S.; software, K.S., E.S., A.G.-J., P.W., A.N., Z.G.-S.; validation, K.S., E.S., A.G.-J., P.W., A.N., Z.G.-S.; formal analysis, K.S., E.S., A.G.-J., P.W., A.N., Z.G.-S.; investigation, K.S., E.S., A.G.-J., P.W., A.N., Z.G.-S.; resources, K.S., E.S., A.G.-J., P.W., A.N., Z.G.-S.; data curation, K.S., E.S., A.G.-J., P.W., A.N., Z.G.-S.; writing—original draft preparation, K.S., E.S., A.G.-J., P.W., A.N., Z.G.-S.; visualization, K.S., E.S., A.G.-J., P.W., A.N., Z.G.-S.; project administration, Z.G.-S. All authors have read and agreed to the published version of the manuscript.

Funding: This publication was financed by a subsidy granted to the Wroclaw University of Economics and Business. This publication was financed by a subsidy granted to the Cracow University of Economics (6/ZZE/2021/POT). The publication was financed by a subsidy for the University of Wroclaw, Funds for the "Excellence Initiative - Research University" program.

Conflicts of Interest: The authors declare no conflict of interest.

References

1. Gardocka-Jałowiec, A.; Szalonka, K. Wybrane aspekty stylu życia w perspektywie odpowiedzialności za zdrowie. In *Zdrowie i Style Życia*; Nowak, W., Szalonka, K., Eds.; Uniwersytet Wrocławski: Wrocław, Poland, 2019; pp. 13–28. [CrossRef]
2. Gardocka-Jałowiec, A. Innowacje wobec wyzwań odpowiedzialności za zdrowie. In *Odpowiedzialność za Zdrowie*; Głowacka, D., Mruk, H., Eds.; Polskie Towarzystwo Nauk o Zdrowiu: Poznań, Poland, 2017; pp. 139–150.
3. Witek, L.; Szalonka, K. Percepcja zdrowej żywności a zachowania konsumentów [Perception of Healthy Food and Consumer Behavior]. *Zesz. Naukowe SGGW—Ekon. Org. Gospod. Żywn.* **2017**, *120*, 159–174. [CrossRef]
4. Niemczyk, A. Turysta wobec koncepcji zrównoważonego rozwoju (wyniki badań ankietowych). In *Zrównoważony Rozwój Turystyki*; Wodejko, S., Ed.; Szkoła Główna Handlowa: Warszawa, Poland, 2008; pp. 71–74.
5. Niemczyk, A.; Seweryn, R.; Smalec, A. Sustainable tourism in the activities of city authorities: Poland—Case Study. *Eur. J. Serv. Manag.* **2018**, *28*, 283–297. [CrossRef]
6. Oleksy-Gębczyk, A.; Niemczyk, A. *Wykorzystanie Instrumentów Marketingu Relacji w Jednostkach Samorządu Terytorialnego*; Wydawnictwo, C.H., Ed.; Beck: Warszawa, Poland, 2020; pp. 36–38.
7. Edinger, W.H. Promoting educated consumer choices: Has EU food information legislation finally matured? *J. Consum. Policy* **2016**, *39*, 9–22. [CrossRef]
8. Balcombe, K.; Fraser, I.; Lowe, B.; Monteiro, D.M.D.S. Information Customization and Food Choice. *Am. J. Agric. Econ.* **2015**, *98*, 54–73. [CrossRef]
9. Segovia, M.S.; Palma, M.A.; Nayga, R.M. Can episodic future thinking affect food choices? *J. Econ. Behav. Organ.* **2020**, *177*, 371–389. [CrossRef] [PubMed]
10. Zhu, C.; Lopez, R.A.; Liu, X. Information cost and consumer choices of healthy foods. *Am. J. Agric. Econ.* **2015**, *98*, 41–53. [CrossRef]
11. Bal-Domańska, B.; Stańczyk, E. Regionalna charakterystyka wybranych aspektów ekoaktywności przedsiębiorstw przemysłowych. *Prace Naukowe Uniw. Ekon. Wrocławiu* **2017**, *477*, 24–39. [CrossRef]
12. Cosentino, C.; Paolino, R.; Freschi, P.; Calluso, A. Short communication: Jenny milk production and qualitative characteristics. *J. Dairy Sci.* **2012**, *95*, 2910–2915. [CrossRef]
13. Adduci, F.; Elshafie, H.; Labella, C.; Musto, M.; Freschi, P.; Paolino, R.; Ragni, M.; Cosentino, C. Abatement of the clostridial load in the teats of lactating cows with lysozyme derived from donkey milk. *J. Dairy Sci.* **2019**, *102*, 6750–6755. [CrossRef]
14. Witek, L.; Szalonka, K. Alergie pokarmowe i ich wpływ na rozwój rynku żywności funkcjonalnej i ekologicznej [Food allergies and their impact on the development of the functional and organic food market]. *Polityki Eur. Finans. Mark.* **2016**, *16*, 128–140.
15. Gardocka-Jałowiec, A.; Szalonka, K.; Stańczyk, P. The determinants of shopping place selection in Poland—The survey results. *Optimum. Econ. Stud.* **2018**, *4*, 274–285. [CrossRef]
16. Griffith, R.; O'Connell, M. Public policy towards food consumption. *Fisc. Stud.* **2010**, *31*, 481–507. [CrossRef]
17. Steiner, J.E. Facial expressions of the neonate infant indicating the hedonics of food related chemical stimuli. In *Taste and Development: The Genesis of Sweet Preference*; US Government Publishing Office: Washington, DC, USA, 1977.
18. Elliott, S.; Bowen, S. Defending motherhood: Morality, responsibility, and double binds in feeding children. *J. Marriage Fam.* **2018**, *80*, 499–520. [CrossRef]
19. Clark, J.E. Taste and flavour: Their importance in food choice and acceptance. *Proc. Nutr. Soc.* **1998**, *57*, 639–643. [CrossRef]
20. Lusk, J.L.; McCluskey, J. Understanding the impacts of food consumer choice and food policy outcomes. *Appl. Econ. Perspect. Policy* **2018**, *40*, 5–21. [CrossRef]
21. World Health Organization (WHO). Obesity and Overweight. Available online: https://www.who.int/news-room/fact-sheets/detail/obesity-and-overweight (accessed on 18 June 2021).
22. Gródek-Szostak, Z.; Malik, G.; Kajrunajtys, D.; Szeląg-Sikora, A.; Sikora, J.; Kuboń, M.; Niemiec, M.; Kapusta-Duch, J. Modeling the dependency between extreme prices of selected agricultural products on the derivatives market using the linkage function. *Sustainability* **2019**, *11*, 4144. [CrossRef]
23. McMichael, A.J.; Powles, J.W.; Butler, C.D.; Uauy, R. Food, livestock production, energy, climate change and health. *Lancet* **2007**, *370*, 1253–1263. [CrossRef]
24. Chen, P.-J.; Antonelli, M. Conceptual models of food choice: Influential factors related to foods, individual differences, and society. *Foods* **2020**, *9*, 1898. [CrossRef]
25. Gedrich, K. Determinants of nutritional behaviour: A multitude of levers for successful intervention? *Appetite* **2003**, *41*, 231–238. [CrossRef] [PubMed]
26. Bryła, P. Organic food consumption in Poland: Motives and barriers. *Appetite* **2016**, *105*, 737–746. [CrossRef] [PubMed]
27. Altheimer, G.; Urry, H.L. Do emotions cause eating? The role of previous experiences and social context in emotional eating. *Curr. Dir. Psychol. Sci.* **2019**, *28*, 234–240. [CrossRef]
28. Cummings, J.R.; Ackerman, J.M.; Wolfson, J.A.; Gearhardt, A.N. COVID-19 stress and eating and drinking behaviors in the United States during the early stages of the pandemic. *Appetite* **2021**, *162*, 105163. [CrossRef]
29. Lusk, J.L. Income and (Ir) rational food choice. *J. Econ. Behav. Organ.* **2019**, *166*, 630–645. [CrossRef]
30. Szeląg-Sikora, A.; Sikora, J.; Niemiec, M.; Gródek-Szostak, Z.; Kapusta-Duch, J.; Kuboń, M.; Komorowska, M.; Karcz, J. Impact of integrated and conventional plant production on selected soil parameters in carrot production. *Sustainability* **2019**, *11*, 5612. [CrossRef]

31. Kapusta-Duch, J.; Szeląg-Sikora, A.; Sikora, J.; Niemiec, M.; Gródek-Szostak, Z.; Kuboń, M.; Leszczyńska, T.; Borczak, B. Health-promoting properties of fresh and processed purple cauliflower. *Sustainability* **2019**, *11*, 4008. [CrossRef]
32. Hertwich, E.I. *Assessing the Environmental Impacts of Consumption and Production: Priority Products and Materials*; A Report of the Working Group on the Environmental Impacts of Products and Materials to the International Panel for Sustainable Resource Management, Hertwich; International Panel on the Sustainable Use of Natural Resources: Paris, France; UNEP: Nairobi, Kenya, 2010.
33. Sikora, J.; Niemiec, M.; Szeląg-Sikora, A.; Gródek-Szostak, Z.; Kuboń, M.; Komorowska, M. The impact of a controlled-release fertilizer on greenhouse gas emissions and the efficiency of the production of chinese cabbage. *Energies* **2020**, *13*, 2063. [CrossRef]
34. Muller, L.; Lacroix, A.; Ruffieux, B. Environmental labelling and consumption changes: A food choice experiment. *Environ. Resour. Econ.* **2019**, *73*, 871–897. [CrossRef]
35. Gardocka-Jałowiec, A. *Zmiany Konsumpcji a Kreowanie Innowacji*; University of Bialystok: Białystok, Poland, 2015.
36. Buzby, J.C.; Farah-Wells, H.; Hyman, J. *The Estimated Amount, Value, and Calories of Postharvest Food Losses at the Retail and Consumer Levels in the United States*; Economic Research Service, Economic Information Bulletin; Department of Agriculture: Washington, DC, USA, 2014.
37. Breiman, L.; Friedman, J.H.; Olshen, R.A.; Stone, C.J. *Classification and Regression Trees*; Brooks/Cole Publishing: London, UK, 1984.
38. Lamborn, K.R.; Chang, S.M.; Prados, M.D. Prognostic factors for survival of patients with glioblastoma: Recursive partitioning analysis. *Neuro-Oncology* **2004**, *6*, 227–235. [CrossRef] [PubMed]
39. Hess, K.R.; Abbruzzese, M.C.; Lenzi, R.; Raber, M.N.; Abbruzzese, J.L. Classification and regression tree analysis of 1000 consecutive patients with unknown primary carcinoma. *Clin. Cancer Res.* **1999**, *5*, 3403–3410. [PubMed]
40. Lemon, S.C.; Roy, J.; Clark, M.A.; Friedmann, P.D.; Rakowski, W. Classification and regression tree analysis in public health: Methodological review and comparison with logistic regression. *Ann. Behav. Med.* **2003**, *26*, 172–181. [CrossRef] [PubMed]
41. Kuhn, L.; Page, K.; Ward, J.; Worrall-Carter, L. The process and utility of classification and regression tree methodology in nursing research. *J. Adv. Nurs.* **2013**, *70*, 1276–1286. [CrossRef] [PubMed]
42. Adriyendi, A. Classification using naïve Bayes and decision tree on food addiction. *Int. J. Database Theory Appl.* **2016**, *9*, 161–180. [CrossRef]
43. Ionuţ, C.; Popa, M.; Laza, V.; Sîrbu, D.; Curşeu, D.; Ionut, R. *Compendiu de Igienă*; Editura Medicală Universitară "Iuliu Haţieganu": Cluj-Napoca, Romania, 2004; pp. 432–436.
44. Laza, V. Contradictions and controversies in contemporary nutrition. *Palestrica Third Millenn. Civiliz. Sport* **2017**, *18*, 220–224. [CrossRef]
45. Raggio, L.; Berrondo, V. Knowledge and perceptions on the reading of food labels: A case study in Uruguay. *Demetra Food Nutr. Health* **2018**, *13*, 891–900.
46. Shamim, K.; Khan, S.A.; Ahmad, S. Consumers' understanding of nutrition labels for ultra-processed food products. *J. Public Aff.* **2020**, e2398. [CrossRef]
47. Bialkova, S.; Sasse, L.; Fenko, A. The role of nutrition labels and advertising claims in altering consumers' evaluation and choice. *Appetite* **2016**, *96*, 38–46. [CrossRef]
48. Kumar, N.; Kapoor, S. Do labels influence purchase decisions of food products? Study of young consumers of an emerging market. *Br. Food J.* **2017**, *119*, 218–229. [CrossRef]
49. Pliner, P.; Mann, N. Influence of social norms and palatability on amount consumed and food choice. *Appetite* **2004**, *42*, 227–237. [CrossRef] [PubMed]
50. Collins, J. Generational change in nutrition and dietetics: The millennial dietitian. *Nutr. Diet.* **2019**, *76*, 369–372. [CrossRef]
51. McCamley, J.; Vivanti, A.; Edirippulige, S. Dietetics in the digital age: The impact of an electronic medical record on a tertiary hospital dietetic department. *Nutr. Diet.* **2019**, *76*, 480–485. [CrossRef]
52. Prestin, A.; Vieux, S.N.; Chou, W.-Y.S. Is online health activity alive and well or flatlining? Findings from 10 years of the health information national trends survey. *J. Health Commun.* **2015**, *20*, 790–798. [CrossRef] [PubMed]
53. Özkaya, F.T.; Durak, M.; Doğan, O.; Bulut, Z.; Haas, R. Sustainable consumption of food: Framing the concept through turkish expert opinions. *Sustainability* **2021**, *13*, 3946. [CrossRef]
54. Kymäläinen, T.; Seisto, A.; Malila, R. Generation Z food waste, diet and consumption habits: A Finnish social design study with future consumers. *Sustainability* **2021**, *13*, 2124. [CrossRef]
55. Binkley, J.K. Nutrition and food choice: Home vs. restaurants. *J. Consum. Aff.* **2018**, *53*, 1146–1166. [CrossRef]
56. Finkelstein, E.A.; Trogdon, J.G.; Cohen, J.W.; Dietz, W. Annual medical spending attributable to obesity: Payer-and service-specific estimates. *Health Aff.* **2009**, *28*, w822–w831. [CrossRef]

energies

MDPI

Article

Prioritizing Energy-Intensive Machining Operations and Gauging the Influence of Electric Parameters: An Industrial Case Study

Ardamanbir Singh Sidhu [1], Sehijpal Singh [2], Raman Kumar [2,*], Danil Yurievich Pimenov [3] and Khaled Giasin [4]

1 Department of Mechanical Engineering, I. K. Gujral Punjab Technical University, Jalandhar 144603, India; aol.ardaman@gmail.com
2 Department of Mechanical and Production Engineering, Guru Nanak Dev Engineering College, Ludhiana 141006, India; sehijpalsingh@yahoo.co.in
3 Department of Automated Mechanical Engineering, South Ural State University, Lenin Prosp. 76, 454080 Chelyabinsk, Russia; danil_u@rambler.ru
4 School of Mechanical and Design Engineering, University of Portsmouth, Portsmouth PO1 3DJ, UK; khaled.giasin@port.ac.uk
* Correspondence: sehgal91@yahoo.co.in; Tel.: +91-9855100530

Citation: Sidhu, A.S.; Singh, S.; Kumar, R.; Pimenov, D.Y.; Giasin, K. Prioritizing Energy-Intensive Machining Operations and Gauging the Influence of Electric Parameters: An Industrial Case Study. *Energies* **2021**, *14*, 4761. https://doi.org/10.3390/en14164761

Academic Editor: Dimitrios A. Georgakellos

Received: 1 July 2021
Accepted: 3 August 2021
Published: 5 August 2021

Abstract: Increasing the energy efficiency of machining operations can contribute to more sustainable manufacturing. Therefore, there is a necessity to investigate, evaluate, and optimize the energy consumed during machining operations. The research highlights a method employed to prioritize the most energy-intensive machining operation and highlights the significance of electric parameters as predictors in power estimation of machining operations. Multi regression modeling with standardized regression weights was used to identify significant power quality predictors for active power evaluation for machining operations. The absolute error and the relative error both decreased when the active power was measured by the power analyzer for each of the identified machining operations, compared to the standard power equation and that obtained from the modeled regression equations. Furthermore, to determine energy-intensive machining operation, a hybrid decision-making technique based on TOPSIS (a technique for order preference by similarity to ideal solution) and DoM (degree of membership) was utilized. Allocation of weights to energy responses was carried out using three methods, i.e., equal importance, entropy weights, and the AHP (analytical hierarchy process). Results revealed that a drilling process carried out on material ST 52.3 is energy-intensive. This accentuates the significance of electric parameters in the assessment of active power during machining operations.

Keywords: machining operations; electric parameters; active power; active energy; specific energy consumption; energy efficiency; TOPSIS; entropy weight; AHP

1. Introduction

Manufacturing activities can negatively impact the environment. It is one of the major consumers of electricity. Electricity production and heat contribute to CO_2 emissions, especially the production of electricity through fossils fuels like coal. Due to Covid-19, there was an unprecedented decline in emissions in 2020. Worldwide CO_2 emissions from the electricity sector reduced by almost 450 million tons in 2020 [1]. This was mainly due to the reduction in industrial production. Worldwide energy demand dropped by 3.8% in the first quarter of 2020 compared with the first quarter of 2019. As a result, annual energy demand in 2020 decreased by 6%. The reduced use of coal amounted to 1.1 Gt of reductions in CO_2 emissions. This highlights that the increase in manufacturing activity and its decrease in the lockdown period due to the coronavirus pandemic directly affected the growth and decline of CO_2 emissions [2,3].

The IEA (International Energy Agency) Global Energy Review 2021 assessed that emissions of CO_2 are likely to increase by 5% to 33 billion tons. Global energy demand is expected to increase by 4.6% in 2021. The electricity sector is expected to contribute to 75% of this increase. Unless tangible steps are taken to curb emissions of CO_2, the situation in 2022 may become alarming [2,3]. It is a clear warning that enough is not being done to introduce clean energy technologies. We face an immense challenge in reforming the worldwide energy system. Therefore, there is a need to inculcate machining practices that limit electrical energy consumption and promote a green environment, thus highlighting the need for this research.

Review papers provide great insight into the work done to improve the energy efficiency of machining processes. For example, Zhang et al. [4] focused on theoretical and experimental models in the preview of energy efficiency of machine tools, Yoon et al. [5] reported on energy-saving strategies in the case of machine tools, Zhou et al. [6] researched in the field of cutting energy models based on machining processes, and Zhao et al. [7] concentrated on the optimization of energy components, process constraints, and improvements in auxiliary system efficiency.

The literature review indicates that research in energy conservation of machining processes started way back in 1994 when Bayoumi and Hutton [8] applied specific cutting energy to measure energy efficiency in the milling process. Draganescu et al. [9] stipulated numerical models for specific energy consumption and energy efficiency at the spindle level. Gutowski et al. [10] reported that the energy consumed by the cutting process accounts only for 20% of the total energy consumed by the machine tool. A theoretical model for the power consumption of a machine tool was put forward and the concept of specific energy consumption (SEC). Li and Kara [11] proposed an empirical SEC model in the turning process on a CNC lathe. The model encompassed coefficients related to workpiece materials and machine tools, along with material removal rate (MRR). He et al. [12] stated that machine tools usually function at an efficiency of less than 30% and have a high possibility for energy saving and efficiency improvement.

The SEC model introduced the milling process and considered the effect of change in spindle energy due to different cutting conditions [13]. A generalized SEC put up a model for automated machine tools in the milling process [14]. The production of machine tools and energy efficiency were solved with the help of scheduling and electricity utilization with a hybrid genetic algorithm. This method could diminish prices, minimize greenhouse gas emissions, and save energy [15]. Velchev et al. [16] set up an energy consumption model for the turning process and optimized the machining parameters. The main barrier faced in getting enhanced industrial energy efficiency is the mindset and attitude of manufacturing unit owners toward sustainable development and product manufacturing. The 11 key hurdles to making more sustainable and low-carbon manufacturing units were discussed [17]. Zhao et al. [18] broadened the model proposed by Li et al. [13] for the SEC for the turning process by considering the coolant pump's unloaded spindle power and power accompanied by standby and cutting power. Sealy et al. [19] measured the energy consumption of the machine tools at the cutting, spindle, and machine tool levels. Zhou et al. [7] broadened the Gutowski [10] and Li [11] models by considering the outcome of the speed of the spindle upon power utilization in the milling process. Other research focused on machining process-level energy. For example, Hu et al. [20] examined the machine tool's variable and fixed energy utilization states. They recommended an online method for checking the energy efficiency and energy utilization ratio of the machine tool.

Many researchers set up empirical models based upon the different working states of the machine tool. For example, an energy model was presented while considering toolset time, tool change, MRR, and embodied energy of the cutting tool [21]. The energy consumption of different machine tools was observed on the basis of each machine tool's size and technical features [22]. The total energy in the milling process was estimated, taking into consideration different tool paths. The total energy consumed was the aggregate of energy consumed for each state, e.g., basic state, tool change spindle rotation, coolant, feed motion, and the cutting state [23]. The machining energy was modeled on the basis of rapid transverse, spindle acceleration, and material removal states [24]. The variation in specific cutting energy was studied using undeformed chip thickness, tool wear, cutting tool nose radius, and dry and flood coolant, assuming that the tip energy was 25% of the total energy demand. The optimum feed rate level helped the study achieve a 72% reduction in tip energy, amounting to about an 18% reduction in total direct energy. It was pointed out that tool wear increases the specific energy coefficient, and the flood-cutting environment decreases the specific energy coefficient. Nose radii do not significantly alter the specific energy demand [25]. A model for energy consumption was suggested in the milling process based on machine tool constituents such as the spindle, feed axes, coolant pump, ATC, and chip conveyor. The spindle power was further modeled as a linear function of spindle speed. The axis feed power was modeled as a linear function of the feed rate. Lastly, the energy consumed in the cutting process was estimated as a difference between the total energy consumed and the energy consumed during the air cut [26].

The machine tool's total energy utilization was computed considering the power model and operation time for each machine tool component such as the spindle, tool changer, axis, coolant, chip conveyor, and clamping [27]. A specific energy consumption model was presented for material removal during the milling process, considering actual cutting energy and air-cutting energy [28]. A therblig-based value stream model was proposed. The therblig approach is built on the micro motions in the machine tool. Thus, the machining operations are split into series of small energy-consuming machine tool motions. This approach helps to analyze the energy consumption of the basic motions of the machine tool [29]. The idle, cutting, and tool change states were considered and put forward a model of direct machining energy. The model included embodied energies of the cutting tool and the coolant as indirect energy [30].

Few studies focused on improving energy efficiency. Instead, experiments were performed to conserve energy considering a weight reduction of moving parts of the machine tool by introducing lotus-type porous carbon steel, energy-saving by reducing standby time, use of an optimum-capacity coolant pump, and the influence of tilting angle on machining energy [31]. In addition, many researchers worked on the optimization of the machining parameters for a decrease in energy consumption. The modification of machining variables led to improved energy efficiency and reduced energy consumption. A higher cutting value led to diminished power of drilling, face, and end milling, but constraints of surface quality and life of tool must be considered. An adaptive pecking cycle also led to the lower power consumption of deep hole machining [32]. The dry milling process was performed on medium carbon steel C45 with input responses of cutting speed, feed, depth of cut, and radial depth of cut, and optimized output responses such as carbon emissions, surface roughness, and MRR [33]. The dry turning process was performed on AISI 1045 steel and optimized with the help of grey relational analysis for cutting power and surface roughness (Ra) [34]. The machining process of grooving was observed under dry conditions on AISI 4340 steel and considered an additional input response factor of the hardness of the material; the MRR and tool wear carbon emissions were optimized with the help of the fuzzy method [35]. Turning experiments were performed on alloy steel with input parameters of cutting speed, feed, depth of cut and nose radius, optimized energy efficiency, active energy consumption, and power factor during the machining process with the help of Taguchi and ANOVA [36]. Kumar et al. [37] researched the wet turning of EN 353 alloy and optimized the output response parameters of energy efficiency, active power consumption, active energy consumption, MRR, Ra, and power factor with the help of the Taguchi and TOPSIS methods. The effect of longitudinal ultrasonic vibrations and minimum quantity lubrication on the drilling force, burr height, and Ra was identified with the help of the RSM technique, showing that the most influential factors were the feed rate, vibration amplitude, and spindle speed [38]. A multi-objective optimization was performed on Ra, MRR, and SEC based on grey relational analysis while turning AISI 304 austenitic steel [39].

The literature review reveals specific terms used to express the energy efficiency of machine tool and machining processes. The energy utilization ratio is the ratio of the energy utilized for actual cutting or machining the workpiece to the total energy consumed by the machine tool. Researchers [19,28,37] utilized the concept of instantaneous energy efficiency in their research. This is the ratio of the power used during the cutting or machining process to the total power consumed by the machine tool at that instant. Researchers [9,20] also used SEC in their work, which can be described in three stages at the process, spindle, and machine tool levels. At the process level, it is defined as the ratio of energy consumed for material removal or the machining process to the volume of the material removed; alternatively, it is the ratio of the power consumed to the material removal rate. At the spindle level, it is the energy consumed by the spindle motor during machining to the volume of the material removed. At the machine tool level, it is the ratio of the volume of the material removed. A few authors used the concept of relative energy efficiency in their works [30,40] for energy benchmarking. This is the ratio of the minimum energy required to remove the material to the actual energy consumed to remove the material.

1.1. Research Gaps Based on Literature Review

The literature survey revealed that the research in the field of energy efficiency of machining processes has mainly been dedicated to building up empirical models for evaluating energy consumption. The empirical models obtained use coefficients and constants that depend on the machine tool's nature, the nature and composition of the workpiece, and the cutting tool. Such constants can be determined from experimental data only. However, there is a current need to evaluate energy or power consumption directly with the help of devices such as power loggers or power quality analyzers rather than empirical models. The literature review also revealed that most research has been conducted on machining processes such as turning or milling for parametric optimization. However, it is better to investigate, evaluate, and optimize machining processes that are energy-intensive. Moreover, the studies conducted previously revealed that electric parameters and their impact on power consumption have not been analyzed. Hence, there is a need to investigate the electric parameters and prioritize the most energy-intensive machining processes. The "most energy-intensive" term refers to the machining operation consuming maximum energy. It is judged on the basis of not only a single condition of energy consumption but also criteria such as specific energy consumption and energy efficiency.

1.2. Research Questions and Intended Contribution of the Study

Considering the research gaps mentioned above, the study addressed the following research question:

> *"Do we need to identify the electric parameters and prioritize the most energy-intensive machining processes?"*

The answer to this query is yes; this study can fill the gap of nonavailability of a systematic methodology to prioritize the most energy-intensive machining process. Research on the shop floor of an industry poses a significant challenge. However, it is one way to get hands-on knowledge and awareness about the industry's real challenges. This increases the chances of implementation of the results of the research by the industry. The time restrictions for conducting experiments represent one of the challenges faced. The production time of the industry may be affected. Halting one machine tool to perform experiments may stop the entire production line. However, actual conditions can be monitored and improved only when a real situation is observed on the shop floor. This helps to identify the impact on the quality of power being supplied to the given machine tool due to the working of other machine tools in the vicinity and the impact of electric parameters on power consumption. An empirical study or theoretical research is of no use unless it is meaningful and practical for the industry. This study intends to contribute the methodology to any industry to find significant electric parameters and identify the most energy-intensive machining operations. Hence, the investigation laid down several research objectives, described below.

1.3. Research Objectives

Considering the research gap and intended contribution, the study based its investigation upon the premises of the following research objectives:

- To investigate the significant electric parameters and to analyze their impact on power consumption.
- To ascertain a methodology to identify and prioritize the most energy-intensive machining processes.

The rest of the paper is organized as follows: Section 2 describes the materials and methods. Then, it identifies methods used to achieve the research objectives, followed by explaining the materials and machine tools used for experimentation. Section 3 deals with an investigation of significant electric parameters with the help of multiple regression modeling and determination of their impact on active power consumption. Section 4 includes the hybrid decision-making methodology and its application to prioritize energy-intensive machining operation, as well as a discussion of the results, followed by the conclusions in Section 5.

2. Materials and Methods

A case study was conducted to prioritize energy-intensive machining operations and gauge the influence of electric parameters. Firstly, the research team identified the manufacturing industry for performing experiments. As a result, Auto International was selected in the industrial hub of Ludhiana, Punjab, India. It is located in Kohara, a place near the industrial town of Ludhiana. The main reason for choosing this industry was the willingness and cooperation of the management of the industry to extend their facility for experimentation despite the busy schedule of the shop floor. In addition, the industry management was interested in the outcome of the investigation and looked forward to the energy conservation of their machining operations. Next, the research team assessed the various machining operations in the concerned industry's machine shop. The identified operation consisted of different machining operations to accomplish the final product. The details of the specified machining operation are shown in Figure 1.

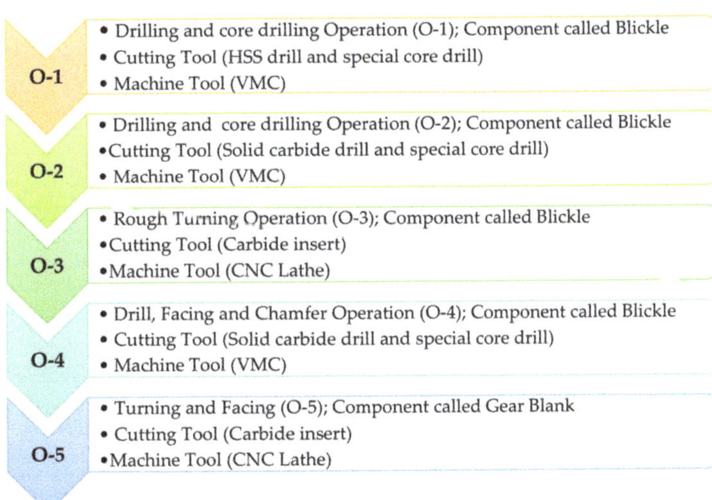

Figure 1. Machining operation details identified for study in the machine shop of the industry.

A component called a Blickle (shown in Figure 2a, used as a component in airport trolleys) was machined on a vertical milling machine (VMC). The first four operations (O-1 to O-4) shown in Figure 1 were performed on the Blickle. Operation O-1 involved drilling with an HSS drill and special core drill. Operation O-2 consisted of drilling with a solid carbide drill and special core drill. Operation O-3 involved rough turning with a carbide insert. Lastly, O-4 consisted of drilling, facing, and chamfering with a solid carbide drill and special core drill. A Gear Blank (Figure 2b, used as a component in tractor gears) was machined as the workpiece for O-5; its machining was completed on a CNC lathe. Operation O-5 involved the turning and facing of the Gear Blank with a carbide insert. The details of the workpiece, including the Blickle and Gear Blank, are shown in Table 1. The cutting tools utilized for machining operations from O-1 to O-5 and their materials and particulars are shown in Table 2. The details of cutting parameters for each operation are shown in Table 3. The specifications of the machine tools used are presented in Table 4. Figure 3a shows the power logger, and Figure 3b shows the connections of the power logger with the main power supply.

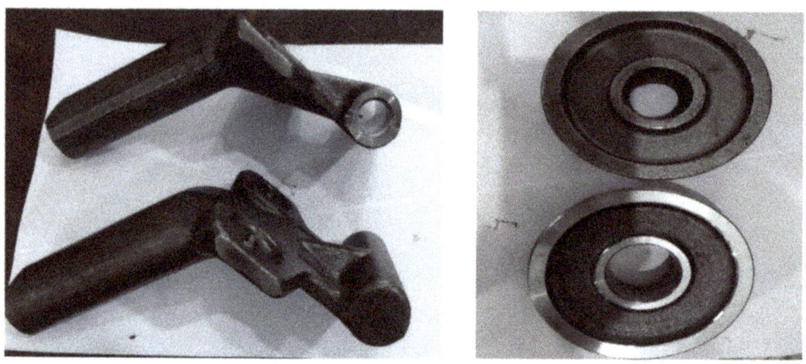

(**a**) Blickle (**b**) Gear Blank for turning

Figure 2. Identified components under study.

Table 1. Details of the workpiece.

Component Blickle for O-1 to O-4	
Workpiece material	Mild steel grade: DIN: ST52.3
Percentage composition	C: 0.207–0.22, Mn: 1.04–1.6, Si: 0.240–0.5, P: 0.033–0.035, Al: 0.038, and rest Fe
Surface Hardness	149/167 BHN
Grain size	6.5 to 7.0
Micro Structure	Pearlite + ferrite
Applications	Manufacturing of automobile parts, airport trolley parts
Component Gear Blank for O-5	
Workpiece material	20MnCr5 steel or EN 10084-2008
Percentage composition	C: 0.17–0.22, Si_{max}: 0.4, Mn: 1.1–1.4, Cr: 1–1.3, P_{max}: 0.035, S_{max}: 0.035
Applications	Auto parts, tractor parts

Table 2. Cutting tools, materials, and particulars.

Operation	Cutting Tool	Technical Particulars
O-1 Drilling	HSS drill; Make: ITM	Diameter Φ 16.0, flute length 110 mm, point angle 140°, number of flutes 2
O-1 Core Drilling	Special core drill (solid carbide) with SECO inserts; Namoh Tooling's	Flute length 85 mm, diameter Φ 17.75, tool with two fine boring inserts SECO make SCGX060204P2
O-2 Drilling	Solid carbide drill; Namoh Tooling's	Diameter Φ 16, point angle of 140°, coating of TiAIN, flute length 100 mm, shank length of 50 mm, external cooling
O-2 Core Drilling	Special core drill (solid carbide) with SECO inserts; Namoh Tooling's	Flute length 85 mm, diameter Φ 17.75, tool with two fine boring inserts SECO make SCGX060204P2
O-2 Chamfer	SECO insert	TCMT110204-F1-TP1501, chamfer at angle 45°
O-2 Facing	SECO insert	SCGX060204P2
O-3 Rough Turning	SECO carbide insert	WNMG060408-M5-TP1501
O-3 Facing	Facing insert; SECO Make	ONMU0900520 ANTN-M13-F40M
O-4 Drilling	Special carbide drill Namoh Tooling	Flute length 50 mm, point angle 140°
O-4 Chamfer	SECO insert	TCMT110204-F1-TP1501 chamfer at angle 45°
O-4 Facing	SECO insert	SCGX060204P2
O-5 Turning and Facing	SECO carbide insert	WNMG060408-M5-TP1501

Table 3. Cutting parameters for machining operations.

Machining Operation	Cutting Parameters Details
O-1	Drilling and Core Drilling-1: (on VMC) Drilling-1 (HSS drill): External diameter of workpiece 30 mm Hole diameter in drilling: 16 mm, Spindle RPM 450 Incremental peck drilling 15 mm of peck length, Feed rate 70 mm/min, the actual depth of the hole 77 mm Core drilling: Core diameter 17.8 mm, Spindle RPM 1200, Feed rate 120 mm/min
O-2	Drilling and Core Drilling-2: (on VMC) Drilling-2 (Solid carbide drill): External diameter of workpiece 30 mm Hole diameter in the drilling 16 mm, Spindle RPM 1050 Incremental peck drilling 10 mm of peck length, Feed rate 125 mm/min, The actual depth of the hole 77 mm Core drilling: Core diameter 17.8 mm, Spindle RPM 1200, Feed rate 120 mm/min
O-3	Rough Turning: (on CNC) External diameter of rough turning 35 mm, Final diameter 31.4 mm, depth of cut of 1.8 mm, feed of 0.18 mm/rev, Cutting speed 31 m/min, Length of cut 105 mm
O-4	Drilling and Chamfer, Facing: (on VMC) Facing: Spindle RPM 1500, Feed rate of 200 mm/min Drilling: Drill diameter 14 mm, Spindle speed 1500 RPM, Incremental peck drilling Peck length 8.2 mm, Hole depth 10.5 mm Chamfer: Spindle RPM 2000, Feed 150 mm/min, 1 × 45°
O-5	Turning and Facing: (on CNC) Turning: Final outer diameter obtained 133.4 mm, Length of cut 19.1 mm, depth of cut 1 mm, Feed 0.18 mm/rev, Cutting speed 168 m/min Facing-1: Outer diameter 133.4 mm and inner diameter 105 mm, faced through the depth of 1 mm. Feed 0.18 mm/rev, cutting speed 150 m/min Facing-2: Outer diameter 68 mm and inner diameter 52.78 mm, faced through the depth 1 mm, Feed 0.18 mm/rev, cutting speed 76 m/min

Table 4. Technical specifications of machine tool used.

Description	Units	VMC BFW/V-4 BT-40	CNC LATHE BFW RHINO (2550)	CNC LATHE JYOTI (DX-100)
CNC System	-	Fanuc 01-MF	Fanuc (B-6i) 828D	Siemens
Spindle Motor Power	kW	7.5 (Cont.) 11 (Int.)	11 15	7 10.5
Spindle Speed	rpm	8000	2000	4000
Table *X*-Axis	mm	600	200 (cross)	360
Saddle *Y*-Axis	mm	450	-	-
Spindle *Z*-Axis	mm	500	625 (longitudinal)	200
Axis Drives Feed Rate	mm/min	1–10,000	20 Rapid feed (*X* and *Z*) axis	24 Rapid feed (*X* and *Z*) axis
Ball screw Día × Pitch	mm	32 × 16	32 × 10 (*X*-axis) 40 × 10 (*Z*-axis)	32 × 10
Table Clamping Area	mm × mm	450 × 900	-	-
ATC (No. of Tools)	-	24	8/12	5
Accuracy Positioning	mm	±0.007	±0.007	±0.007
Accuracy Repeatability	mm	±0.005	±0.005	±0.005
Power Supply		3-Phase, 415 V, 50 Hz	3-Phase, 415 V, 50 Hz	3-Phase, 415 V, 50 Hz
Total Machine Power	KVA	18	16	
Chuck Size	mm	-	250	170
Std. Turning Diameter	mm	-	350 (max)	100
Max. Turning Length	mm	-	200	200

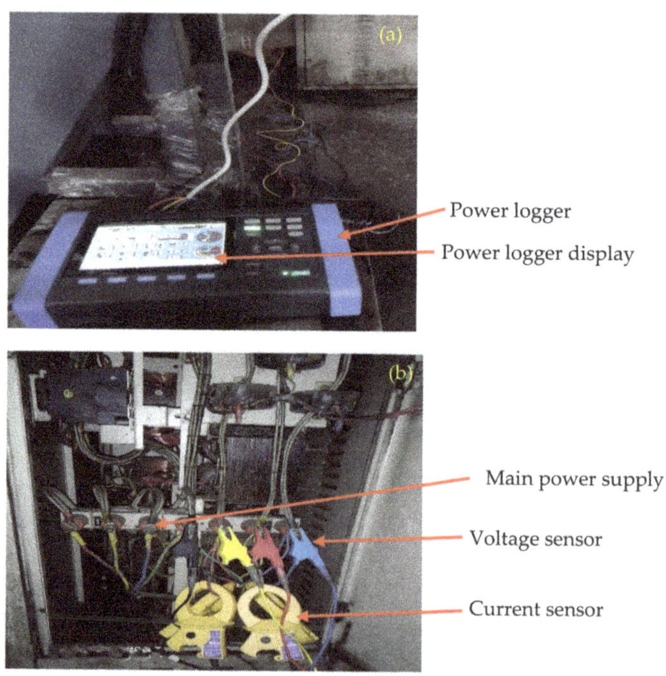

Figure 3. (**a**) Hioki power quality analyzer; (**b**) Hioki power analyzer connections.

The equipment used for making the observations was a Power Quality Analyzer, HIOKI Make, model PQ3100 (HIOKI EE Corporation, Ueda, Nagano 386-1192, Japan). The wiring mode used was 3P4W (3-Phase 4 wire). The voltage measurement was 415 V (line to line), the current sensor used was CT7126, the rated current was 60 A AC, the maximum current was 100 A peak, the maximum rated voltage to earth was 300 V AC, the anticipated transient overvoltage was 4000 V, the active power was 300 W to 9 MW with an accuracy of ±0.3% rdg. ±0.1% f.s. + clamp sensor accuracy, the measurable conductor diameter was Φ15 mm, and the measurement accuracy was as follows: frequency (45 Hz $\leq$ f $\leq$ 66 Hz), for (±0.3% rdg. ±0.1% f.s.), and phase (±0.2°).

2.1. Electric Parameters and Energy Responses

- Active power consumption by machine ($\ddot{A}PC_{m/c}$) in kW

This is the total power consumed by the machine and machining operation during the actual machining of the workpiece [41–43].

- Power factor ($PF_{m/c}$)

The $PF_{m/c}$ is the ratio of active power to apparent power, as shown in Equation (1) [36].

$$PF_{m/c} = \frac{\text{Active power}}{\text{Apparent power}}. \quad (1)$$

Active power is the actual power utilized to do useful work. Reactive power in (Var-volt-ampere reactive) is when the power swings back and forth without any work. It is the product of the apparent power and the sine of the phase difference (sinθ). It results from inductive loads known as lag reactive power and reactive power ensuing from capacitive loads known as lead reactive power. Apparent power in VA is obtained by combining active power and reactive power vectorially. Good PF usually ranges from 1.0 to 0.95, poor PF ranges from 0.95 to 0.85, and bad PF is below 0.85.

- Active energy consumption by machine ($\ddot{A}EC_{m/c}$) in kWh

This is the total energy consumed by the machine and machining operation during the actual machining of the workpiece. Energy is a product of average active power over a complete cycle, and machining time ($t_{m/c}$) is shown in Equation (2) [36,42].

$$\ddot{A}EC_{m/c} = \ddot{A}PC_{m/c} \times t_{m/c}. \quad (2)$$

- Energy efficiency ($\dot{E}\dot{E}_{\acute{\eta}}$)

This gives the ratio of the energy consumed by the machining process to the total energy consumed by the machine. The $\dot{E}\dot{E}_{\acute{\eta}}$ is shown in Equation (3) [36,44–47].

$$\dot{E}\dot{E}_{\acute{\eta}} = \frac{\ddot{A}EC_{m/c} - \ddot{A}EC_{m/c}(AC)}{\ddot{A}EC_{m/c}} \times 100, \quad (3)$$

where $\ddot{A}EC_{m/c}(AC)$ is the energy consumed by the machine during air cutting.

- Specific energy consumption (ȿ) in kJ/cm^3

This is the energy required to remove a unit volume of the material. It is obtained by dividing the total energy consumed by a machine during the machining of a workpiece by the total volume of material removed, as shown in Equation (4) [48–50].

$$\text{Specific energy consumption (ȿ)} = \frac{\ddot{A}EC_{m/c}}{\text{Unit volume}}. \quad (4)$$

- Current RMS (root mean square) value ($\breve{I}_{m/c}$) in ampere

The power quality analyzer records the current at an interval of 1 s. The current registered is the average of the root mean square (RMS) value of current of each of the three phases at that instant. Thus, the magnitude of the current documented and used for calculations is yet again the average of all these observations over the entire machining cycle.

- Voltage RMS value ($\breve{V}_{av}$) in volt

The power quality analyzer logs the voltage at an interval of 1 s. The voltage measured is a line-to-line voltage in a three-phase supply. The voltage noted is the average of the RMS value of voltage across each of the three phases at that instant. Thus, the magnitude of the voltage documented and used for calculations is yet again the average of all these observations over the entire machining cycle [51].

- Current unbalance ($\check{I}_{m/c}$ub)

This is measured as a percentage of the fundamental current, as shown in Equation (5). The three-phase power system is balanced or symmetrical if the three-phase voltages and currents have the same amplitudes and same phase shifting (angular difference) at 120° to each other. If either or both of these conditions are not met, the system is unbalanced or asymmetrical [51].

$$\check{I}_{m/c_{ub}}\,(\%) = \frac{\text{Maximum deviation from } \check{I}_{m/c\ m/c}}{\check{I}_{m/c\,\frac{m}{c}}} \times 100. \quad (5)$$

- Voltage unbalance ($\check{V}$ub)

This is measured as a percentage of the fundamental voltage, as shown in Equation (6). Thus, the current unbalance factor is several times larger than the voltage unbalance factor [51].

$$\check{V}_{ub}\,(\%) = \frac{\text{Maximum deviation from the average voltage}}{\text{average voltage}} \times 100. \quad (6)$$

Thus, the current unbalance causes power and energy losses. International standards such as EN-50160 and IEC 1000-3-series give limits for the unbalance voltage calculated using the ratio of sequences method up to 2% for LV (low voltage) and MV (medium voltage) systems measured as 10 min values with an instantaneous maximum of 4% [51,52].

- Current total harmonic distortion factor ($\check{I}_{m/cthdf}$)

Harmonics are measured in terms of the total harmonic distortion factor. This indicates the extent to which the total harmonic component is distorting the fundamental waveform. The power quality analyzer used for experimentation can measure all types of triplen harmonics. According to the general system's IEEE 519-1992 standard, the harmonic voltage limits the maximum harmonic distortion factor's to 5.0% [41,53].

- Electrical power transients

Electrical transients are brief bursts of energy that occur on power, data, or communication networks. Transients are momentary fluctuations in voltage or current that last less than a millisecond. However, for a fraction of a second, high voltages are utilized to drive large amounts of current into an electrical circuit. The cause of the power transients may be due to utility grid switching, arc welders (arc flash), equipment cycling, grounding, lightning strike, and voltage/current drops [41,53].

2.2. *Experimental Procedure, Observations, and Calculations*

Under the current study, a particular industry was selected. Five separate machining operations on different machine tools were considered. These operations were designated as O-1 to O-5. Details of these machining operations were presented in Section 2.1. The machine tools were placed in the machine shop of the industry. The experiment was performed under the actual operating conditions. Other machine tools in the vicinity of the machine tool under study were simultaneously functioning. Power and electric parameters for each of the machining operations were noted with the help of the Hioki-3100 power quality analyzer attached to the main power supply of the machine tool. Readings were noted at an interval of 1 s. The data obtained were analyzed with the help of Hioki PQ-one version 4.00 software (HIOKI EE Corporation, Ueda, Nagano 386-1192, Japan. The machining time for each operation was carefully noted. Three sets of observations were recorded for each of the machining operations.

Furthermore, two types of observations were made for each machining operation, first under air cutting (AC) and then under an actual cut. In air cutting, observations were recorded without actual cutting taking place, whereas, under the actual cut, observations were recorded when actual cutting or machining was taking place.

The observations were stored in the SD memory card of the power quality analyzer. The memory card was then transferred to the laptop or computer where the Hioki PQ-1 software Ver. 4.00 was installed, and different electric parameters were recorded at an interval of 1 s for the entire duration of the machining operation. The observed data were tabulated under an excel chart. The average values of each electric parameter corresponding to each machining operation were noted. In the study, the machining processes did not have any power transients due to switching, welding, equipment cycling, grounding, and lightning supplies. Consequently, these were not taken into account as factors affecting electric parameters. Furthermore, motor conditions for machines such as VMC/CNC do not change perceptibly with time and remain for a long time. Moreover, the machine tools under study were no more than 3 years old; thus, this effect was negligible. Higher-order harmonics (e.g., fifth and ninth) were seen in the observations, but their percentage values were too minimal to make an impact; hence, they could be safely rejected. Moreover, these harmonics were well within IEEE 519 and even EN50160 harmonic ratings or permissible limits. The experimental observations and calculations for air cutting, i.e., without machining, are shown in Table 5.

Table 5. Experimental observations and calculations of air cutting (without machining).

Operation	$ɿ_{m/c}$	$\ddot{A}PC_{AC}$	$\ddot{A}EC_{AC}$	$\ddot{I}_{m/cm/c}$	$\ddot{I}_{m/cthdf}$	$\ddot{I}_{m/cub}$	$\ddot{V}_{av}$	$\ddot{V}_{ub}$	$\ddot{V}_{thdf}$
O-1	154	1.467	0.062	3.335	26.345	33.645	416.540	2.658	3.132
O-2	104	1.449	0.041	3.277	26.676	31.750	411.223	1.570	5.406
O-3	76	2.205	0.046	5.423	29.082	14.240	415.302	0.410	2.149
O-4	77	1.080	0.023	2.457	41.035	35.515	417.354	0.260	2.255
O-5	63	1.626	0.030	3.594	52.480	12.930	428.041	0.260	5.900

Active energy consumption was calculated as per Equation (2). The experimental observations taken from the power logger for actual cutting, i.e., during machining operations for operations O-1 to O-5, are shown in Table 6. The calculations of active energy, energy efficiency, and specific energy consumption as per Equations (2), (3), and (4) are shown in Table 7. These data were used to evaluate the impact of electric parameters on active power consumption and ascertain a methodology to identify and prioritize the most energy-intensive machining processes.

Table 6. Experimental observations of the actual cut (machining).

Operation	$ɿ_{m/c}$ (sec)	$\ddot{A}PC_{m/c}$ (kW)	$\ddot{A}EC_{m/c}$ (kWh)	V (cm^3)	$PF_{m/c}$	$\ddot{I}_{m/cm/c}$ (A)	$\ddot{I}_{m/cthdf}$ (%)	$\ddot{I}_{m/cub}$ (%)	$\ddot{V}_{av}$ (V)	$\ddot{V}_{ub}$ (%)	$\ddot{V}_{thdf}$ (%)
O-1	154	1.973	0.084	19.759	0.677	4.027	40.898	40.316	415.500	2.58	3.638
O-2	104	2.321	0.067	19.759	0.675	4.719	40.965	35.670	409.706	1.630	6.565
O-3	76	3.706	0.077	19.702	0.686	7.404	41.435	10.94	415.234	0.405	2.26
O-4	77	1.542	0.033	3.760	0.647	3.478	43.673	18.583	416.825	0.225	2.257
O-5	63	2.975	0.053	14.300	0.738	5.691	51.447	19.420	426.704	0.285	6.349

Table 7. Calculations of actual cut during O-1 to O-5.

Operation	$\ddot{A}EC_{m/c}$ (kJ)	$\dot{E}\dot{E}_{\eta}$ (%)	$\$$ (kJ/cm^3)
O-1	302.472	25.637	15.308
O-2	241.200	38.209	12.207
O-3	277.74	40.260	14.097
O-4	118.620	30.303	31.548
O-5	190.800	43.396	13.343

3. Regression Models of Active Power Consumption for Machining Operations

This section describes the multiple regression modeling of active power consumption in each machining operation from O-1 to O-5. Experiments were carried out in the real world at a machine shop in the industry. They were not carried out in a laboratory under controlled settings. Many machine tools and pieces of equipment were used simultaneously, along with the observed machine tools. As a result, the quality of the electric power supplied to the machine tool was affected. Therefore, the main aim of the regression modeling was to identify the significant electric parameters affecting the active power consumed by the machining operations. The predictors or the independent variables considered were the following electric parameters: the average current, power factor, current total harmonic distortion factor, current unbalance, voltage unbalance, average voltage, and voltage total harmonic distortion factor.

The regression models for the five machining operations were developed using the backward elimination method. This involved eliminating nonsignificant terms from the model. The elimination of nonsignificant terms is based on the *p*-value of the *t*-statistic test evaluated at $\alpha = 0.05$ for each of the predictors. If the *p*-value was higher than 0.05, the predictor was considered nonsignificant. The coefficient of determination (R^2) and ANOVA were used to assess the fitness of the proposed model. The R-squared value describes the variance in the response data interpreted by the regression model. The model's predicted R-squared (pred) indicates how it might anticipate data. When unnecessary variables are included in the model, the adjusted R-squared (adj) will typically decrease. There is a good chance that significant terms have been introduced into the model if the difference between R-squared and R-squared (adj) is small [54].

The regression models were made with coded regression coefficients. These were obtained by subtracting the mean from each predictor's value and dividing it by the standard deviation. Coded coefficients help to determine the relative importance of each of the significant predictors to the dependent variable. The R-squared Value was noted after the first iteration of the regression modeling process. The R-squared values are shown in Table 8.

Table 8. Coefficient of determination (R-sq.) for O-1 to O-5.

Operation	Iteration-1		Final Iteration	
	R-Sq.	(%)	Non-Significant Terms Removed	R-Sq. (%)
O-1	R-sq.	98.56	$\ddot{V}_{ub}$, $\ddot{V}_{thdf}$, $\ddot{I}_{m/cub}$	98.55
	R-sq. (adj)	98.50		98.51
	R-sq. (pred)	98.19		98.27
O-2	R-sq.	99.52	$\ddot{V}_{ub}$, $\ddot{V}_{thdf}$, $\ddot{I}_{m/cthdf}$	99.50
	R-sq. (adj)	99.48		99.48
	R-sq. (pred)	99.39		99.42
O-3	R-sq.	99.69	$\ddot{V}_{thdf}$, $\ddot{I}_{m/cthdf}$, $\ddot{V}_{av}$, $\ddot{I}_{m/cub}$	99.65
	R-sq. (adj)	99.66		99.64
	R-sq. (pred)	99.46		99.49
O-4	R-sq.	98.60	$\ddot{V}_{ub}$, $\ddot{V}_{av}$, $\ddot{I}_{m/cub}$	98.41
	R-sq. (adj)	98.46		98.33
	R-sq. (pred)	98.07		98.02
O-5	R-sq.	93.96	$\ddot{V}_{thdf}$, $\ddot{V}_{av}$	93.64
	R-sq. (adj)	93.20		93.08
	R-sq. (pred)	89.57		89.30

The developed regression equations for machining operations in uncoded regression coefficients from O-1 to O-5 are shown in Equations (7)–(11).

$$\ddot{A}PC_{m/c}\ (O\text{-}1) = -12{,}133 + 23.8\ \ddot{V}_{av} + 7.893\ \ddot{I}_{m/cthdf} + 3964.8\ PF_{m/c} + 301.4\ \ddot{I}_{m/cm/c}. \tag{7}$$

$$\ddot{A}PC_{m/c}\ (O\text{-}2) = -17{,}763 + 13.25\ \ddot{I}_{m/cub} + 34.97\ \ddot{V}_{av} + 4494\ PF_{m/c} + 478.76\ \ddot{I}_{m/cm/c}. \tag{8}$$

$$\ddot{A}PC_{m/c}\ (O\text{-}3) = -1621 - 1441\ \ddot{V}_{ub} + 2316\ PF_{m/c} + 582.94\ \ddot{I}_{m/cm/c}. \tag{9}$$

$$\ddot{A}PC_{m/c}\ (O\text{-}4) = -1618 - 418\ \ddot{V}_{thdf} + 6.466\ \ddot{I}_{m/cthdf} + 4002\ PF_{m/c} + 355.57\ \ddot{I}_{m/cm/c}. \tag{10}$$

$$\ddot{A}PC_{m/c}\ (O\text{-}5) = -5019 - 11296\ \ddot{V}_{ub} + 68.4\ \ddot{I}_{m/cub} + 24.7\ \ddot{I}_{m/cthdf} + 6853\ PF_{m/c} + 629.2\ \ddot{I}_{m/cm/c}. \tag{11}$$

After removing nonsignificant terms (p-values greater than 0.05), final models of active power for each operation were accepted with electric parameters having significant effects (p-value less than 0.05). The results of ANOVA are shown in Table 9. The coded coefficients for machining operations from O-1 to O-5 are shown in Table 10.

Table 9. ANOVA Table for O-1 to O-5.

Source	DF	Adj SS	Adj MS	F-Value	*p*-Value
			O-1		
Regression	4	41,924,033	10,481,008	2526.98	0.000
Error	149	617,999	4148		
Total	153	42,542,032			
			O-2		
Regression	4	103,689,348	25,922,337	4895.57	0.000
Error	99	524,211	5295		
Total	103	104,213,559			
			O-3		
Regression	3	68,657,447	22,885,816	6859.26	0.000
Error	72	240,227	3336		
Total	75	68,897,674			
			O-4		
Regression	4	19,636,536	4,909,134	1117.05	0.000
Error	72	316,419	4395		
Total	76	19,952,955			
			O-5		
Regression	5	143,722,619	28,744,524	167.81	0.000
Error	57	9,763,666	171,292		
Total	62	153,486,286			

Table 10. Coded regression coefficients for O-1 to O-5.

Term	Coef	SE Coef	T-Value	*p*-Value	VIF
			O-1		
Constant	1942.20	5.19	374.24	0.000	
$\ddot{V}av$	12.04	5.39	2.23	0.000	1.07
$\ddot{I}_{m/c}thdf$	82.22	8.76	9.39	0.000	2.83
$PF_{m/c}$	369.48	7.55	48.93	0.000	2.10
$\ddot{I}_{m/cm/c}$	239.48	8.97	26.71	0.000	2.95

Table 10. *Cont.*

Term	Coef	SE Coef	T-Value	*p*-Value	VIF
			O-2		
Constant	2302.78	7.14	322.73	0.000	
$\breve{I}_{m/cub}$	96.0	11.7	8.19	0.000	2.67
$\ddot{V}_{av}$	30.05	8.04	3.74	0.000	1.26
$PF_{m/c}$	386.7	13.8	28.03	0.000	3.70
$\breve{I}_{m/cm/c}$	755.8	10.8	69.77	0.000	2.28
			O-3		
Constant	3685.24	6.63	556.20	0.000	
$\ddot{V}_{ub}$	−30.80	6.91	-4.46	0.000	1.07
$PF_{m/c}$	158.1	13.2	11.94	0.000	3.94
$\breve{I}_{m/cm/c}$	816.1	13.1	62.42	0.000	3.84
			O-4		
Constant	1528.34	7.55	202.30	0.000	
$\ddot{V}_{thdf}$	−17.43	8.09	−2.15	0.035	1.13
$\breve{I}_{m/cthdf}$	105.1	13.8	7.60	0.000	3.30
$PF_{m/c}$	431.4	13.0	33.31	0.000	2.90
$\breve{I}_{m/cm/c}$	415.02	8.96	46.34	0.000	1.39
			O-5		
Constant	2942.9	52.1	56.44	0.000	
$\ddot{V}_{ub}$	−183.9	62.2	−2.96	0.005	1.40
$\breve{I}_{m/cub}$	367	180	2.04	0.046	11.68
$\breve{I}_{m/cthdf}$	218.9	98.9	2.21	0.031	3.54
$PF_{m/c}$	1108.4	61.2	18.12	0.000	1.35
$\breve{I}_{m/cm/c}$	1211	133	9.11	0.000	6.40

3.1. Analysis of the Regression Model for Operation-1 (O-1)

In machining operation O-1, the overall regression model as per Equation (7) was statistically the best fit because the coefficient of determination as per Table 8 after the final iteration had values of R-sq. 98.50% and R-sq. (adj) 98.47%. This means that the regression model explained the maximum variation of the dependent variable, which was active power, up to 98.47%. A look at the residual plots for $\ddot{A}PC_{m/c}$ for (O-1) in Figure 4b indicates that the residuals lie close to the diagonal line representing an ideal normal distribution. The points of the residual plots were not skewed, and they were randomly distributed. Therefore, it seems that the data were normally distributed. Furthermore, a look at the histogram of residuals gives evidence that our residuals were normally distributed. The distribution of residuals along the straight horizontal line was similar for all significant parameters, suggesting equality of variance. Therefore, the conditions of normality of residuals and equality of variance were fulfilled. The ANOVA results for (O-1) revealed that there was significance between-group variance according to the value of the F-statistic (see Table 9). At $\alpha = 0.05$, the F-value was equal to 2526.98 with a p-value < 0.001. This indicates evidence of a regression relationship between the dependent variable $\ddot{A}PC_{m/c}$ and the independent variables $PF_{m/c}$, $\breve{I}_{m/cm/c}$, $\ddot{V}_{av}$, and $\breve{I}_{m/cthdf}$ combined.

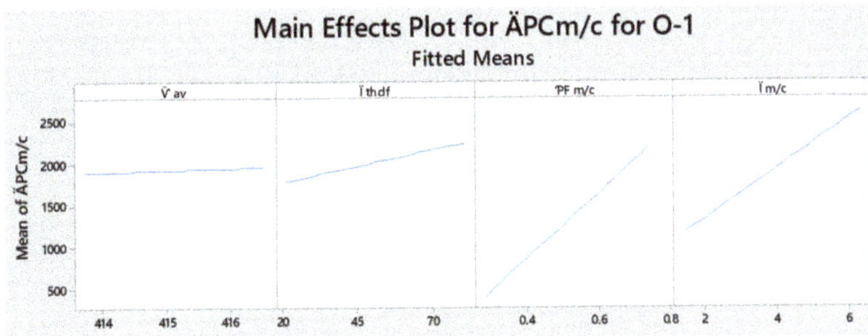

(a)

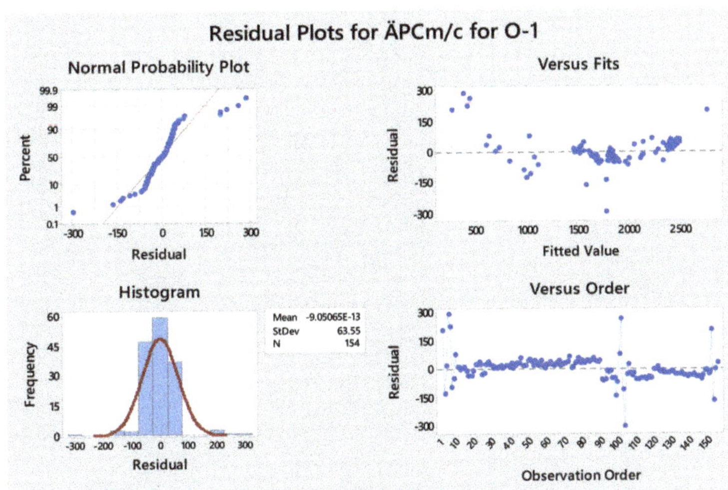

(b)

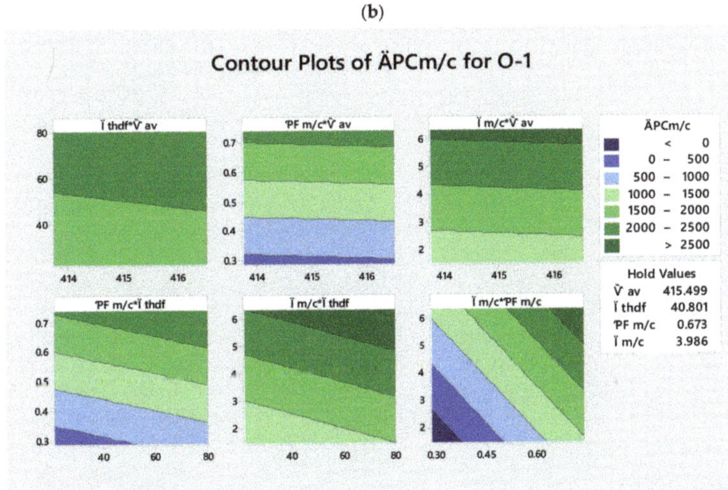

(c)

Figure 4. (**a**) Main effects plot for O-1. (**b**) Residuals plot for O-1. (**c**) Contour plots for O-1.

Individual coefficients contributed meaningful information in the prediction of $\ddot{A}PC_{m/c}$ (see Table 10 for O-1). The test statistics showed significant *t*-values of $PF_{m/c}$, $\breve{I}_{m/cm/c}$, $\breve{V}_{av}$, and $\breve{I}_{m/cthdf}$ at $\alpha = 0.05$. Therefore, $PF_{m/c}$, $\breve{I}_{m/cm/c}$, $\breve{V}_{av}$, and $\breve{I}_{m/cthdf}$ were individually useful in the prediction of $\ddot{A}PC_{m/c}$. The coded coefficients in Table 10 reveal that, for machining operation O-1, the predictor with the highest impact on the dependent variable $\ddot{A}PC_{m/c}$ was $PF_{m/c}$ (369.48), followed by $\breve{I}_{m/cm/c}$ (239.48), $\breve{I}_{m/cthdf}$ (82.22), and $\breve{V}_{av}$ (12.04). The value of the VIF (variance inflation factor) in Table 10 for O-1 of all predictors was less than 5. This indicates that there was no significant multicollinearity between the significant predictors.

A look at the main effects plots for $\ddot{A}PC_{m/c}$ in Figure 4a also validates the above result. The slope of significant predictors versus $\ddot{A}PC_{m/c}$ indicates that, in this machining operation (O-1), $PF_{m/c}$ had a maximum impact on $\ddot{A}PC_{m/c}$, followed by $\breve{I}_{m/cm/c}$, $\breve{I}_{m/cthdf}$, and $\breve{V}_{av}$.

The contour plots of $\breve{I}_{m/cm/c}$, $PF_{m/c}$, and $\ddot{A}PC_{m/c}$ in Figure 4c indicate the maximum impact of $\breve{I}_{m/cm/c}$ and $PF_{m/c}$ on $\ddot{A}PC_{m/c}$ as the number of contour lines representing different ranges of $\ddot{A}PC_{m/c}$ consumption was greater. This suggests that any variation in $\breve{I}_{m/cm/c}$ and $PF_{m/c}$ significantly affects the power consumption. The contour plot between $PF_{m/c}$ and $\breve{I}_{m/cthdf}$ and between $\breve{I}_{m/cm/c}$ and $\breve{I}_{m/cthdf}$ indicate that both parameters were significant, but their impact on $\ddot{A}PC_{m/c}$ was not as substantial as $\breve{I}_{m/cm/c}$ and $PF_{m/c}$, whereas the plot between $\breve{I}_{m/cthdf}$ and $\breve{V}_{av}$ exhibited the least significance.

3.2. *Analysis of the Regression Model for Operation-2 (O-2)*

In the machining operation O-2, the overall regression model as per Equation (8) was statistically the best fit because the coefficient of the determination as per Table 8 after the final iteration had values of R-sq. 99.50% and R-sq. (adj) 99.48%. This means that the regression model explained the maximum variation of the active power up to 99.48%. A look at the residual plots for $\ddot{A}PC_{m/c}$ for (O-2) in Figure 5b indicates that the residuals lie close to the diagonal line, representing an ideal normal distribution. The points of the residual plots were not skewed, and they were randomly distributed. Therefore, it seems that the data were normally distributed. Furthermore, a look at the histogram of residuals gives evidence that our residuals were normally distributed. The distribution of residuals along the straight horizontal line was similar for all significant parameters, suggesting equality of variance. Therefore, the conditions of normality of residuals and equality of variance were fulfilled. The ANOVA results for O-2 revealed that there was significant between-group variance based on the value of the F-statistic (see Table 9). At $\alpha = 0.05$, the F-value was equal to 4895.57 (*p*-value < 0.001). This indicates evidence of a regression relationship between the dependent variable $\ddot{A}PC_{m/c}$ and the independent variables $PF_{m/c}$, $\breve{I}_{m/cm/c}$, $\breve{V}_{av}$, and $\breve{I}_{m/cub}$ combined.

Individual coefficients contributed meaningful information to the prediction of $\ddot{A}PC_{m/c}$ (see Table 10 for O-2). The test statistics showed *t*-values of $PF_{m/c}$, $\breve{I}_{m/cm/c}$, $\breve{V}_{av}$, and $\breve{I}_{m/cub}$ at $\alpha = 0.05$. Therefore, $PF_{m/c}$, $\breve{I}_{m/cm/c}$, $\breve{V}_{av}$, and $\breve{I}_{m/cub}$ were individually useful in the prediction of $\ddot{A}PC_{m/c}$. The coded coefficients in Table 10 reveal that, for machining operation O-2, the predictor with the highest impact on the dependent variable $\ddot{A}PC_{m/c}$ was $\breve{I}_{m/cm/c}$ (755.8) followed by $PF_{m/c}$ (386.7), $\breve{I}_{m/cub}$ (96.0), and $\breve{V}_{av}$ (30.05). The value of the VIF (variance inflation factor) in Table 10 for O-2 of all predictors was less than 5. This indicates that there was no significant multicollinearity between the significant predictors.

A look at the main effects plots for $\ddot{A}PC_{m/c}$ in Figure 5a also validates the above result. The slope of significant predictors versus $\ddot{A}PC_{m/c}$ indicates that, in this machining operation (O-2), $\breve{I}_{m/cm/c}$ had the maximum impact on $\ddot{A}PC_{m/c}$, followed by, $PF_{m/c}$, $\breve{I}_{m/cub}$, and $\breve{V}_{av}$.

The contour plots of $\breve{I}_{m/cm/c}$, $PF_{m/c}$, and $\ddot{A}PC_{m/c}$ in Figure 5c indicate the maximum impact of $\breve{I}_{m/c}m/c$ and $PF_{m/c}$ on $\ddot{A}PC_{m/c}$ as the number of contour lines representing different ranges of $\ddot{A}PC_{m/c}$ consumption was greater. This suggests that any variation in $\breve{I}_{m/cm/c}$ and $PF_{m/c}$ significantly affected the power consumption. The contour plots between $PF_{m/c}$ and $\breve{I}_{m/cub}$ and between $\breve{I}_{m/cm/c}$ and $\breve{I}_{m/cub}$ indicate that both parameters were significant, but their impact on ÄPC$_{m/c}$ was not as substantial as Ï$_{m/cm/c}$ and ƤF$_{m/c}$, whereas the plot between Ï$_{m/cub}$ and Ṽ$_{av}$ exhibited the least significance.

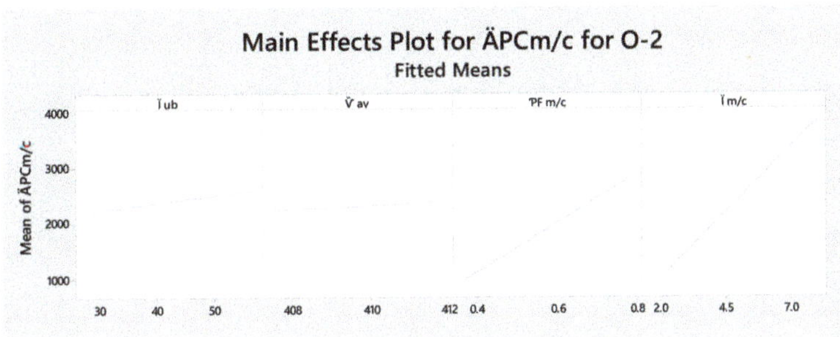

(**a**)

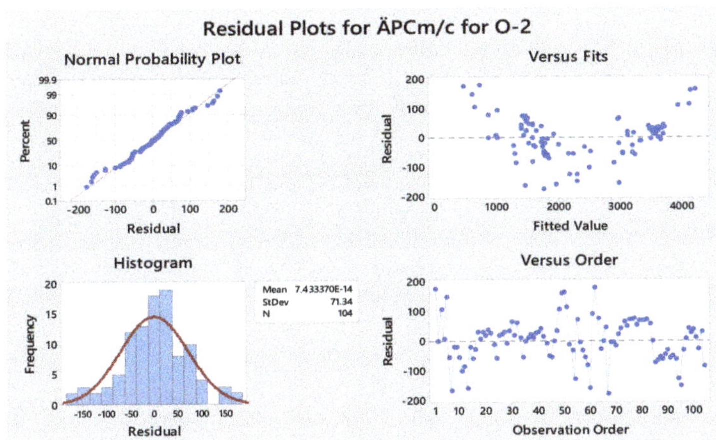

(**b**)

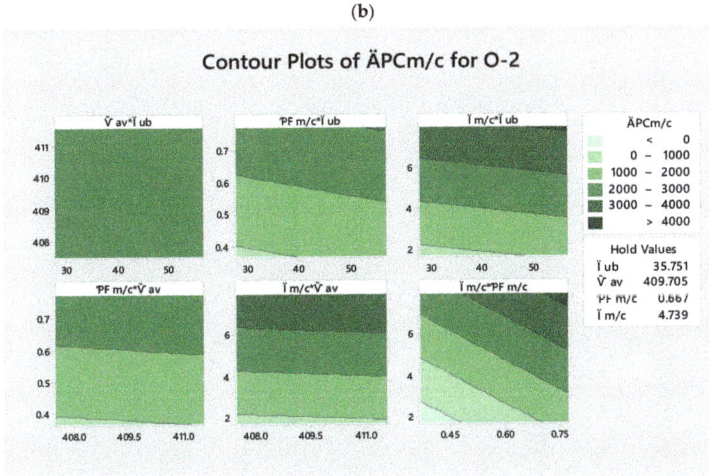

(**c**)

Figure 5. (**a**) Main effects plot for O-2. (**b**) Residuals plot for O-2. (**c**) Contour plots for O-2.

3.3. Analysis of the Regression Model for Operation-3 (O-3)

In the machining operation O-3, the overall regression model as per Equation (9) was statistically the best fit because the coefficient of determination as per Table 8 after the final iteration had values of R-sq. 99.65% and R-sq. (adj) 99.64%. This means that the regression model explained the maximum variation of the active power up to 99.64%. A look at the residual plots for $\ddot{A}PC_{m/c}$ for (O-3) in Figure 6b indicates that the residuals lie close to the diagonal line representing an ideal normal distribution. The points of the residual plots were not skewed, and they were randomly distributed. Therefore, it seems that the data were normally distributed. Furthermore, a look at the histogram of residuals gives evidence that our residuals were normally distributed. The distribution of residuals along the straight horizontal line was similar for all significant parameters, suggesting equality of variance. Therefore, the conditions of normality of residuals and equality of variance were fulfilled. The ANOVA results for O-3 revealed that there was significant between-group variance based on the value of the F-statistic (see Table 9). At $\alpha = 0.05$, the F-value was equal to 6859.26 with a p-value < 0.001. This indicates evidence of a regression relationship between the dependent variable $\ddot{A}PC_{m/c}$ and the independent variables $\text{ƤF}_{m/c}$, $\ddot{I}_{m/cm/c}$, and $\breve{V}_{ub}$ combined.

Individual coefficients contributed meaningful information in the prediction of $\ddot{A}PC_{m/c}$ (see Table 10 for O-3). The test statistics showed t-values of $\text{ƤF}_{m/c}$, $\ddot{I}_{m/cm/c}$, and $\breve{V}_{ub}$ at $\alpha = 0.05$. Therefore, $\text{ƤF}_{m/c}$, $\ddot{I}_{m/cm/c}$, and $\breve{V}_{ub}$ were individually useful in the prediction of $\ddot{A}PC_{m/c}$. The coded coefficients in Table 10 reveal that, for machining operation O-3, the predictor with the highest impact on the dependent variable $\ddot{A}PC_{m/c}$ was $\ddot{I}_{m/cm/c}$ (816.1), followed by $\text{ƤF}_{m/c}$ (158.1) and $\breve{V}_{ub}$ (-30.80). The negative sign indicates that the predictor $\breve{V}_{ub}$ had the highest negative effect on the consumption of $\ddot{A}PC_{m/c}$. The value of the VIF (variance inflation factor) inTable 10 for O-3 of all predictors was less than 5. This indicates that there was no significant multicollinearity between the significant predictors.

A look at the main effects plots for $\ddot{A}PC_{m/c}$ in Figure 6a also validates the above result. The slope of significant predictors versus $\ddot{A}PC_{m/c}$ indicates that, in this machining operation (O-3), $\ddot{I}_{m/cm/c}$ had the maximum impact on $\ddot{A}PC_{m/c}$, followed by $\text{ƤF}_{m/c}$, whereas $\breve{V}_{ub}$ had a negative slope.

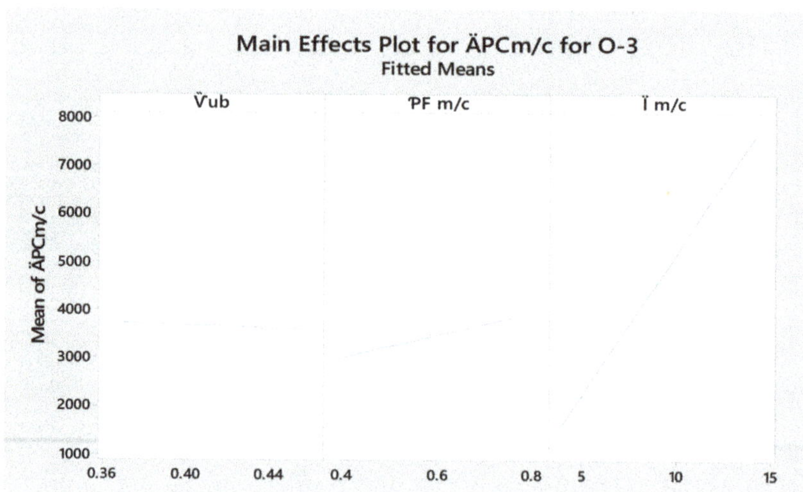

(a)

Figure 6. *Cont.*

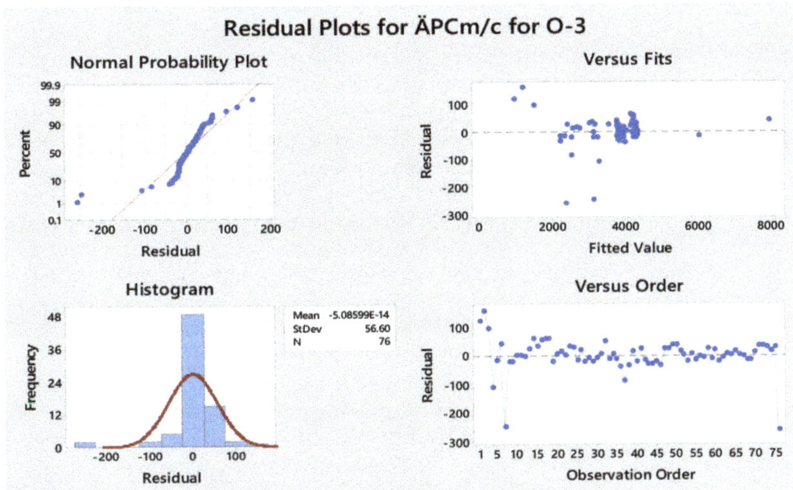

(b)

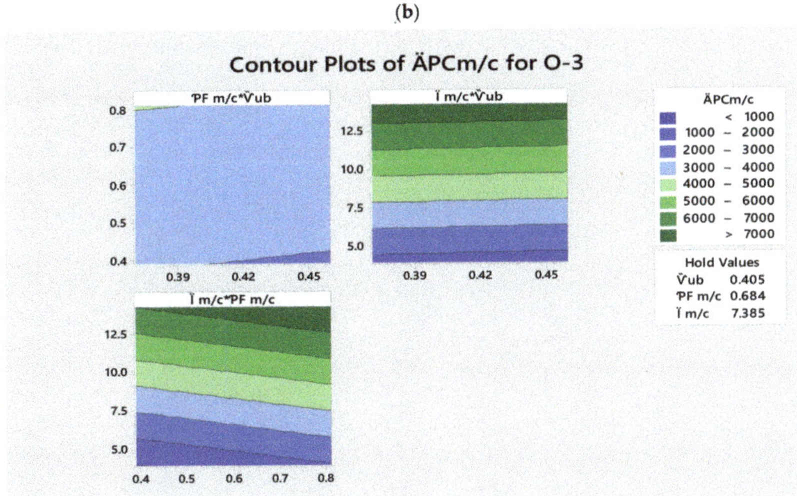

(c)

Figure 6. (**a**) Main effects plot for O-3. (**b**) Residuals plot for O-3. (**c**) Contour plots for O-3.

The contour plots of $\breve{I}_{m/cm/c}$, $\text{ƤF}_{m/c,}$ and $\ddot{A}PC_{m/c}$ in Figure 6c indicate the maximum impact of $\breve{I}_{m/c}m/c$ and $\text{ƤF}_{m/c}$ on $\ddot{A}PC_{m/c}$ as the number of contour lines representing different ranges of $\ddot{A}PC_{m/c}$ consumption was greater. This suggests that any variation in $\breve{I}_{m/cm/c}$ and $\text{ƤF}_{m/c}$ significantly affected the power consumption. The contour plot between $\breve{I}_{m/cm/c}$ and $\tilde{V}_{ub}$ indicates that both parameters were significant, but their impact on $\ddot{A}PC_{m/c}$ was not as substantial as $\breve{I}_{m/cm/c}$ and $\text{ƤF}_{m/c}$, whereas the plot between $\text{ƤF}_{m/c}$ and $\tilde{V}_{ub}$ exhibited the least significance.

3.4. Analysis of the Regression Model for Operation-4 (O-4)

In the machining operation O-4, the overall regression model as per Equation (10) was statistically the best fit because the coefficient of determination as per Table 8 for O-4, after the final iteration, had values of R-sq. 98.41% and R-sq. (adj) 98.33%. This means that, for O-4, the regression model explained the maximum variation of the active power up to 98.33%. A look at the residual plots for $\ddot{A}PC_{m/c}$ for (O-4) in Figure 7b indicates that the residuals lie close to the diagonal line, representing an ideal normal distribution. The points of the residual plots were not skewed, and they were randomly distributed. Therefore, it seems that the data were normally distributed. Furthermore, a look at the histogram of residuals gives evidence that our residuals were normally distributed. The distribution of residuals along the straight horizontal line was similar for all significant factors, suggesting equality of variance. Therefore, the conditions of normality of residuals and equality of variance were fulfilled. The ANOVA results for O-4 revealed that there was significant between-group variance based on the value of the F-statistic (see Table 9). At $\alpha = 0.05$, the *F*-value was equal to 1117.05 (*p*-value < 0.001). This indicates evidence of a regression relationship between the dependent variable $\ddot{A}PC_{m/c}$ and the independent variables $PF_{m/c}$, $\breve{I}_{m/cm/c}$, $\breve{I}_{m/cthdf}$, and $\ddot{V}_{thdf}$ combined.

Individual coefficients contributed meaningful information in the prediction of $\ddot{A}PC_{m/c}$ (see Table 10 for O-4). The test statistics showed *t*-values of $PF_{m/c}$, $\breve{I}_{m/cm/c}$, $\breve{I}_{m/cthdf}$, and $\ddot{V}_{thdf}$ at $\alpha = 0.05$. Therefore, $PF_{m/c}$, $\breve{I}_{m/cm/c}$, $\breve{I}_{m/cthdf}$, and $\ddot{V}_{thdf}$ were individually useful in the prediction of $\ddot{A}PC_{m/c}$. The coded coefficients in Table 10 reveal that, for machining operation O-4, the predictor with the highest impact on the dependent variable $\ddot{A}PC_{m/c}$ was $PF_{m/c}$ (431.4), followed by $\breve{I}_{m/cm/c}$ (415.02), $\breve{I}_{m/cthdf}$ (105.1), and $\ddot{V}_{thdf}$ (−17.43). The negative sign indicates that the predictor $\ddot{V}_{ub}$ had the highest negative effect on the consumption of $\ddot{A}PC_{m/c}$. The value of the VIF (variance inflation factor) in Table 10 for O-4 of all predictors was less than 5. This indicates that there was no significant multicollinearity between the significant predictors. A look at the main effects plots for $\ddot{A}PC_{m/c}$ in Figure 7a also validates the above result. The slope of significant predictors versus $\ddot{A}PC_{m/c}$ indicates that, in this machining operation (O-4), $PF_{m/c}$ had the maximum impact on $\ddot{A}PC_{m/c}$, followed by $\breve{I}_{m/cm/c}$, and $\breve{I}_{m/cthdf}$. The plots also show that $\ddot{V}_{thdf}$ had a slight negative slope.

For O-4, the contour plots of $\breve{I}_{m/cm/c}$, $PF_{m/c}$, and $\ddot{A}PC_{m/c}$ in Figure 7c indicate the maximum impact of $\breve{I}_{m/c}m/c$ and $PF_{m/c}$ on $\ddot{A}PC_{m/c}$ as the number of contour lines representing different ranges of $\ddot{A}PC_{m/c}$ consumption was greater. This suggests that any variation in $\breve{I}_{m/cm/c}$ and $PF_{m/c}$ significantly affected the power consumption. The contour plot between $\breve{I}_{m/cm/c}$ and $\breve{I}_{m/cthdf}$ indicates that both parameters were significant, but their impact on $\ddot{A}PC_{m/c}$ was not as substantial as $\breve{I}_{m/cm/c}$ and $PF_{m/c}$. The impact of $\breve{I}_{m/cthdf}$ and $\ddot{V}_{thdf}$ on $\ddot{A}PC_{m/c}$ was the least significant.

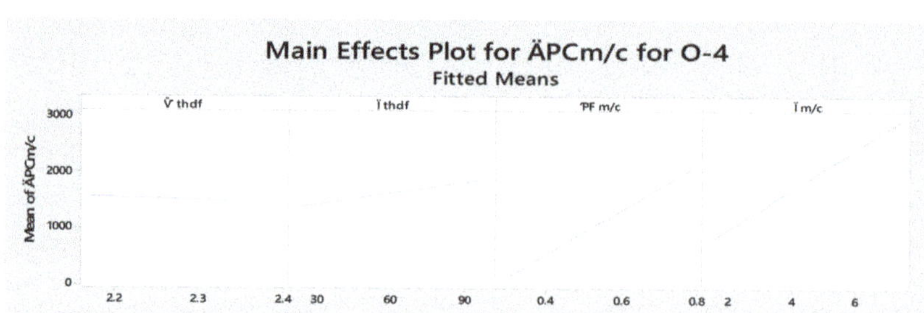

(a)

Figure 7. *Cont.*

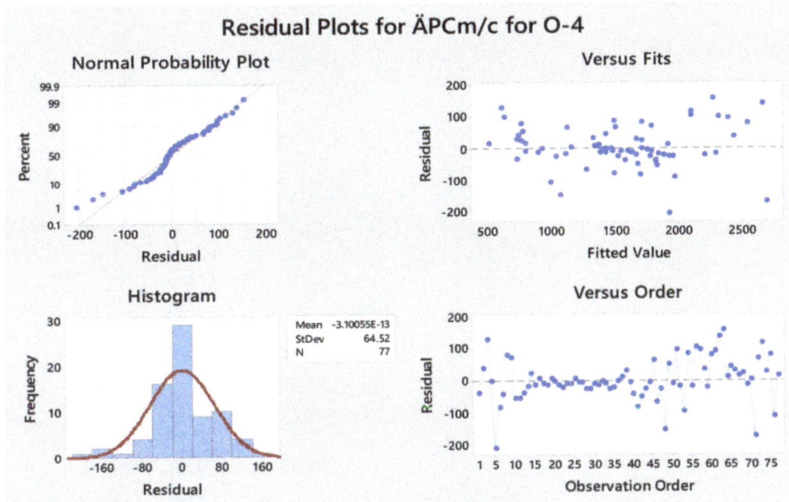

(**b**)

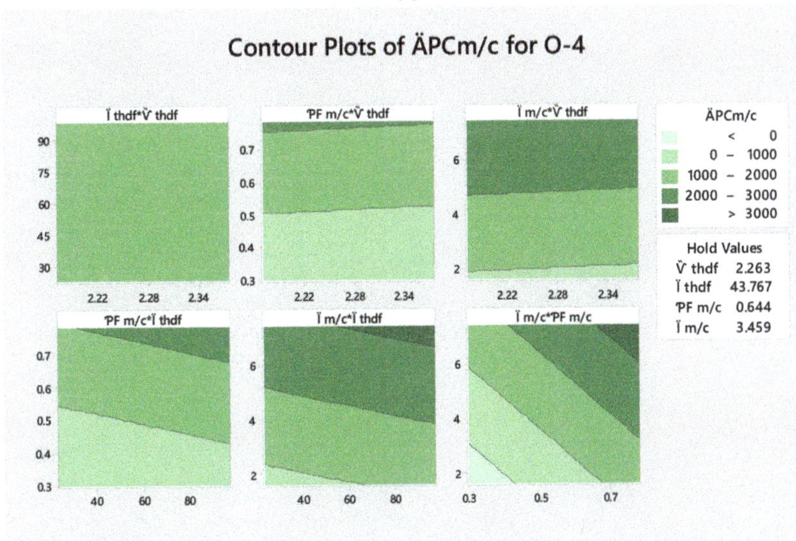

(**c**)

Figure 7. (**a**) Main effects plot for O-4. (**b**) Residuals plot for O-4. (**c**) Contour plots for O-4.

3.5. Analysis of the Regression Model for Operation-5 (O-5)

In the machining operation O-5, the overall regression model as per Equation (11) was statistically the best fit because the coefficient of the determination as per Table 8 for O-5, after the final iteration, had values of R-sq. 93.64% and R-sq. (adj) 93.08%. This means that, for O-5, the regression model explained the maximum variation of the active power up to 93.08%. A look at the residual plots for $\ddot{A}PC_{m/c}$ for (O-5) in Figure 8b indicates that the residuals lie close to the diagonal line, representing an ideal normal distribution. The points of the residual plots were not skewed, and they were randomly distributed. Therefore, it seems that the data were normally distributed. Furthermore, a look at the histogram of residuals gives evidence that our residuals were normally distributed. The distribution of residuals along the straight horizontal line was similar for all significant parameters, suggesting equality of variance. Therefore, the conditions of normality of residuals and equality of variance were fulfilled.

The ANOVA results for O-5 reveal that there was significant between-group variance based on the value of the F-statistic (see Table 9). At $\alpha = 0.05$, the F-value was equal to 167.81 (p-value < 0.001). This indicates evidence of a regression relationship between the dependent variable $\ddot{A}PC_{m/c}$ and the independent variables $PF_{m/c}$, $\breve{I}_{m/cm/c}$, $\breve{I}_{m/cthdf}$, $\breve{I}_{m/cub}$, and $\breve{V}_{ub}$ combined.

Individual coefficients contributed meaningful information in the prediction of $\ddot{A}PC_{m/c}$ (see Table 10 for O-5). The test statistics showed t-values of $PF_{m/c}$, $\breve{I}_{m/cm/c}$, $\breve{I}_{m/cthdf}$, $\breve{I}_{m/cub}$, and $\breve{V}_{ub}$ at $\alpha = 0.05$. Therefore, $PF_{m/c}$, $\breve{I}_{m/cm/c}$, $\breve{I}_{m/cthdf}$, $\breve{I}_{m/cub}$, and $\breve{V}_{ub}$ were individually useful in the prediction of $\ddot{A}PC_{m/c}$. The coded coefficients in Table 10 reveal that, for machining operation O-5, the predictor with the highest impact on the dependent variable $\ddot{A}PC_{m/c}$ was $\breve{I}_{m/cm/c}$ (1211), followed by $PF_{m/c}$ (1108.4), $\breve{I}_{m/cub}$ (367), $\breve{I}_{m/cthdf}$ (218.9), and $\breve{V}_{ub}$ (-183.9). The negative sign indicates that the predictor $\breve{V}_{ub}$ had the highest negative effect on consumption of $\ddot{A}PC_{m/c}$. The value of the VIF (variance inflation factor) in Table 10 for O-5 for predictors $\breve{V}_{ub}$, $\breve{I}_{m/cthdf}$, and $PF_{m/c}$ was less than 5. This indicates that there was no significant multicollinearity between these significant predictors. However, for the predictors $\breve{I}_{m/cub}$ and $\breve{I}_{m/cm/c}$, the value of VIF was greater than 5, which indicates slight multicollinearity for these predictors.

A look at the main effects plots for $\ddot{A}PC_{m/c}$ in Figure 8a also validates the above result. The slope of significant predictors versus $\ddot{A}PC_{m/c}$ indicates that, in this machining operation (O-5), $\breve{I}_{m/cm/c}$ had a maximum impact on $\ddot{A}PC_{m/c}$, followed by $PF_{m/c}$, $\breve{I}_{m/cub}$, and $\breve{I}_{m/cthdf}$. The plots also show that $\breve{V}_{ub}$ had a negative slope.

For O-5, the contour plots of $\breve{I}_{m/cm/c}$, $PF_{m/c}$, and $\ddot{A}PC_{m/c}$ in Figure 8c indicate the maximum impact of $\breve{I}_{m/c}$m/c and $PF_{m/c}$ on $\ddot{A}PC_{m/c}$ as the number of contour lines representing different ranges of $\ddot{A}PC_{m/c}$ consumption was greater. The contour plot between $\breve{I}_{m/cm/c}$ and $\breve{I}_{m/cthdf}$ indicates that both parameters were significant, but their impact on $\ddot{A}PC_{m/c}$ was not as substantial as $\breve{I}_{m/cm/c}$ and $PF_{m/c}$. The impact of $\breve{I}_{m/cthdf}$ and $\breve{V}_{ub}$ on $\ddot{A}PC_{m/c}$ was the least significant.

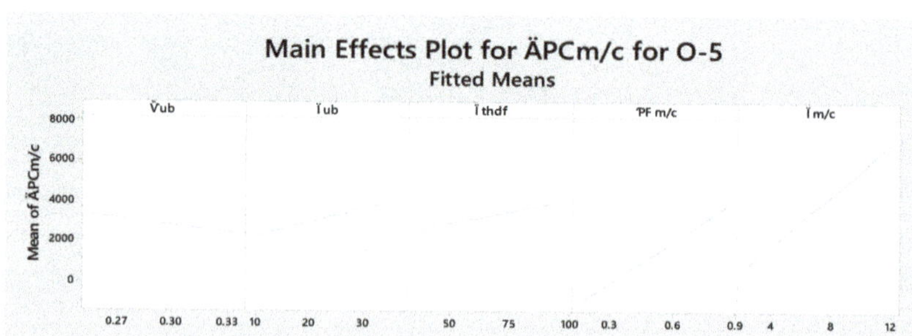

(a)

Figure 8. *Cont.*

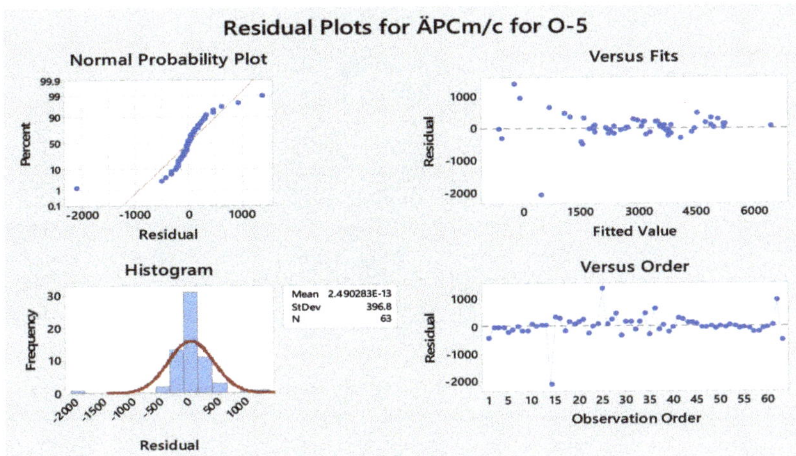

(b)

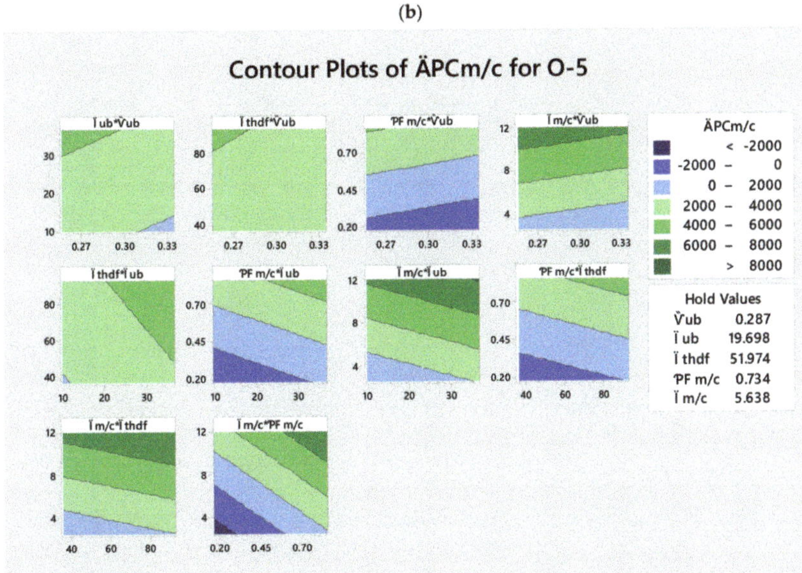

(c)

Figure 8. (**a**) Main effects plot for O-5. (**b**) Residuals plot for O-5. (**c**) Contour plots for O-5.

3.6. Comparative Analysis of Modeling

In order to establish the impact of significant electric parameters of estimated active power consumption of machining operations, the predicted results of equations developed with regression modeling for machining operations O-1 to O-5 were compared in this section with a standard equation of power for three-phase supply, as shown in Equation (12). Lastly, the experimental observations and predicted results from regression models and standard Equation (12) were compared in terms of absolute error calculated using Equation (13) and relative error (%) calculated using Equation (14) [55].

$$P = \sqrt{3}\breve{V}_{av}\breve{I}_{m/c_{m/c}}\mathrm{Cos}\phi. \tag{12}$$

$$\text{Absolute error} = |\text{Experimental value} - \text{Predicted value}|. \tag{13}$$

$$\text{Relative error } (\%) = \frac{|\text{Experimental value} - \text{Predicted value}|}{\text{Experimental value}} \times 100. \quad (14)$$

The absolute error and relative error for experimental (Exp.) active power and that predicted by standard Equation (12) (i.e., Std. Equation) are shown in Table 11 for machining operations O-1 to O-5. The absolute error and relative error for experimental active power and that predicted by regression models (Reg. Mod.) using Equations (7)–(11) are shown in Table 12 for machining operations O-1 to O-5. The average active power consumed during each of the machining operations was noted with the help of a power quality analyzer, denoted as $\ddot{A}PC_{m/c}$ (Exp.). This experimental value was also considered the true value of the power. The average active power was then calculated using the standard Equation (12), denoted as $\ddot{A}PC_{m/c}$ (Pred.). The average active power was also estimated from the regression model for each machining operation, denoted as $\ddot{A}PC_{m/c}$ (Pred.) Reg. Mod.

Table 11. Active power determined using standard equation and measured by power analyzer.

Machining Operation	Avg. $\ddot{V}_{av}$	Avg. $\mathcal{P}F_{m/c}$	Avg. $\ddot{I}_{m/c m/c}$	$\ddot{A}PC_{m/c}$ (W) (Exp.)	$\ddot{A}PC_{m/c}$ (Pred.) Std. Equation	Abs. Error Std. Equation	Relative Error
O-1	415.500	0.673	3.986	1942.203	1930.828	11.375	0.586
O-2	409.706	0.667	4.739	2302.782	2243.197	59.585	2.588
O-3	415.234	0.684	7.386	3685.243	3633.658	51.585	1.400
O-4	416.825	0.645	3.459	1528.342	1610.101	81.759	5.350
O-5	426.704	0.735	5.638	2942.897	3061.838	118.941	4.042

Table 12. Active power determined using regression equation and measured by power analyzer.

Machining Operation	Avg. $\ddot{V}_{ub}$	Avg. $\ddot{V}_{thdf}$	Avg. $\ddot{I}_{m/c ub}$	Avg. $\ddot{V}_{av}$	Avg $\ddot{I}_{m/c thdf}$	Avg. $\mathcal{P}F_{m/c}$	Avg. $\ddot{I}_{m/c m/c}$	$\ddot{A}PC_{m/c}$ (Exp.)	$\ddot{A}PC_{m/c}$ (Pred.) Reg. Mod.	Abs. Error Reg. Mod.	Relative Error
O-1	2.600	3.626	39.887	415.500 *	40.801 *	0.673 *	3.986 *	1942.203	1941.442	0.761	0.039
O-2	1.655	6.869	35.751 *	409.706 *	42.887	0.667 *	4.739 *	2302.782	2302.557	0.225	0.010
O-3	0.405 *	2.260	10.958	415.234	40.800	0.684 *	7.386 *	3685.243	3685.243	0.000	0.000
O-4	0.238	2.263 *	19.060	416.825	43.767 *	0.645 *	3.459 *	1528.342	1528.332	0.010	0.001
O-5	0.288 *	6.350	19.698 *	426.704	51.975 *	0.735 *	5.638 *	2942.897	2942.847	0.050	0.002

* Significant predictors as per developed regression models (Equations (7) to (11)).

Figure 9a represents the machining operation along the x-axis, relative error along the primary y-axis, and the absolute error along the secondary y-axis. The relative error between the values of active power obtained from the power analyzer and those calculated using the standard Equation (12) is illustrated using a bar chart. The minimum relative error was 0.586 for machining operation O-1, and the maximum relative error was 5.350 for O-4.

Figure 9b shows that the relative error was minimum for O-3 (0.000) and maximum for O-1 (0.011). Figure 9 indicates the comparison between the relative errors obtained for power estimation using the standard Equation and regression models for each machining operation. It is visible that both the absolute error and the relative error were more significant in magnitude when the standard Equation (12) was used to estimate the active power. However, they become negligible when power was computed using the regression models considering the significant electric parameters. This indicates that electric parameters influence the active power and play a decisive role in gauging the magnitude of the active power consumed during the machining operation.

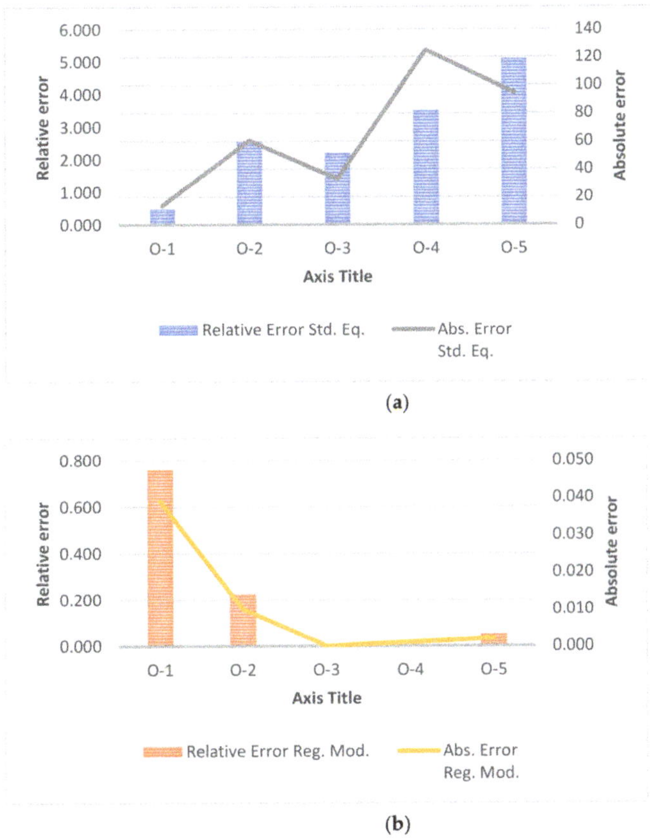

Figure 9. (**a**). Relative and absolute error of experimental and standard equation values. (**b**) Relative and absolute error of experimental and regression equation values.

The regression models validate that electric parameters significantly impact the power consumption of a machining operation. It was observed that the average current had the most significant impact on electrical power consumption, followed by the power factor. Therefore, the average current consumption and power factor need to be monitored to reduce electric power consumption during the machining process.

Other electric parameters impacting power consumption are the current total harmonic distortion factor and current unbalance. Their impact was less as compared to the average current and power factor. These parameters have not been researched much. Their impact is comparatively new. There is a need to study the impact of these parameters on electrical power consumption. As a result, they were considered in the analysis. It was also observed that the average value of the power factor for different machining processes varied between 0.61 and 0.71. This value is low and needs to be improved.

It was further observed that, even for the same machining process and repetitive cuts, the significance of electric parameters concerning power consumption changed. This could have been because the electrical power being consumed by machine tools is a dynamic quantity. The working of other machine tools on the shop floor of an industry may affect the electric parameters from time to time. It is not easy to insulate a particular machine tool from the effect of other machines consuming power.

4. Hybrid Decision-Making Methodology

This section is dedicated to prioritizing the considered machining operations on the basis of energy consumption. The term "prioritization" refers to ranking the five different machining operations used in the industry. The ranks were assigned to the machining operations while considering energy consumption responses, as well as electric parameters. As a result, the user, engineer, or manager can quickly identify the machining operation consuming maximum energy. Consequently, the machining operation consuming maximum energy can be optimized on a priority basis.

Earlier regression modeling was completed to find the significant electric parameters affecting the active power consumption in the machining operation. The significant energy consumption responses and electric parameters were defined in Section 2.1. The energy responses active power, active energy, specific energy consumption, and energy efficiency were considered for decision making to prioritize the different machining operations studied in the industry. The significant electric parameters considered in the decision matrix for prioritizing machining operations were the power factor, average current, current total harmonic distortions, and current unbalance. These parameters were considered because of their significant impact on active power consumption. This was evident from the values of the coded coefficients observed during regression modeling of machining operations O-1 to O-5 (see Table 10). The electric parameters such as the average voltage, voltage unbalance, and voltage total harmonic distortion factor had a negligible impact on power consumption. In some cases, it was also negative. Moreover, these parameters were found to be nonsignificant in all the considered machining operations. As a result, they were not considered in the decision matrix.

A hybrid decision-making technique was utilized, based on the Technique for Order Preference by Similarity to Ideal Solution (TOPSIS). It was employed to convert multiple performances into a single score called the multiple composite score (MCS). The TOPSIS assumes that the chosen alternative will have the shortest Euclidean divergence from the ideal positive solution and the most divergence from the ideal negative solution. The steps described below are usually employed in this technique [37,56,57].

Step 1: Identification of study objectives and responses. The decision pattern was considered as per Equation (15). Every row of the decision matrix (DM) was assigned to each experiment number and column to one response, i.e., $\ddot{A}PC_{m/c}$, $\ddot{A}EC_{m/c}$, $\dot{E}\dot{E}_{\acute{\eta}}$, and $\mathcal{P}F_{m/c}$. q_{ij} is an element of the decision matrix 'DM' [q_{ij}, i = 1, 2, ... , a number of experiments (n), j = 1, 2, ..., number of responses (m)] which is input.

$$DM = \begin{bmatrix} q_{11} & q_{12} & -- & q_{1j} & -- & q_{1m} \\ q_{21} & q_{22} & -- & q_{2j} & -- & q_{2m} \\ -- & -- & -- & -- & -- & -- \\ q_{i1} & q_{i2} & -- & q_{ij} & -- & q_{im} \\ -- & -- & -- & -- & -- & -- \\ q_{n1} & q_{n2} & -- & q_{nj} & -- & q_{nm} \end{bmatrix}. \quad (15)$$

Step 2: The vector normalization obtained the normalized DM (M_{ij}) as per Equation (16).

$$M_{ij} = \left[\frac{q_{ij}}{\left[\sum_{i=1}^{n} q_{ij}^2\right]^{\frac{1}{2}}} \right]. \quad (16)$$

Step 3: The weights (w_j) of significance were assigned to the responses, w_j; j = 1, 2 ... ... m, such that $\sum w_{j_j} = 1$. In the present study, three methods of weight assignment were used: identical, objective, and subjective preferences.

- Equal-weights method

In this technique, weights were attained, dividing one by the total number of responses, as per Equation (17).

$$w_{j_j} = \frac{1}{m}. \quad (17)$$

Since there eight responses in the present case, the weight assigned to each response was 0.125 ($w_{j_j} = 12.5\%$).

- Entropy-weights method

Weights are established without considering the influence of decision-makers by using probability and measuring uncertain information. The fundamental principle of weight estimation with entropy is that a higher weight index value is more efficient than a lower index value [58]. The DM listed in Equation (15) is normalized using Equation (18) for a beneficial response, e.g., power factor, and Equation (19) for a nonbeneficial reaction, e.g., energy consumption by the linear normalization technique; it is noticeable that the normalized decision matrix $NDM_{ij} \in [0, 1]$ [59].

The probability of the response (Pr_{ij}) to happen is computed suing Equation (20), and Equation (21) is utilized to attain the entropy (En_j) of the jth response. In Equation (21), $Y = \frac{1}{\log_e (n)}$ is a stable expression, n is the number of experiments, and the value of En_j lies between zero and one.

Equation (22) is utilized to compute the degrees of divergence (Div_j), and Equation (23) obtains the entropy weight (E_w) of the jth response.

$$NDM_{ij} = \frac{q_{ij}}{Max q_{ij}} \text{ (Beneficial)}. \quad (18)$$

$$NDM_{ij} = \frac{Min q_{ij}}{q_{ij}} \text{ (Nonbeneficial)}. \quad (19)$$

$$Pr_{ij} = \frac{NDM_{ij}}{\sum_{i=1}^{n} NDM_{ij}}. \quad (20)$$

$$En_j = -Y \sum_{i=1}^{n} Pr_{ij} \log_e (Pr_{ij}). \quad (21)$$

$$Div_j = |1 - En_j|. \quad (22)$$

$$Ew_j = \frac{Div_j}{\sum_{j=1}^{m} Div_j}. \quad (23)$$

- Analytic Hierarchy Process (AHP) weights

This method attains weights for responses with the consent of the decision-maker. In the end, a nine-point inclination scale is used to obtain the relative importance of responses by building a pairwise comparison matrix [60].

Assuming m responses (R_m), a pairwise comparison matrix ($R_{m\times m}$) can be built, Equation (24), for the pairwise comparison of response (R_i) with a response (R_j). To attain the elements of $R_{m\times m}$, the suitable comparative significance can be assigned to every row response ($R_1, R_2, \ldots, R_m$) by comparing the value with the response from every column ($R_1, R_2, \ldots, R_m$). In $R_{m\times m}$, the value of $r_{ij} = 1$ for i = j, i.e., a response compared with itself is always equal to 1, and other leftover elements of the $R_{m\times m}$ (for $i \neq j$) are the reciprocal of the equivalent component, e.g., $r_{12} = 1/r_{21}$. Consequently, the first diagonal entries of $R_{m\times m}$ are equal answers of responses. Half of the remaining entries are reciprocal of the equivalent elements to 1, and the other half are agreed by contrasting the corresponding elements. The relative normalized weight (w_j) of every response is obtained by taking the ratio of the geometric mean [?]M of the equivalent row in the $B_{m\times m}$ comparison to the sum of the geometric means of all the rows (see Equations (25) and (26)).

Equation (27) yields the consistency index (ĆÍ), where λ_{max} is the maximum eigenvalue of $B_{m\times m}$. λ_{max} is obtained by taking the average value of the sum of the matrix product of the pairwise $B_{m \times m}$ comparison and w_j vectors and dividing by the relative normalized weight of the equivalent response. A lower value of ĆÍ, denotes a lower variation from the consistency. Equation (28) provides the consistency ratio (ĆŔ). In Equation (28), ŔÍ is the random index value (see [60]). Generally, a ĆŔ of 0.10 or less is considered adequate, and it imitates an informed opinion attributable to the acquaintance of a market analyst concerning the problem of study.

$$\begin{array}{c} \text{Response (j)} \; R_1 \; R_2 \; -- \; R_j \; -- R_m. \\ \text{(i)} \end{array}$$

$$R_{m \times m} = \begin{matrix} R_1 \\ R_2 \\ -- \\ R_i \\ -- \\ R_m \end{matrix} \begin{bmatrix} 1 & r_{12} & -- & r_{1j} & -- & r_{1m} \\ r_{21} & 1 & -- & r_{2j} & -- & r_{2m} \\ -- & -- & -- & -- & -- & -- \\ r_{i1} & r_{i2} & -- & r_{ij} & -- & r_{im} \\ -- & -- & -- & -- & -- & -- \\ r_{m1} & r_{m2} & -- & r_{mj} & -- & 1 \end{bmatrix} \tag{24}$$

$$ǴM = \left[\prod_{j=1}^{m} r_{ij} \right]^{\frac{1}{m}}. \tag{25}$$

$$w_{ij} = \frac{[Ǵ]M}{\sum_{i=1}^{m} [Ǵ]M}. \tag{26}$$

$$ĆÍ = \frac{\lambda_{max} - m}{m - 1}. \tag{27}$$

$$ĆŔ = \frac{ĆÍ}{ŔÍ}. \tag{28}$$

Step 4: The weighted normalized matrix ($\widehat{W}\dot{Z}_{ij}$) was obtained by multiplying the columns of M_{ij} with their respective assigned weight, w_j. Subsequently, $\widehat{W}\dot{Z}_{ij}$ was attained using Equation (29).

Step 5: In this stage, it was necessary to find out the ideal best (Z^+) and ideal worst (Z^-) solutions with the help of Equations (30) and (31), respectively. Here, Z^+ and Z^- solutions were the largest and smallest values amongst all response values, respectively. In Equation (30), j and j′ are related to the beneficial (m) and nonbeneficial attributes (m′), respectively.

Step 6: Separation measures (Sep_m) were prepared on the basis of Euclidean distance (see Equations (32) and (33)).

Step 7: The relative closeness or 'MCS' of all experiments was computed, i.e., alternatives representing the ideal resolution using Equation (34).

$$\widehat{W}\dot{Z}_{ij} = \left[w_j \times M_{ij} \right]. \tag{29}$$

$$\begin{aligned} Z_j^+ &= \{\text{best}\,(\widehat{W}\,\dot{Z}ij)\}_{i=1}^{n}. \\ Z^+ &= \left\{ Z_1^+, Z_2^+, \ldots, Z_j^+, \ldots Z_m^+ \right\}. \end{aligned} \tag{30}$$

$$\begin{aligned} Z_{j'}^- &= \{\text{worst}\,(\widehat{W}\dot{Z}_{ij'})\}_{i=1}^{n}. \\ Z^- &= \left\{ Z_1^-, Z_2^-, \ldots, Z_{j'}^-, \ldots Z_{m'}^- \right\}. \end{aligned} \tag{31}$$

$$Sep_i^+ = \left\{ \sum_{j=1}^{m} \left(Z_{ij} - Z_j^+ \right)^2 \right\}^{0.5}. \tag{32}$$

$$\mathrm{Sep}_i^- = \left\{ \sum_{j'=1}^{m'} \left(Z_{ij} - Z_{j'}^- \right)^2 \right\}^{0.5}. \quad (33)$$

$$\mathrm{MCS} = \frac{\mathrm{Sep}_i^-}{\mathrm{Sep}_i^+ + \mathrm{Sep}_i^-}. \quad (34)$$

Step 8: The methods to select the final ranks of alternatives based on individual results from different MCDM weights are described below.

Step 9: Degree of Membership (DoM).

Let $\mathcal{R}_{xy}$ be the rank matrix of the yth alternative using the xth MCDM method (x = 1, 2, ..., k, y = 1, 2, ..., t), where k is the number of MCDM methods and t is the number of alternatives.

Step 9.1: Constitute the rank matrix $\mathcal{R} = (r_{xy})$ k × t.

Step 9.2: Calculate the values of the rank state variables; x = 1, 2, ..., k, y = 1, 2, ..., t, z = 1, 2, ..., t from the rank matrix $\mathcal{R} = (r_{xy})$ k × t, using Equation (35).

Step 9.3: Constitute rank frequency number matrix F = (f_{yz}) txt, where f_{yz} is the rank frequency number that the rank of the yth alternative is the zth place according to different MCDM methods, and f_{yz} is expressed as Equation (36).

Step 9.4: Constitute the membership degree matrix $\varphi = (\varphi_{yz})$ txt, where φ_{yz} is the membership degree that the rank of the yth alternative belongs to the zth place according to different MCDM methods, and φ_{yz} is expressed as Equation (37).

The yth row ($\varphi_{y1}, \varphi_{y1}, \ldots, \varphi_{yt}$) of the membership degree matrix $\varphi = (\varphi_{yz})$ txt represents the degree that the rank of yth alternative belongs to k places, where $0 \leq \varphi_{yz} \leq 1$ and $\sum_{z=1}^{k} \varphi_{yz} = 1$.

Step 9.5: Calculate the final rank index $\mathcal{P}_y$ of the yth alternative ($y = 1, 2, \ldots, t$), where $\mathcal{P}_y$ is calculated using Equation (38).

Step 9.6: Determine final ranks r01, r02, ..., r0t of the operations in ascending order based on the values of $\mathcal{P}_1, \mathcal{P}_2, \mathcal{P}_3, \ldots \mathcal{P}_t$.

$$\delta_{yz}^{(x)} = \begin{cases} 1; & r_{xy} = z \\ 0; & r_{xy} \neq z \end{cases} \quad (x = 1,\ 2,\ \ldots,\ \mathrm{k},\ y = 1,\ 2,\ \ldots,\ \mathrm{t},\ z = 1,\ 2,\ \ldots,\ \mathrm{t}. \quad (35)$$

$$f_{yz} = \sum_{x=1}^{k} \delta_{yz}^{(x)}\ (y = 1,\ 2,\ \ldots, \mathrm{t}, z = 1,\ 2, \ldots, \mathrm{t}). \quad (36)$$

$$\varphi_{yz} = f_{yz}/\mathrm{k}\ (y = 1,\ 2,\ \ldots, \mathrm{t}, z = 1,\ 2, \ldots, \mathrm{t}). \quad (37)$$

$$\mathcal{P}_y = \sum_{z=1}^{t} z \cdot \varphi_{yz}. \quad (38)$$

Table 13 describes the active power consumption by the machine ($\ddot{A}PC_{m/c}$) in kW, active energy consumption by the machine ($\ddot{A}EC_{m/c}$) in kWh, energy efficiency ($\dot{E}\dot{E}_{\acute{\eta}}$) as a percentage, specific energy consumption (ʂ) in kJ/cm^3 for the power factor ($\mathcal{P}F_{m/c}$), the average current of the actual or machining cut ($\check{I}_{m/cm/c}$) in rms, the current total harmonic distortion factor of the actual or machining cut ($\check{I}_{m/cthdf}$), and the current unbalance of the actual cut ($\check{I}_{m/cub}$) using a decision matrix as per Equation (15). The normalized decision matrix was calculated as per Equation (16) and is shown in Table A1 (Appendix A). $\ddot{A}PC_{m/c}$, $\ddot{A}EC_{m/c}$, ʂ, $\check{I}_{m/cm/c}$, $\check{I}_{m/cthdf}$, and $\check{I}_{m/cub}$ were "the lower, the better" energy responses, and $\dot{E}\dot{E}_{\acute{\eta}}$ and $\mathcal{P}F_{m/c}$ were "the higher, the better" energy responses; the calculation was done to four significant decimal places.

Estimation of weights for responses

- Equal-Weights Method

The weights for responses were estimated using the equal-weights method as per Equation (3), revealing 12.5% or 0.125 each.

- Entropy-Weights Method

Equations (18) and (19) yielded the normalized decision matrix (NDM_{ij}) for computation of entropy weights for beneficial and by the nonbeneficial responses, respectively. The NDM_{ij} for entropy-weights is shown in Table A1. The probability of the response (Pr_{ij}) was computed using Equation (20), and the result is shown in Table A2. The entropy (En_j) of the response was calculated using Equation (21), while the degrees of divergence (Div_j) and entropy-weights (Ew_j) were calculated using Equations (22) and (23), respectively. The attained results are shown in Table 14.

Table 13. Decision matrix of energy responses.

Operation	ÄPC$_{m/c}$	ÄEC$_{m/c}$	ĖĖ$_{ή}$	ȿ	ƤF$_{m/c}$	Ĭ$_{m/cm/c}$	Ĭ$_{m/cthdf}$	Ĭ$_{m/cub}$
O-1	1.973	0.084	25.619	15.308	0.677	4.027	40.898	40.316
O-2	2.321	0.067	38.209	12.207	0.675	4.719	40.965	35.67
O-3	3.707	0.077	40.260	14.097	0.687	7.405	41.435	10.94
O-4	1.542	0.033	30.303	31.548	0.647	3.478	43.673	18.583
O-5	2.975	0.053	43.396	13.343	0.738	5.691	51.447	19.42

Table 14. Entropy-weights computation and weights.

	ÄPC$_{m/c}$	ÄEC$_{m/c}$	ĖĖ$_{ή}$	ȿ	ƤF$_{m/c}$	Ĭ$_{m/cm/c}$	Ĭ$_{m/cthdf}$	Ĭ$_{m/cub}$
En_j	0.9718	0.9600	0.9890	0.9741	0.9994	0.9798	0.9978	0.9316
Div_j	0.0282	0.0400	0.0110	0.0259	0.0006	0.0202	0.0022	0.0684
Ew_j	0.1437	0.2034	0.0561	0.1321	0.0030	0.1028	0.0111	0.3481
Ew_j(%)	14.371	20.343	5.6058	13.207	0.2957	10.279	1.1134	34.806

- AHP Method

To attain AHP weights, a pairwise comparison matrix of responses was established as per Equation (24). The pairwise comparison matrix was a collective decision of the research group, as well as people from academia and industry. A questionnaire was drafted on the basis of eight energy consumption responses to get opinions from the local industry experts and people from academia. The questionnaire content was decided by selecting energy responses from the literature and assigning electric parameters based on regression modeling. Lastly, an event was arranged, and 12 experts were invited for a brainstorming session to complete a pairwise comparison matrix of energy responses. This event comprised six experts from academia and six from the industry. The final agreed pairwise comparison matrix is shown in Table 15.

Table 15. Pairwise comparison matrix.

Responses	ÄPC$_{m/c}$	ÄEC$_{m/c}$	ĖĖ$_{ή}$	ȿ	ƤF$_{m/c}$	Ĭ$_{m/cm/c}$	Ĭ$_{m/cthdf}$	Ĭ$_{m/cub}$
ÄPC$_{m/c}$	1	1/5	1/7	1/5	1/7	1/3	1/5	1/2
ÄPC$_{m/c}$	5	1	1	1	1	1/3	1/3	1
ĖĖ$_{ή}$	7	1	1	1	1	1/3	1/3	1
ȿ	5	1	1	1	1	3	1/2	1
ƤF$_{m/c}$	7	1	1	1	1	1/3	1/3	1
Ĭ$_{m/cm/c}$	3	3	3	1/3	3	1	1/3	3
Ĭ$_{m/cthdf}$	5	3	3	2	3	3	1	3
Ĭ$_{m/cub}$	2	1	1	1	1	1/3	1/3	1
Eigen Vector	0.03	0.093	0.1	0.151	0.1	0.174	0.269	0.083

The relative normalized weight and the weights of responses were computed using Equations (25) and (26), respectively. The consistency index and ratio were calculated using Equations (27) and (28), respectively. The maximum eigenvalue was λ_{max} = 8.816, and the consistency ratio CR was 8.3%. The CR was observed to be below the permitted value of

10%, indicating substantial accuracy in the decision-maker's judgment when assigning values in the matrix for pairwise comparisons. The weights of significance computed using the three different methods are listed in Table 16.

Table 16. Weights of energy responses computed using three methods.

	$\ddot{A}PC_{m/c}$	$\ddot{A}EC_{m/c}$	$\dot{E}\dot{E}_{\acute{\eta}}$	ş	ƤF$_{m/c}$	$\ddot{I}_{m/cm/c}$	$\ddot{I}_{m/cthdf}$	$\ddot{I}_{m/cub}$
Equal	12.5%	12.5%	12.5%	12.5%	12.5%	12.5%	12.5%	12.5%
Entropy	14.37%	20.34%	5.60%	13.21%	0.29%	10.27%	1.11%	34.81%
AHP	3%	9.3%	10%	15.10%	10%	17.4%	26.9%	8.3%

The normalized decision matrix for TOPSIS method calculations was obtained as per Equation (16) and is shown in Table A3. The weighted normalized matrix ($\hat{W}\dot{Z}_{ij}$) was computed using Equation (29) and is tabulated in Table A3 for equal, entropy, and AHP weights. The positive ideal (best) answer was calculated using Equation (30), and the negative ideal (worst) response was calculated using Equation (31); the attained values are depicted in Table A6 for equal, entropy, and AHP weights. Equations (32) and (33) were applied to calculate separation measures (Sep_m) for positive and negative answers, respectively. The evaluated responses are depicted in Table 17. The relative closeness or MCS was calculated using Equation (34), and the computed results are tabulated in Table 17. The final ranks attained using the three different weight methods are shown in Table 17.

Table 17. Separation measures and multiple composite scores (MCSs).

Operation	Equal Weights				Entropy Weights				AHP Weights			
	Sep_i^+	Sep_i^-	MCS	Rank	Sep_i^+	Sep_i^-	MCS	Rank	Sep_i^+	Sep_i^-	MCS	Rank
O-1	0.0806	0.0721	0.0341	5	0.1827	0.2555	0.0208	5	0.0573	0.0702	0.0387	3
O-2	0.0628	0.0752	0.0410	4	0.1502	0.2321	0.0289	4	0.0450	0.0697	0.0423	2
O-3	0.0733	0.0839	0.0448	3	0.0883	0.2649	0.1177	1	0.0660	0.0666	0.0335	5
O-4	0.0636	0.0890	0.0519	2	0.0754	0.2315	0.1052	2	0.0514	0.0770	0.0461	1
O-5	0.0477	0.0829	0.0526	1	0.0688	0.2102	0.0951	3	0.0476	0.0656	0.0380	4

The ranks obtained using the various weight methods were different for each machining operation. Thus, to get the final combined ranks, the degree of membership technique was applied, as shown in Step 9 of the decision-making methodology. First, the constitute rank frequency number of the machining operations from O-1 to O-5 was calculated using Equation (35), and the results are shown in Table A6. Then, a membership degree was constituted using Equation (36). Lastly, the final rank index of each machining operation was obtained using Equation (37), as shown in Table 18. The final ranks of each machining operation were then calculated, and the results are presented in Figure 10.

Table 18. Final ranks of operations using degree of membership (DoM).

Weights	1	2	3	4	5	SUM	Rank
O-1	0.0000	0.0000	1.0000	0.0000	3.3333	4.3333	5
O-2	0.0000	0.6667	0.0000	2.6667	0.0000	3.3333	4
O-3	0.3333	0.0000	1.0000	0.0000	1.6667	3.0000	3
O-4	0.3333	1.3333	0.0000	0.0000	0.0000	1.6667	1
O-5	0.3333	0.0000	1.0000	1.3333	0.0000	2.6667	2

Prioritizing the machining process in terms of energy consumption was conducted systematically. Machining operation O-1 emerged as the most energy-intensive process. Five machining operations O-1 to O-5 were considered, and eight response parameters were measured. The allocation of weights to responses was an important step. To rule out any bias of the researcher, three methods were employed. The first assigned equal weights to all the responses. Under this method, the reciprocal of the total number of the responses was taken, and equal weights were assigned to all the responses. The second was an objective method, i.e., the entropy weight method, in which weights were assigned without considering the researcher's input. First, the probability of the response was worked out, and then the entropy of the response was calculated. This step was followed by working out the degree of divergence and assigning entropy weights. This was a purely analytical method, and the bias of the researcher was ruled out.

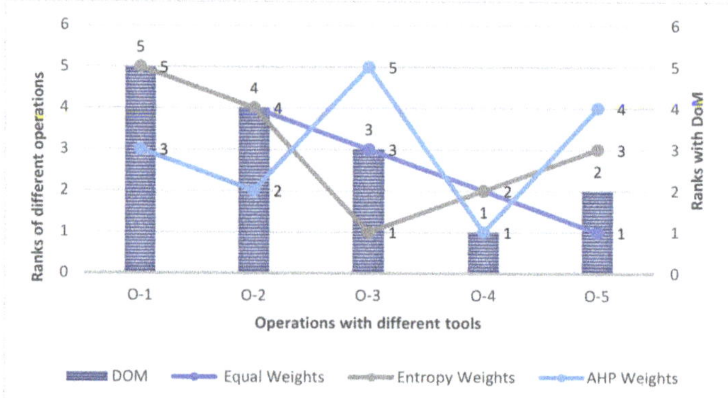

Figure 10. Ranks established using different weight methods, along with the final rank determined using a degree of membership.

The third method involved giving importance to the responses by a process of subjective preference, i.e., the AHP method. In this method, a pairwise comparison was made between identified responses. A nine-point Saaty scale was used to obtain the relative importance of responses [60]. The decision-makers were a group of experts. For example, for the present research, mechanical engineers, electrical engineers, technical experts from the concerned industry, and technical tool suppliers formed the expert group. First, the judgment values of each expert were considered. Then, the relative normalized weight and the weight of responses were calculated. In this case, the consistency ratio worked out to be 8.3%, which was less than the allowed value of 10%. This indicates precision in the decision-maker's judgment in allocating values for pairwise comparison.

Table 16 shows that the equal weights method assigned equal weights to all responses. The entropy method gave the maximum importance to the current unbalance (34.81%), followed by actual energy consumption (20.34%) and the average power consumption (14.37%). The AHP assigned maximum weight to the current total harmonic distortion factor (26.9%). Even in the multiple regression analysis, the current total harmonic distortion factor remained a significant parameter in machining operations O-1, O-2, and O-3. The second parameter to which the entropy method assigned weight was the average current (17.4%). This indicates the extreme impact of current on active power consumption.

Thus, in terms of prioritizing a machining operation based on energy consumption, rank 1 indicated the most energy-efficient machining operation, and rank 5 showed the most energy-intensive operation. It was observed that, under TOPSIS, the various weights methods graded the machining operations differently. The method of equal weights and the entropy weights method assigned rank 5 to the machining operation O-5. This indicates that machining operation 5 was the most energy-intensive process. At the same time, the method of AHP ranked machining operation O-3 as rank 5. To bring uniformity to the results and get a final combined rank, the degree of membership (DOM) was implemented. The results are presented in Table 18 and highlighted in Figure 10.

Machining operation O-1 emerges as the most energy-intensive process. O-1 is a machining process in which drilling occurs with the help of a high-speed steel (HSS) drill of diameter 16 mm and flute length 110 mm. The drilled hole depth is 77 mm. The spindle RPM is 450, and the feed rate is 70 mm/min (see Table 3). The process takes 154 s to complete, which is a lot of time. Hence, the energy consumed is more. This process will be looked into, and possibilities for a reduction in energy consumption will be suggested with the help of experimentation. This becomes a scope for future research.

5. Conclusions

This study presented a practical technique for enhancing the sustainability of machining operations in an industry. The article emphasized the significance of electric parameters in the active power consumption of machining operations. It encompassed a systematic approach for the identification of the most energy-intensive machining operation. Based on the critical outcome, the following conclusions were drawn:

- The quality of the power supplied to the machine tool seems to be affected by the concurrent functioning of the other machine tools in the surrounding area of the machine shop. This is evident from the fact that different electric parameters become significant at different times in the power consumption of a machining operation examined on the same shop floor of the industry. Therefore, the quality of power being supplied to the machine tool needs to be monitored and corrected to lower power consumption.
- Multiple regression analysis revealed that, out of the seven electric parameters considered, the rms values of the current and the power factor emerged as significant in all five machining operations. The current total harmonic distortion factor appeared significant in the three machining operations (O-1, O-4, and O-5). Current unbalance, the rms value of voltage, and voltage unbalance were significant in two machining operations each (O-2 and O-5, O-1 and O-2, and O-3 and O-5, respectively). The voltage total harmonic distortion factor was significant in only a single machining operation (O-4).
- The values of the coded coefficients of regression models revealed the relative impact of significant electric parameters on active power consumption. It was observed that the rms value of the current had the maximum direct impact, followed by the power factor. Therefore, their optimization would lead to a maximum reduction in electric power and energy. These factors were followed by the current unbalance, which also had a direct impact on power consumption. The rms value of voltage ranked last, with a small positive impact, whereas the voltage unbalance and the total harmonic distortion factor negatively affected the power consumption. To reduce the power consumption of a machining operation, it is imperative to assess the significance of electric parameters and evaluate their relative importance.
- The maximum absolute error in the estimation of active power using the standard power equation was 118.941 for machining operation O-5, but that with the developed regression model was 0.05. Similarly, the maximum relative error using the standard power equation was 5.350 for O-4 compared to 0.001 using the developed regression model. The R-squared value was more than 98% for O-1 to O-4 and 93% for O-5 for the developed models. This proves the results are theoretically correct.
- The TOPSIS equal-weights method identified machining operation O-1 as the most energy-intensive and O-5 as the least energy-intensive. With the assistance of the entropy weights without the decision maker's input, the same technique classified machining operation O-1 as the most energy-intensive but O-3 as the least energy-intensive process. The AHP weights method with the decision-maker's input ranked O-5 as the maximum and O-4 as the minimum energy consumption. Furthermore, the degree of membership (DoM) approach was employed to establish the final conjoined ranks. Machining operation O-1 was the most energy-intensive, followed by O-2, O-3, O-5, and O-4.

- The awareness of the importance of electric parameters in the active power consumption of machining processes and the identification of the energy-intensive machining operation benefit researchers and the industry in reducing energy consumption and minimizing the impact of carbon dioxide emissions in the industry environment.
- The most energy-intensive machining operation identified, i.e., O-1, a drilling process, needs to be optimized for minimum energy consumption. In addition, research is required to investigate or explore why some electric parameters become significant and others do not for the machine tools and machining processes conducted on the same shop floor. Furthermore, some electric parameters have positive and small negative impacts on the machining operation's power consumption and need to be investigated.

Author Contributions: Conceptualization, A.S.S. and R.K.; methodology, R.K. and S.S.; software, A.S.S., R.K., S.S., D.Y.P. and K.G.; validation, K.G., A.S.S. and R.K.; formal analysis, R.K.; investigation, A.S.S. and R.K.; resources, A.S.S.; data curation, A.S.S.; writing—original draft preparation, A.S.S., R.K., K.G. and D.Y.P.; writing—review and editing, A.S.S., R.K. and S.S.; visualization, S.S., D.Y.P. and K.G.; supervision, S.S.; project administration, A.S.S. All authors have read and agreed to the published version of the manuscript.

Funding: This research received no external funding.

Institutional Review Board Statement: Not applicable.

Informed Consent Statement: Not applicable.

Data Availability Statement: Not applicable.

Acknowledgments: The authors express deep gratitude to I. K. Gujral Punjab Technical University, Kapurthala-Jalandhar (Punjab), India, for providing the facilities to carry out this research work. Furthermore, the authors earnestly thank the joint MD of Auto International, Village Mangarh on Kohara, Machhiwara Road, Ludhiana-Punjab (India), Rajan Mittal, and Rohit Gupta for permitting observations and experimentation in the machine shop of their industry. The authors also thank Kuldeep Singh, the production supervisor.

Conflicts of Interest: The authors declare no conflict of interest.

Nomenclature

Abbreviations	**Acronym**	**Units**
Machining Time	$\text{ l}_{m/c}$	seconds (s)
Active power consumption air cut	$\ddot{A}PC_{air}$	kilowatt (kW)
Active power consumption by machine	$\ddot{A}PC_{m/c}$	kilowatt (kW)
Active energy consumption air cut	$\ddot{A}EC_{air}$	kilowatt hour (kWh)
Active energy consumption by machine	$\ddot{A}EC_{m/c}$	kilowatt hour (kWh)
Energy efficiency	$\dot{E}\dot{E}_{\acute{\eta}}$	No units
Specific energy consumption	ʂ	kilojoule/cm^3
Power factor	$PF_{m/c}$	No units
Average current (rms)	$\breve{I}_{m/cm/c}$	ampere (A)
Current total harmonic distortion factor	$\breve{I}_{m/cthdf}$	(%)
Current unbalance	$\breve{I}_{m/cub}$	(%)
Average voltage (rms)	$\ddot{V}_{av}$	volt
Voltage total harmonic distortion factor	$\ddot{V}_{thdf}$	(%)
Voltage unbalance	$\ddot{V}_{ub}$	(%)
Air cut	AC	No unit
Actual cut	ACT	No unit
p-Value	Ƥ	No unit
Volume of material removed	V	cm^3
Cutting speed	V_c	m/min
Depth of cut	ɗ	mm
Feed rate	F_d	mm/min

Abbreviations	**Acronym**	**Units**
Specific cutting energy	Śe	kJ/cm^3
Tool wear	$Ŧ_w$	mm
Material removal rate	ω	cm^3/min

Appendix A

Table A1. Normalized decision matrix (entropy method).

Operation	$ÄPC_{m/c}$	$ÄEC_{m/c}$	$ĖĖ_{ή}$	ş	$ƤF_{m/c}$	$Ĭ_{m/cm/c}$	$Ĭ_{m/cthdf}$	$Ĭ_{m/cub}$
O-1	0.7814	0.3929	0.5904	0.7974	0.9169	0.8638	1.0000	0.2714
O-2	0.6644	0.4925	0.8805	1.0000	0.9146	0.7370	0.9984	0.3067
O-3	0.4160	0.4286	0.9277	0.8659	0.9309	0.4697	0.9870	1.0000
O-4	1.0000	1.0000	0.6983	0.3869	0.8767	1.0000	0.9365	0.5887
O-5	0.5183	0.6226	1.0000	0.9149	1.0000	0.6111	0.7950	0.5633

Table A2. Probability of responses (entropy method).

Operation	$ÄPC_{m/c}$	$ÄEC_{m/c}$	$ĖĖ_{ή}$	ş	$ƤF_{m/c}$	$Ĭ_{m/cm/c}$	$Ĭ_{m/cthdf}$	$Ĭ_{m/cub}$
O-1	0.2312	0.1338	0.1441	0.2011	0.1977	0.2346	0.2120	0.0994
O-2	0.1966	0.1677	0.2149	0.2522	0.1972	0.2002	0.2117	0.1123
O-3	0.1231	0.1459	0.2264	0.2184	0.2007	0.1276	0.2093	0.3663
O-4	0.2959	0.3405	0.1704	0.0976	0.1890	0.2716	0.1985	0.2156
O-5	0.1533	0.2120	0.2441	0.2307	0.2156	0.1660	0.1685	0.2063

Table A3. Normalized decision matrix (TOPSIS).

Operation	$ÄPC_{m/c}$	$ÄEC_{m/c}$	$ĖĖ_{ή}$	ş	$ƤF_{m/c}$	$Ĭ_{m/cm/c}$	$Ĭ_{m/cthdf}$	$Ĭ_{m/cub}$
				Equal weights				
O-1	0.3372	0.5746	0.3168	0.3654	0.4415	0.3430	0.4169	0.6592
O-2	0.3966	0.4583	0.4725	0.2914	0.4404	0.4020	0.4176	0.5833
O-3	0.6334	0.5267	0.4979	0.3365	0.4483	0.6308	0.4224	0.1789
O-4	0.2635	0.2257	0.3748	0.7530	0.4222	0.2963	0.4452	0.3039
O-5	0.5083	0.3625	0.5367	0.3185	0.4815	0.4848	0.5245	0.3176
				Entropy weights				
O-1	0.3372	0.5746	0.3168	0.3654	0.4415	0.3430	0.4169	0.6592
O-2	0.3966	0.4583	0.4725	0.2914	0.4404	0.4020	0.4176	0.5833
O-3	0.6334	0.5267	0.4979	0.3365	0.4483	0.6308	0.4224	0.1789
O-4	0.2635	0.2257	0.3748	0.7530	0.4222	0.2963	0.4452	0.3039
O-5	0.5083	0.3625	0.5367	0.3185	0.4815	0.4848	0.5245	0.3176
				AHP weights				
O-1	0.3372	0.5746	0.3168	0.3654	0.4415	0.3430	0.4169	0.6592
O-2	0.3966	0.4583	0.4725	0.2914	0.4404	0.4020	0.4176	0.5833
O-3	0.6334	0.5267	0.4979	0.3365	0.4483	0.6308	0.4224	0.1789
O-4	0.2635	0.2257	0.3748	0.7530	0.4222	0.2963	0.4452	0.3039
O-5	0.5083	0.3625	0.5367	0.3185	0.4815	0.4848	0.5245	0.3176

Table A4. Weighted normalized matrix ($\hat{W}\dot{Z}_{ij}$) TOPSIS.

Operation	ÄPC$_{m/c}$	ÄEC$_{m/c}$	ĖĖ$_{ή}$	ȿ	ƤF$_{m/c}$	Ĭ$_{m/cm/c}$	Ĭ$_{m/cthdf}$	Ĭ$_{m/cub}$
				Equal weights				
O-1	0.0421	0.0718	0.0396	0.0457	0.0552	0.0429	0.0521	0.0824
O-2	0.0496	0.0573	0.0591	0.0364	0.0551	0.0502	0.0522	0.0729
O-3	0.0792	0.0658	0.0622	0.0421	0.0560	0.0789	0.0528	0.0224
O-4	0.0329	0.0282	0.0468	0.0941	0.0528	0.0370	0.0557	0.0380
O-5	0.0635	0.0453	0.0671	0.0398	0.0602	0.0606	0.0656	0.0397
				Entropy weights				
O-1	0.0485	0.1169	0.0178	0.0483	0.0013	0.0353	0.0046	0.2295
O-2	0.0570	0.0932	0.0265	0.0385	0.0013	0.0413	0.0046	0.2030
O-3	0.0910	0.1071	0.0279	0.0444	0.0013	0.0648	0.0047	0.0623
O-4	0.0379	0.0459	0.0210	0.0994	0.0012	0.0305	0.0050	0.1058
O-5	0.0731	0.0737	0.0301	0.0421	0.0014	0.0498	0.0058	0.1105
				AHP weights				
O-1	0.0101	0.0534	0.0317	0.0365	0.0667	0.0597	0.1122	0.0547
O-2	0.0119	0.0426	0.0473	0.0291	0.0665	0.0699	0.1123	0.0484
O-3	0.0190	0.0490	0.0498	0.0336	0.0677	0.1098	0.1136	0.0148
O-4	0.0079	0.0210	0.0375	0.0753	0.0637	0.0516	0.1198	0.0252
O-5	0.0152	0.0337	0.0537	0.0318	0.0727	0.0844	0.1411	0.0264

Table A5. Ideal positive and negative results.

	ÄPC$_{m/c}$	ÄEC$_{m/c}$	ĖĖ$_{ή}$	ȿ	ƤF$_{m/c}$	Ĭ$_{m/cm/c}$	Ĭ$_{m/cthdf}$	Ĭ$_{m/cub}$
				Equal weights				
Ideal positive	0.0329	0.0282	0.0671	0.0364	0.0602	0.0370	0.0521	0.0224
Ideal negative	0.0792	0.0718	0.0396	0.0941	0.0528	0.0789	0.0656	0.0824
				Entropy weights				
Ideal positive	0.0379	0.0459	0.0301	0.0385	0.0014	0.0305	0.0046	0.0623
Ideal negative	0.0910	0.1169	0.0178	0.0994	0.0012	0.0648	0.0058	0.2295
				AHP weights				
Ideal positive	0.0079	0.0210	0.0537	0.0291	0.0727	0.0516	0.1122	0.0148
Ideal negative	0.0190	0.0534	0.0317	0.0753	0.0637	0.1098	0.1411	0.0547

Table A6. Frequency of ranks according to DoM.

Ranks	1	2	3	4	5
O-1	0	0	1	0	2
O-2	0	1	0	2	0
O-3	1	0	1	0	1
O-4	1	2	0	0	0
O-5	1	0	1	1	0

References

1. Aktar, M.A.; Alam, M.M.; Al-Amin, A.Q. Global economic crisis, energy use, CO_2 emissions, and policy roadmap amid COVID-19. *Sustain. Prod. Consum.* **2021**, *26*, 770–781. [CrossRef]
2. Sutrisno, A.; Nomaler, Ö.; Alkemade, F. Has the global expansion of energy markets truly improved energy security? *Energy Policy* **2021**, *148*, 111931. [CrossRef]
3. Bogdanov, D.; Ram, M.; Aghahosseini, A.; Gulagi, A.; Oyewo, A.S.; Child, M.; Caldera, U.; Sadovskaia, K.; Farfan, J.; De Souza Noel Simas Barbosa, L.; et al. Low-cost renewable electricity as the key driver of the global energy transition towards sustainability. *Energy* **2021**, *227*, 120467. [CrossRef]

4. Zhang, Y. Review of recent advances on energy efficiency of machine tools for sustainability. *Proc. Inst. Mech. Eng. Part B J. Eng. Manuf.* **2014**, *229*, 2095–2108. [CrossRef]
5. Yoon, H.-S.; Kim, E.-S.; Kim, M.-S.; Lee, J.-Y.; Lee, G.-B.; Ahn, S.-H. Towards greener machine tools—A review on energy saving strategies and technologies. *Renew. Sustain. Energy Rev.* **2015**, *48*, 870–891. [CrossRef]
6. Zhou, L.; Li, J.; Li, F.; Meng, Q.; Li, J.; Xu, X. Energy consumption model and energy efficiency of machine tools: A comprehensive literature review. *J. Clean. Prod.* **2016**, *112*, 3721–3734. [CrossRef]
7. Zhou, L.; Li, J.; Li, F.; Xu, X.; Wang, L.; Wang, G.; Kong, L. An improved cutting power model of machine tools in milling process. *Int. J. Adv. Manuf. Technol.* **2017**, *91*, 2383–2400. [CrossRef]
8. Bayoumi, A.E.; Yücesan, G.; Hutton, D.V. On the closed form mechanistic modeling of milling: Specific cutting energy, torque, and power. *J. Mater. Eng. Perform.* **1994**, *3*, 151–158. [CrossRef]
9. Draganescu, F.; Gheorghe, M.; Doicin, C.V. Models of machine tool efficiency and specific consumed energy. *J. Mater. Process. Technol.* **2003**, *141*, 9–15. [CrossRef]
10. Gutowski, T.; Dahmus, J.; Thiriez, A. Electrical Energy Requirements for Manufacturing Processes. In Proceedings of the 13th CIRP International Conference of Life Cycle Engineering, Lueven, Belgium, 31 May–2 June 2006; pp. 1–5.
11. Li, W.; Kara, S. An empirical model for predicting energy consumption of manufacturing processes: A case of turning process. *Proc. Inst. Mech. Eng. Part B J. Eng. Manuf.* **2011**, *225*, 1636–1646. [CrossRef]
12. He, Y.; Liu, B.; Zhang, X.; Gao, H.; Liu, X. A modeling method of task-oriented energy consumption for machining manufacturing system. *J. Clean. Prod.* **2012**, *23*, 167–174. [CrossRef]
13. Li, L.; Yan, J.; Xing, Z. Energy requirements evaluation of milling machines based on thermal equilibrium and empirical modelling. *J. Clean. Prod.* **2013**, *52*, 113–121. [CrossRef]
14. Li, T.; Yuan, C. Numerical Modeling of Specific Energy Consumption in Machining Process. In Proceedings of the ASME 2013 International Manufacturing Science and Engineering Conference collocated with the 41st North American Manufacturing Research Conference, Madison, WI, USA, 10–14 June 2013.
15. Moon, J.-Y.; Shin, K.; Park, J. Optimization of production scheduling with time-dependent and machine-dependent electricity cost for industrial energy efficiency. *Int. J. Adv. Manuf. Technol.* **2013**, *68*, 523–535. [CrossRef]
16. Velchev, S.; Kolev, I.; Ivanov, K.; Gechevski, S. Empirical models for specific energy consumption and optimization of cutting parameters for minimizing energy consumption during turning. *J. Clean. Prod.* **2014**, *80*, 139–149. [CrossRef]
17. O'Rielly, K.; Jeswiet, J. Strategies to Improve Industrial Energy Efficiency. *Procedia CIRP* **2014**, *15*, 325–330. [CrossRef]
18. Zhao, G.; Hou, C.; Qiao, J.; Cheng, X. Energy consumption characteristics evaluation method in turning. *Adv. Mech. Eng.* **2016**, *8*, 1687814016680737. [CrossRef]
19. Sealy, M.P.; Liu, Z.Y.; Zhang, D.; Guo, Y.B.; Liu, Z.Q. Energy consumption and modeling in precision hard milling. *J. Clean. Prod.* **2016**, *135*, 1591–1601. [CrossRef]
20. Hu, S.; Liu, F.; He, Y.; Hu, T. An on-line approach for energy efficiency monitoring of machine tools. *J. Clean. Prod.* **2012**, *27*, 133–140. [CrossRef]
21. Mativenga, P.T.; Rajemi, M.F. Calculation of optimum cutting parameters based on minimum energy footprint. *CIRP Ann.* **2011**, *60*, 149–152. [CrossRef]
22. Behrendt, T.; Zein, A.; Min, S. Development of an energy consumption monitoring procedure for machine tools. *CIRP Ann.* **2012**, *61*, 43–46. [CrossRef]
23. Aramcharoen, A.; Mativenga, P.T. Critical factors in energy demand modelling for CNC milling and impact of toolpath strategy. *J. Clean. Prod.* **2014**, *78*, 63–74. [CrossRef]
24. Guo, Y.; Duflou, J.R.; Qian, J.; Tang, H.; Lauwers, B. An operation-mode based simulation approach to enhance the energy conservation of machine tools. *J. Clean. Prod.* **2015**, *101*, 348–359. [CrossRef]
25. Balogun, V.A.; Gu, H.; Mativenga, P.T. Improving the integrity of specific cutting energy coefficients for energy demand modelling. *Proc. Inst. Mech. Eng. Part B J. Eng. Manuf.* **2015**, *229*, 2109–2117. [CrossRef]
26. Altıntaş, R.S.; Kahya, M.; Ünver, H.Ö. Modelling and optimization of energy consumption for feature based milling. *Int. J. Adv. Manuf. Technol.* **2016**, *86*, 3345–3363. [CrossRef]
27. Albertelli, P.; Keshari, A.; Matta, A. Energy oriented multi cutting parameter optimization in face milling. *J. Clean. Prod.* **2016**, *137*, 1602–1618. [CrossRef]
28. Ma, F.; Zhang, H.; Cao, H.; Hon, K.K.B. An energy consumption optimization strategy for CNC milling. *Int. J. Adv. Manuf. Technol.* **2017**, *90*, 1715–1726. [CrossRef]
29. Jia, S.; Yuan, Q.; Lv, J.; Liu, Y.; Ren, D.; Zhang, Z. Therblig-embedded value stream mapping method for lean energy machining. *Energy* **2017**, *138*, 1081–1098. [CrossRef]
30. Chen, X.; Li, C.; Jin, Y.; Li, L. Optimization of cutting parameters with a sustainable consideration of electrical energy and embodied energy of materials. *Int. J. Adv. Manuf. Technol.* **2018**, *96*, 775–788. [CrossRef]
31. Fujishima, M.; Mori, M.; Oda, Y. Energy-efficient manufacturing on machine tools by machining process improvement. *Prod. Eng.* **2014**, *8*, 217–224. [CrossRef]
32. Mori, M.; Fujishima, M.; Inamasu, Y.; Oda, Y. A study on energy efficiency improvement for machine tools. *CIRP Ann.* **2011**, *60*, 145–148. [CrossRef]

33. Yan, J.; Li, L. Multi-objective optimization of milling parameters—The trade-offs between energy, production rate and cutting quality. *J. Clean. Prod.* **2013**, *52*, 462–471. [CrossRef]
34. Kant, G.; Sangwan, K.S. Prediction and optimization of machining parameters for minimizing power consumption and surface roughness in machining. *J. Clean. Prod.* **2014**, *83*, 151–164. [CrossRef]
35. Iqbal, A.; Zhang, H.-C.; Kong, L.L.; Hussain, G. A rule-based system for trade-off among energy consumption, tool life, and productivity in machining process. *J. Intell. Manuf.* **2015**, *26*, 1217–1232. [CrossRef]
36. Bilga, P.S.; Singh, S.; Kumar, R. Optimization of energy consumption response parameters for turning operation using Taguchi method. *J. Clean. Prod.* **2016**, *137*, 1406–1417. [CrossRef]
37. Kumar, R.; Bilga, P.S.; Singh, S. Multi objective optimization using different methods of assigning weights to energy consumption responses, surface roughness and material removal rate during rough turning operation. *J. Clean. Prod.* **2017**, *164*, 45–57. [CrossRef]
38. Teimouri, R.; Amini, S.; Lotfi, M.; Alinaghian, M. Sustainable drilling process of 1045 steel plates regarding minimum energy consumption and desired work quality. *Int. J. Lightweight Mater. Manuf.* **2019**, *2*, 397–406. [CrossRef]
39. Su, Y.; Zhao, G.; Zhao, Y.; Meng, J.; Li, C. Multi-Objective Optimization of Cutting Parameters in Turning AISI 304 Austenitic Stainless Steel. *Metals* **2020**, *10*, 217. [CrossRef]
40. Warsi, S.S.; Jaffery, H.I.; Ahmad, R.; Khan, M.; Akram, S. Analysis of power and specific cutting energy consumption in orthogonal machining of al 6061-T6 alloys at transitional cutting speeds. In Proceedings of the ASME 2015 International Mechanical Engineering Congress and Exposition, IMECE 2015, Houston, TX, USA, 13–19 November 2015.
41. Bustillo, A.; Pimenov, D.Y.; Mia, M.; Kapłonek, W. Machine-learning for automatic prediction of flatness deviation considering the wear of the face mill teeth. *J. Intell. Manuf.* **2021**, *32*, 895–912. [CrossRef]
42. Khan, A.M.; Anwar, S.; Gupta, M.K.; Alfaify, A.; Hasnain, S.; Jamil, M.; Mia, M.; Pimenov, D.Y. Energy-Based Novel Quantifiable Sustainability Value Assessment Method for Machining Processes. *Energies* **2020**, *13*, 6144. [CrossRef]
43. Singh, G.; Sharma, V.S.; Gupta, M.K.; Nguyen, T.-T.; Królczyk, G.M.; Pimenov, D.Y. Parametric optimization of multi-phase MQL turning of AISI 1045 for improved surface quality and productivity. *J. Prod. Syst. Manuf. Sci.* **2021**, *2*, 5–16.
44. Markopoulos, A.P.; Karkalos, N.E.; Mia, M.; Pimenov, D.Y.; Gupta, M.K.; Hegab, H.; Khanna, N.; Aizebeoje Balogun, V.; Sharma, S. Sustainability Assessment, Investigations, and Modelling of Slot Milling Characteristics in Eco-Benign Machining of Hardened Steel. *Metals* **2020**, *10*, 1650. [CrossRef]
45. Jamil, M.; Zhao, W.; He, N.; Gupta, M.K.; Sarikaya, M.; Khan, A.M.; R, S.M.; Siengchin, S.; Pimenov, D.Y. Sustainable milling of Ti–6Al–4V: A trade-off between energy efficiency, carbon emissions and machining characteristics under MQL and cryogenic environment. *J. Clean. Prod.* **2021**, *281*, 125374. [CrossRef]
46. Pimenov, D.Y.; Abbas, A.T.; Gupta, M.K.; Erdakov, I.N.; Soliman, M.S.; El Rayes, M.M. Investigations of surface quality and energy consumption associated with costs and material removal rate during face milling of AISI 1045 steel. *Int. J. Adv. Manuf. Technol.* **2020**, *107*, 3511–3525. [CrossRef]
47. Karim, M.R.; Tariq, J.B.; Morshed, S.M.; Shawon, S.H.; Hasan, A.; Prakash, C.; Singh, S.; Kumar, R.; Nirsanametla, Y.; Pruncu, C.I. Environmental, Economical and Technological Analysis of MQL-Assisted Machining of Al-Mg-Zr Alloy Using PCD Tool. *Sustainability* **2021**, *13*, 7321. [CrossRef]
48. Kumar, R.; Singh, S.; Sidhu, A.S.; Pruncu, C.I. Bibliometric Analysis of Specific Energy Consumption (SEC) in Machining Operations: A Sustainable Response. *Sustainability* **2021**, *13*, 5617. [CrossRef]
49. Sen, B.; Gupta, M.K.; Mia, M.; Pimenov, D.Y.; Mikołajczyk, T. Performance Assessment of Minimum Quantity Castor-Palm Oil Mixtures in Hard-Milling Operation. *Materials* **2021**, *14*, 198. [CrossRef] [PubMed]
50. Abbas, A.T.; Gupta, M.K.; Soliman, M.S.; Mia, M.; Hegab, H.; Luqman, M.; Pimenov, D.Y. Sustainability assessment associated with surface roughness and power consumption characteristics in nanofluid MQL-assisted turning of AISI 1045 steel. *Int. J. Adv. Manuf. Technol.* **2019**, *105*, 1311–1327. [CrossRef]
51. Markiewicz, H.; Klajn, A. Voltage disturbances standard en 50160-voltage characteristics in public distribution systems. *Wroc. Univ. Technol.* **2004**, *21*, 215–224.
52. Pimenov, D.Y.; Bustillo, A.; Mikolajczyk, T. Artificial intelligence for automatic prediction of required surface roughness by monitoring wear on face mill teeth. *J. Intell. Manuf.* **2018**, *29*, 1045–1061. [CrossRef]
53. Association, C.D. Voltage Disturbances. Standard EN 50160 Voltage Characteristics in Public Distribution Systems. In *Power Quality Application Guide*; European Copper Institute: Brussels, Belgium, 2004; pp. 7–21.
54. Anderson, M.J.; Whitcomb, P.J. *RSM Simplified: Optimizing Processes Using Response Surface Methods for Design of Experiments*, 2nd ed.; CRC Press: Boca Raton, FL, USA, 2016; p. 311.
55. Zhang, S. Analysis of some measurement issues in bushing power factor tests in the field. *IEEE Trans. Power Deliv.* **2006**, *21*, 1350–1356. [CrossRef]
56. Kumar, R.; Kaur, S. Multi Attribute Decision Making Approach to Select Microwave Oven with TOPSIS Method. In Proceedings of the 7th International Conference on Advancements in Engineering and Technology (ICAET-2019), Bhai Gurdas Institute of Engineering & Technology, Sangrur, India, 15–16 March 2019; pp. 357–361.
57. Kumar, R.; Dubey, R.; Singh, S.; Singh, S.; Prakash, C.; Nirsanametla, Y.; Królczyk, G.; Chudy, R. Multiple-Criteria Decision-Making and Sensitivity Analysis for Selection of Materials for Knee Implant Femoral Component. *Materials* **2021**, *14*, 2084. [CrossRef]

58. Kumar, R.; Singh, S.; Bilga, P.S.; Jatin, K.; Singh, J.; Singh, S.; Scutaru, M.-L.; Pruncu, C.I. Revealing the benefits of entropy weights method for multi-objective optimization in machining operations: A critical review. *J. Mater. Res. Technol.* **2021**, *10*, 1471–1492. [CrossRef]
59. Chodha, V.; Dubey, R.; Kumar, R.; Singh, S.; Kaur, S. Selection of industrial arc welding robot with TOPSIS and Entropy MCDM techniques. *Mater. Today Proc.* **2021**, in press. [CrossRef]
60. Saaty, T.L. Decision making—The Analytic Hierarchy and Network Processes (AHP/ANP). *J. Syst. Sci. Syst. Eng.* **2004**, *13*, 1–35. [CrossRef]

energies MDPI

Article

The Effect of Using Social Media in the Modern Marketing Communication on the Shaping an External Employer's Image

Agnieszka Izabela Baruk [1,*] and Grzegorz Wesołowski [2]

1 Department of Management Systems and Innovation, Faculty of Management and Production Engineering, Lodz University of Technology, 90-924 Lodz, Poland
2 Department of Organizational Personnel Management, Faculty of Social Sciences, John Paul II Catholic University of Lublin, 20-950 Lublin, Poland; grzegorz.wesolowski@kul.lublin.pl
* Correspondence: agnieszka.baruk@poczta.onet.pl

Abstract: The aim of this article was to determine the significance of modern marketing communication channels used in the process of shaping the external image of an enterprise as an employer. An analysis of the world literature on marketing, management, marketing communication and human resource management was used to prepare the theoretical part. The results of the analysis indicate a cognitive and research gap regarding the use of modern communication channels for building the external image of an enterprise in the role of an employer. In order to reduce the gap, empirical studies were conducted among young Polish potential employees, in which the survey method was used to gather primary data. The collected data were subjected to statistical analysis, during which the following methods and statistical tests were applied: the analysis of average values, exploratory factor analysis, Kruskal–Wallis test (KW), Pearson chi-square independence test and V-Cramer coefficient analysis. The results of the analyses conducted indicate, inter alia, that statistically significant diversity was identified in the case of non-professional media in terms of respondents' opinions on whether the employer's image created by modern media is better than the employer's image created on the basis of classical marketing communication channels. In the case of professional and non-professional media, the age of the respondents was not a differentiating feature. Moreover, neither for professional media nor for non-professional media were statistically significant dependencies identified between respondents' opinions on the impact of actions undertaken by enterprises on shaping their positive external image as an employer and respondents' opinions on whether the employer's image created on the basis of modern marketing communication channels is more beneficial than the employer's image created on the basis of classical marketing communication channels. The results obtained on the basis of the research have a cognitive and applicability value, characterized by originality. Until now, the importance of using modern marketing communication channels in shaping the employer's external image has not been analysed. This also applies to enterprises operating on the energy market.

Keywords: marketing communication; social media; image; enterprise; employer; employer's image; employee; employer's image of energy enterprise

Citation: Baruk, A.I.; Wesołowski, G. The Effect of Using Social Media in the Modern Marketing Communication on the Shaping an External Employer's Image. *Energies* **2021**, *14*, 4177. https://doi.org/10.3390/en14144177

Academic Editors: Wen-Hsien Tsai and Dimitrios A. Georgakellos

Received: 23 June 2021
Accepted: 8 July 2021
Published: 10 July 2021

1. Introduction

In order to develop or at least maintain their current market position, contemporary enterprises are forced to constantly search for sources of a relatively long-lasting competitive advantage [1,2]. Extremely important, unique sources of such an advantage are intangible assets, such as employees [3], knowledge [4] and positive image of an enterprise [5–7], including the image as an employer. These sources are very important for each enterprise, including energy companies. Competing for talented employees has become almost as important for enterprises as competing for external customers [8–10] leading to sustainable development.

A positive image of an enterprise as an employer may significantly facilitate attracting the most valuable employees [11,12] and keeping them. Investing in human capital positively affects the value of an enterprise and its organizational efficiency [10,13]. Therefore, enterprises should undertake activities aimed at shaping their positive image as employers not only among present but also among potential employees, as well as among other stakeholders of the organization [14–16], through, for example, the use of social media.

The inclusion of the Internet technology in activities allows an enterprise to survive and compete [17]. Especially, the use of social media in marketing communication in the 21st century has clearly gained importance [18]. This results among other things, from the fact that thanks to the development of the Internet, including social media, enterprises have gained the ability to communicate marketing messages to a large number of customers at the same time and to prompt their immediate interaction [19,20]. This process also brings some threats to the image of an enterprise. They are related to, among others, the possibility of expressing opinions about the enterprise and its offer, including the personnel offer, by an unlimited number of stakeholders, also by those whose opinions are negative [21].

The results of an analysis of the literature on the subject indicate that the concept of shaping the image of an enterprise as an employer is often wrongly identified with the creation of a brand. Many researchers discussing these issues refer to broadly understood employer branding [8,9,15,22–25]. Analysing this process and indicating the factors which may have an impact on employer branding, researchers rather do not take into account the nature of marketing communication channels used by an enterprise as a factor which could potentially influence its perception as an employer. It is true that some researchers in their works refer to modern marketing communication channels but mainly or only in the context of their importance in the process of the recruitment of employees [26–28]. This approach is widely used regardless of the industry, including energy one (for example Heilmann, Saarenketo and Liikkanen [29]). Becoming aware of the key importance of shaping the employer's image in a comprehensive way involving the use of social media can definitely enhance the market potential of energy enterprises. That is why the proposed approach brings many valuable implications for this group of companies because each of them plays the role of employer. It is one of the key market roles for energy enterprises.

The results of the analysis of the literature presented in the next section of this article clearly indicate the existence of a cognitive gap and research gap in terms of the possibility of shaping the image of an enterprise as an employer through the use of modern marketing communication channels. These aspects have not been investigated either in the case of energy companies or other ones. Therefore, the proposed approach has the feature of novelty especially in the case of enterprises on the energy market.

In order to reduce the identified gaps, an attempt was made to solve the following research problem: how important is the use of modern marketing communication channels when it comes to the perception of an enterprise as an employer by potential employees?

The aim of the article is therefore to determine the importance of using modern marketing communication channels in the process of shaping the external image of an enterprise as an employer.

The article is a theoretical and empirical study. The theoretical part presents the results of the cognitive and critical analysis of the literature on the subject presenting the issues of the image of an enterprise as an employer and its shaping, as well as marketing communication and the use of modern channels, especially social media. The empirical part presents the results of primary research aimed at identifying the role of using modern communication channels in building the external image of an employer. In the further parts of the article, the results are analysed, and conclusions, limitations of the research conducted and the directions of future research are presented.

2. Literature Review

2.1. Marketing Communication

Communication is a process of transferring ideas and sharing meanings with particular organizations or individuals [30]. One type of communication is marketing communication, used for transferring specific information to recipients in order to simplify marketing processes carried out as part of satisfying indirect and/or final demand [31]. It may also refer to the strategy applied by an offeror, allowing him/her to reach the target market participants [32].

According to the definition proposed by the American Marketing Association, marketing communication should be understood as "coordinated promotional messages and related media used for communication with the market" [33]. Marketing messages can be transmitted by means of modern digital media and/or classical media, including radio, press and TV, as well as by means of personal sales. Thus, the content of the message is exposed in this approach, which distinguishes it from the process approach, in which attention is focused on the sequence of actions taken to make this message available to the recipients. Such an approach in defining marketing communication is presented by, among others Smith, Gopalakrishna and Chatterjee [34].

In the context of the subject of this article, an integrated approach to marketing communication (the so-called IMC) seems particularly important. According to Batra and Keller, integrated marketing communication is "coordinated, consistent means by which companies try to inform, encourage, convince and remind consumers—directly or indirectly—about products and brands they sell" [35]. The main premise of integrated marketing communication for an organization is to centralize and skilfully coordinate all methods, tools and communication channels with entities from its environment, so that they are effective in the long term [36]. The effectiveness of an enterprise's communication may influence the formation of positive relationships with clients [37], which, in turn, may determine the enterprise's image [38]. Mentioned above, IMC includes the following elements: advertising, direct marketing, sales promotion, public relations, personal sales and digital/internet marketing [39]. Each of those elements can also be used for influencing potential and current employees.

Effective marketing communication is one of the factors that may potentially affect the attitudes and behaviour of employees, including the degree of their involvement in the work performed or the attitude towards the employer. Mutual employee-employer relations are based on emotions generated in the psychological and cognitive spheres. Those emotions determine the level of employees' trust in the employer, their willingness to accept changes [40] as well as their openness towards changes [41], etc.

It can therefore be assumed that enterprises wishing to consciously shape their image as employers should pay special attention to the way of communicating with individual participants of the external and internal labour market, as well as with other stakeholders. The employer's image is determined by other sub-images and by the general picture of a particular enterprise.

2.2. The Image of an Enterprise as an Employer

The most general definition of an image is a way of perceiving someone or something or a picture created in a recipient's mind [42]. In the literature on management, company image is treated in terms of a derivative of an organization's identity and the ways in which the organization communicates with its stakeholders [43]. Sometimes the image is wrongly identified with identity or even with a brand. In addition, new "brand image" thinking constructs are created [44,45], whose introduction to the academic discussion only introduces conceptual chaos.

According to Alvesson [46], company image is a holistic picture of an organization in the minds of entities from its surroundings. This picture is a result of shared values of members of a given organization, which are passed on to entities from its surroundings in the form of intentionally designed information. A similar way of defining company image is also presented by Hatch, and Schultz [47].

In turn, Dowling defines image as a "set of meanings that allow a given subject to be identified, described, remembered and referred to" [48]. Image is a direct picture of a given enterprise shaped in the minds of recipients [49].

Therefore, company image can be identified with a set of information and opinions about the company as an employer, a client, a community, a supplier and a participant of a corporation. It is a mental picture which arises in the minds of stakeholders, based on specific activities of the organization [50].

Company image can also be considered as a way of receiving a message of the nature of self-presentation by recipients interpreting the message in a broader context and in relation to their own reference framework [51].

According to Fombrun and Van Riel [52], company image is a result of the assessment of an enterprise by other entities, and the assessment refers to the way the entities perceive the previous and planned activities of the organization and compare them with the activities of its leading market rivals.

Ferrell and Hartline [53] point out that company image may include positive and negative impressions, which result from the company's past and present activity and from the perception of the actions it may take in the future.

The overall company image is the sum of its many sub-images, which include the image as an employer [54]. Minchington [55], defining the essence of the image of an enterprise as an employer, focuses on the perception of an organization as a workplace by its current employees, as well as other entities interested in this issue.

According to Jenner and Taylor [56], the image of an enterprise as an employer reflects the efforts of the organization oriented towards communication with internal and external recipients, as a result of which the enterprise is perceived as a desirable and distinctive employer. Martin [57] also narrows the definition of employer's image to the so-called desired image; according to him, the image is what senior management wants to convey to public opinion about the package of functional, economic and psychological benefits offered by a given company. Martin also emphasizes that this image is characterized by a large dose of subjectivism. This results, among other things, from the fact that the information provided can be received in a different way depending on how important it is to its particular addressees [57]. The analyses so far have focused mainly on a formal shaping of the image by an enterprise as an employer.

A characteristic feature of the image of an enterprise as an employer is the fact that it can be shaped both by internal and external recipients. This bidirectionality evidently affects the effects of shaping the image of an enterprise as an employer. Current employees (internal recipients) perceive an enterprise as an employer on the basis of their personal experience related to horizontal interpersonal relationships (between employees of the same level of organization) and vertical interpersonal relationships (between subordinates and superiors). Positive relations occurring between the organization's members in both of these systems have a positive impact on shaping the image of the enterprise as an employer among current employees. Their opinions reach participants of the external labour market.

Therefore, another area in which an enterprise should shape its image as an employer is its environment, including people appearing in the role of potential employees and people having direct relationships with them [54]. Baruk defines the image of an enterprise as an employer as "a picture (reflection) shaped in the awareness of current and potential employees based on their experience (in the case of people employed in the company) or information reaching potential members of the organization, whose source are people creating a given organization and all types of mass media" [54]. This image, both in the formal and informal dimension, can be shaped by means of classical and modern marketing communication channels, including social media.

2.3. Modern Marketing Communication Channels

People responsible for planning communication marketing activities in enterprises have been facing a serious challenge since the beginning of the 21st century. On the one hand, they have at their disposal a significant number of marketing communication channels, and on the other hand, they must skilfully coordinate the channels, creating their proper composition based on constant verification of their applicability. According to many researchers, traditional marketing communication channels, such as radio, television, newspapers, etc., will gradually lose their importance [58,59] for modern communication channels. That is why it is extremely important to define the channels unequivocally.

In the literature on the subject, one can come across many definitions of modern channels and modern media. Some of the definitions emphasize that a group of modern media consists of websites and other forms of digital communication, as well as information channels, in which active consumers are involved [60]. According to Pratt [61], modern (or new) media are inextricably linked with multimedia, and he claims that new media include all multimedia systems on-line, multimedia systems on the disk and multimedia systems connected with the development of technology for broadcasting or recording text, sound and image.

In turn, Manovich [62] points out that in most cases new media are identified with the dissemination and sharing of information by means of a computer. According to him, this way of defining new media is too simplistic and does not fully present the complexity of the issue, which cannot be argued with. New media are distinguished by the following features [62]: (1) numerical coding (the object of new media has a digital record that allows the object to be changed for example by using an appropriate algorithm); (2) modularity (the object of new media is created by a huge number of independent parts, able to function in an autonomous way); (3) automation (the use of an appropriate algorithm allows the entire object to be diametrically changed); (4) volatility (the object of new media may occur almost in an unlimited number of versions); (5) cultural transcoding.

On the other hand, some authors (including Lister, et al. [63]) believe that new media can be seen through the prism of adjectives that allow them to be described. New media are digital, interactive, hypertext, virtual, network and simulated. It is, however, worth emphasizing that not all of the above features must be present at the same time and with the same intensity.

In turn, according to Winer [64], the characteristics of new media can be limited to two features that distinguish them from traditional media. According to this author, these characteristics are interactivity and their digital character.

An interesting definition of new media was proposed by Witczak [65]. According to him, new media should be defined as a digital process of creating and disseminating as well as processing, exchanging and storing some information which is subject to social communication of an individual and mass character. The key technologies that dynamize the development of new media are the Internet and mobile telephony.

As can be seen, taking into account the approaches applied, all definitions of modern (new) media can be divided into the following three basic groups: technological definitions, attribute definitions and process definitions.

Rapid technological development has meant that nowadays practically every recipient of communication activities can post their opinions about a given offeror, including an employer, on the Internet network platforms. The combination of these platforms together with applications enabling the collection of information has made it possible to obtain specific information about the employer immediately, at all times and from almost anywhere on earth [66], being able to respond to them without deferring in time.

Before the emergence of new media, employees were able to express their opinions and share their experience about an employer usually within internal communication systems, which were created and controlled by the employer himself/herself [67]. This meant that the possibility for free expression was limited.

The dynamic development of new media has made marketing communication methods of enterprises with the environment undergo quick changes. More and more organizations are taking advantage of modern marketing communication channels in their activities. This results, among other things, in the increased interest of researchers in issues related to social media marketing, digital marketing and mobile marketing [68–74].

The results of an analysis of the literature on the subject indicate that research is being conducted on the use of modern marketing communication channels in human resource management, yet it is mainly limited to recruitment activities [75,76]. However, the use of modern marketing communication channels in the process of shaping the image of an enterprise as an employer, especially from the perspective of employees, is still not being analysed.

Marketing channels based on the media include social media. According to Kaplan, and Haenlien, media include "a group of Internet applications based on ideological and technological principles of Web 2.0 and enabling the creation and exchange of user-generated content" [77]. At this point, it is worth emphasizing that in Web 2.0, unlike Web 1.0, platform users are no longer just recipients of information, but also co-creators [78]. Social media are similarly defined by other authors (including [79]), who identify them with Internet applications based on Web 2.0. Some authors defining social media, however, exhibit more aspects related to information resources that enable interaction and creation of network communities (inter alia [80,81]).

Social media are characterized by openness, fast information exchange and high level of user involvement. They create a virtual environment where one can express and exchange opinions, share experiences, distribute and control other information anywhere and anytime [82,83]. Social media enable social interactions to take place between people who would not make contact in any other way [84]. It means that they change individual and group behaviours [85] in each area of life.

In the case of personnel activity, employees, using social media and running their own blog, can express not only positive but also negative opinions about their employer. This may have a significant adverse effect on company image, both among potential employees and other stakeholders [7]. Despite the fact that an increasing number of enterprises are deciding to take actions aimed at shaping their image as an employer [86], it is worth remembering that social media are much more difficult to manage, as well as control [87,88]. Therefore, enterprises should carefully select social media in which they publish information about themselves, especially in relation to fulfilling the role of employer [89].

The most well-known and widely used social media include Facebook (social networking site), Twitter (microblog), YouTube (site for posting videos), Instagram (portal for posting photos) and specialized social media platforms, such as LinkedIn and GoldenLine (portals with a recruitment and professional-business profile). Research conducted by O'Connor, Schmidt and Drouin [90] showed that 86% of employees have co-workers among friends in social media, and 77% of people use social media while working. From the point of view of the subject matter of this article, it is also important that 44% of people provide information about their work via social media [90]. Additionally, during the financial crisis, Facebook became the main social platform where users sought business contacts and new places of employment [91].

Mobile marketing is another marketing channel based on new media [92]. According to Kaplan, mobile marketing can be described as "any marketing activity carried out by means of the ubiquitous network, with which consumers are constantly connected through a personal mobile device" [93]. Mobile marketing may include actions within enterprise communication with customers using short text messages (SMS), QR codes and websites displayed on mobile devices [94] but also the transmission of picture messages (MMS) and image advertising on the screens of mobile devices.

While SMSs and MMSs have not found wider use as image-creating instruments, mobile social marketing has started to play a significant role in shaping company image as an employer. The creation of mobile marketing was related to the fact that the proliferation of the mobile Internet resulted in a specific combination of mobile communication with the classical Internet. However, the increase in the interest in social media meant that people using portals and social applications, wanting to be in constant contact with other users, started using mobile devices for this purpose [73]. An advantage of this combination was the ceasing of restrictions on the time and place of the receipt of a marketing message. Internet access via mobile devices has meant that enterprises can communicate with their employees not only during their work and travel but also during leisure [38]. The specific inseparable link between an employee and an employer can be treated as a negative consequence of the use of social media in marketing communication, which has a clear image overtone.

However, in practice, social media are still used rather as marketing communication channels regarding an offer of a given company [95]. This is clearly evident on the energy market. Social media are used in this case to identify consumers' attitudes towards renewable energy [96], level of environmental consumers' awareness [97], stakeholders' opinions [98], etc.

Application of social media in personnel policy is very limited, although social media can successfully be used for activities aimed at creating a positive image of an enterprise as an employer. This statement applies to enterprises functioning on different markets, inter alia, energy market where often no attention is paid to employer's image [99]. It is clearly visible even though each energy company acts also as an employer. That is why the aspects connected with creating the employer's image, especially external one, are as important for these firms as for the other companies. In the case of enterprises on the energy market, shaping their image in the role of employer is especially significant because of the importance of this industry for the economy on a regional, national and global scale.

In order to reduce the identified cognitive and research gap, the following research hypotheses were verified:

Hypothesis 1. *Respondents' attitude towards the use of modern communication channels for image-related purposes is a feature that differentiates opinions regarding activities that affect the shaping of a positive external image of an employer.*

Hypothesis 2. *There is a dependence between organizing promotional campaigns on non-professional social media (Facebook, Instagram) showing an enterprise as an attractive employer and respondents' opinions regarding the use of modern communication channels for image-related purposes.*

Hypothesis 3. *There is a dependence between organizing promotional campaigns on professional social media (LinkedIn, GoldenLine) showing an enterprise as an attractive employer and respondents' opinions regarding the use of modern communication channels for image-related purposes.*

Hypothesis 4. *The age of respondents is a feature that differentiates opinions regarding activities that affect the shaping of a positive external image of an employer.*

3. Research Methodology

Primary empirical research was conducted using the socio-psychological method of a questionnaire developed by the authors of the article. The research was carried out at the end of 2020. Its subjective scope included 430 people aged between 18 and 45 years old (the so-called mobile age) representing Polish users of social media. The selection of the research sample was conducted using quota-sampling and targeting. The population frame was data provided by the Central Statistical Office. Most of the respondents, like in the case of the structure of the general population, were women (58%). In terms of age structure, the majority of respondents had 36–45 years (40%). The rest of the respondents had, respectively, 26–35 years (38%) and 25 years or less (22%). This structure corresponded to the structure of the general population of adult Poles aged 18–45.

The research was Internet-based. The choice of the form of the research was conditioned by the adopted assumption that only active users of the Internet would be included in the study. The Computer Assisted Web Interview (CAWI) method was used, using the option of creating online questionnaires made available on the Google Forms platform.

In order to determine the importance of using modern forms of marketing communication in the context of the perception of an enterprise as an employer by the participants of the external labour market, the research included fifteen activities undertaken by employers in relation to employees. The actions were selected based on the results of a cognitive-critical analysis of the literature on the subject and on the results of unstructured interviews conducted among 25 people before the creation of the authors' original questionnaire. For each activity, the respondents were asked to determine its impact on shaping a positive image of an employer, using the Likert scale, which is one of the most fundamental psychometric tools in the social sciences [100]. In this article, a five-stage version of the Likert scale was used, in which the mark 1 meant that a given activity definitely has no impact; 2—it is unlikely to have an impact; 3—it is hard to say; 4—it is likely to have some impact; 5—it definitely has an impact. In accordance with the approach presented in the literature, among the analysed activities undertaken by employers, professional social media (such as LinkedIn) and non-professional social media (such as Facebook) were treated as two separate categories [101].

The primary data collected were subjected to quantitative analysis including methods and tests of statistical analysis, such as the analysis of average values, exploratory factor analysis, Kruskal–Wallis test (KW), Pearson chi-square independence test and V-Cramer analysis. The application of the Likert scale was a prerequisite for the application of the first two methods. The chi-square test was applied to determine whether there are dependencies between the variables analysed, and the V-Cramer coefficient was applied to determine the strength of the dependencies between the analysed variables.

The Kruskal–Wallis test allows an answer to be found as to whether the diversity in terms of separating individual groups (e.g., respondents' opinions, respondents' age) is statistically significant enough to say that the respondents' opinion determined by the analysed response is significantly different. It is a non-parametric equivalent of ANOVA [102]. It consists in checking whether the number of independent results from a group comes from the same population or from a population with the same median. Individual samples do not have to be of the same number. The input data are an n-element statistical sample divided into "k" of disjointed groups with numbers ranging from n1 to nk.

The explorative factor analysis allows an in-depth analysis of the collected primary data. It is used to reduce the number of variables constituting primary data obtained from surveys and to detect structures in relationships between these variables, that is to classify them. In order to determine the number of common factors (the so-called main components), the Kaiser criterion method was used to leave only those factors with eigenvalues greater than 1. Each factor explains a certain level of overall variability of the system under consideration, determined by a percentage of variance that can be interpreted as a measure of explanation of the problem. Within individual factors, the variables with the highest factor loadings against given factors were distinguished (value $\geq$0.7 was assumed) [103,104].

Factor analysis does not enable the answer to be found as to whether the diversity in terms of separating individual groups is statistically significant enough to say that the respondents' answers determined by the analysed response are significantly different. The Kruskal–Wallis test is used to answer this question. The higher the value with a satisfactory level of significance, the greater the diversity.

The statistical analysis was performed using the IBM SPSS Statistics Ver. 22 package.

4. Research Results

4.1. Respondents' Opinions about Communication Activities Undertaken by Enterprises on Shaping Their Positive External Image as an Employer

For all of the fifteen actions analysed, in total more than half of the respondents believed that they exert a greater or lesser impact on the beneficial perception of an employer, and more than 50% of people clearly expressed such an opinion in relation to the organization of paid internships (Table 1). In turn, the highest percentage of the respondents in total claimed that the creation of a positive image of an employer is not affected by organizing promotional campaigns on the radio showing a company as an attractive employer. It was the only activity for which the percentage exceeded 20%, amounting to 23.1%. At the same time, it was the only activity whose average rating was less than 3.50.

Table 1. Respondents' opinions on the impact of activities undertaken by enterprises on shaping enterprises' positive external image as an employer.

Activity	Indications (%)					Average Rating	Position
	1	2	3	4	5		
active participation in job fairs	3.3	5.5	8.8	53.0	29.4	**4.00**	4
activities carried out at universities and high schools aimed at increasing the knowledge about a company as an employer	1.5	4.5	14.5	48.5	30.9	**4.03**	3
offering internship programmes	1.8	2.1	9.1	39.7	47.3	**4.28**	1
organizing paid work experience	1.8	3.9	11.2	32.7	50.3	**4.26**	2
organizing free work experience	4.8	10.6	27.3	36.4	20.9	3.58	12
organizing training	2.1	1.2	9.4	42.7	44.5	**4.26**	2
maintaining contacts and cooperation with university environments through student representatives of a given enterprise	1.2	2.7	18.8	53.3	23.9	3.96	5
participation in competitions/rankings for the best employer	2.1	9.1	20.3	42.4	26.1	3.81	6
supporting activities undertaken by student organizations	1.8	6.1	26.7	47.6	17.9	3.74	8
organizing competitions for potential employees	1.5	7.9	24.8	42.7	23.0	3.78	7
organizing promotional campaigns on television, showing an enterprise as an attractive employer	3.9	11.5	22.7	42.4	19.4	3.62	10
organizing promotional campaigns on the radio, showing an enterprise as an attractive employer	6.1	17.0	25.5	37.3	14.2	3.37	14
organizing promotional campaigns on social media (Facebook, Instagram), showing an enterprise as an attractive employer	3.6	10.9	21.2	44.2	20.0	3.66	9
organizing promotional campaigns on social media of a professional nature (LinkedIn, GoldenLine), showing an enterprise as an attractive employer	4.2	10.3	25.5	43.9	16.1	3.57	13
offering employment to people aged 50+	4.8	7.9	30.0	37.0	20.3	3.60	11

1—definitely has no impact; 2—is unlikely to have an impact; 3—hard to say; 4—is likely to have an impact; 5—definitely has an impact.

It is worth noting that both activities involving the use of social media in image-related campaigns also obtained relatively low average ratings not exceeding the value of 3.70. They took respectively the 13th and 9th positions in the overall hierarchy of the activities analysed. It should be added that the use of professional social media gained a clearly lower average rating.

Within the fifteen activities analysed, five of them obtained an average rating exceeding the limit of 4.00, although none of the activities gained an assessment of at least 4.50, despite the five-point scale. Three activities with the largest average ratings had a common feature, which is the creation of conditions for acquiring new skills and/or knowledge by an employer. They were connected with internships programmes, work experience (but only paid) and training. Therefore, it can be assumed that the respondents paid a lot of attention to offering employees opportunities for professional development, noticing the close relationship between the employer's care for those aspects and the employer's image. The two remaining activities, which received average scores of at least 4.00, also referred to an active increase in knowledge, yet with respect to potential employees who are the participants of the external labour market.

On the other hand, the organization of promotional campaigns aimed at shaping the image of a company as an attractive employer obtained average ratings with much lower values, regardless of whether it was about using traditional or modern forms of communication in these activities. In addition, targeting via TV received an average rating with a slightly higher value than the use of professional social media for this purpose, which is of great significance considering the subject matter of this article. It can be concluded that the creation of opportunities for development turned out to be relatively more important from the point of view of the perception of an employer by participants of the external labour market than undertaking typical promotional activities.

The respondents were also asked whether an employer's image created by modern marketing communication channels (including the Internet and mobile applications) is more beneficial than the image built by means of classical channels (including press, television and the radio). It turns out that the majority of respondents in (53.64%) believed that the effect of using modern channels is a more favourable image, although only 7.58% expressed such an opinion. In total, almost every fourth person (23.94%) was of the opposite opinion, claiming that the image built thanks to modern channels is not better than the image shaped using traditional channels. It is worth adding that 4.55% of people categorically expressed such an opinion. It is also characteristic that as many as 22.42% of respondents did not have an opinion on this matter.

The answers to this question only apparently do not confirm the results presented in the previous considerations. It should be remembered that this question concerned the overall image of an employer, without distinguishing the external and internal image, and modern communication channels, without limiting the analysis to social media. Therefore, the question was more general than the question concerning actions taken by an employer analysed before. However, it seems important to identify whether respondents' opinions regarding the comparison of the image obtained thanks to modern and classical communication channels are significant when it comes to the role of individual actions of employers in shaping their external image as offerors of personal values.

4.2. Respondents' Structure Depending on Their Opinions on Whether the Employer's Image Created on the Basis of Modern Marketing Communication Channels Is More Beneficial Than the Employer's Image Created on the Basis of Classical Marketing Communication Channels

In order to identify the structure of the phenomenon studied, in particular, to compare the significance attributed to individual activities by the total of the respondents by the respondents who believe that the image shaped by modern channels is more beneficial than the image constructed using classical channels and by the respondents with a different opinion in this regard, an exploratory factor analysis was carried out for each of the three groups of the respondents. In order to conduct the analysis, the indications "definitely yes" and "rather yes" were merged, treating the respondents who gave such responses as one group agreeing that the image constructed by using modern communication channels is more advantageous than the image formed by traditional channels, whereas the indications "definitely not", "rather not" and "hard to say" were merged, treating people who gave such responses as one group of respondents disagreeing with this statement. The Cronbach Alpha test for the total number of the respondents had the value of 0.896, indicating high reliability. For respondents who agreed with the statement presented above, the value was 0.814, whereas for respondents with a different opinion it amounted to 0.799, which also indicates high reliability.

On the basis of the Kaiser criterion, for each of the three groups of respondents mentioned, three factors with eigenvalues greater than 1 were distinguished. They explain, in total, over 60% of the total variability of the system (Table 2). The first factor, separated for all respondents, has an eigenvalue of 4.423 and explains almost 30% of variability. It includes 4 activities undertaken by employers, in the case of which the factor loading exceeded the 0.7 limit value adopted during the factor analysis (Table 3). These are the same activities which received the highest average ratings, occupying the top positions in the hierarchy of activities, which, according to the respondents, had the greatest impact on shaping the positive external image of an enterprise as an employer. These are internal and external activities based on building knowledge capital.

Table 2. Hierarchy of factors according to their eigenvalues established on the basis of the Kaiser criterion (depending on respondents' opinions on whether the employer's image created on the basis of modern marketing communication channels is more beneficial than the employer's image created on the basis of classical marketing communication channels).

Factor	Eigenvalue			Cumulated Eigenvalue			% of Total Eigenvalues (Variance)			Cumulated % of Eigenvalues		
	tot. *	y #	n ^	tot.	y	n	tot.	y	n	tot.	y	n
1	4.423	4.709	3.277	4.423	4.709	3.277	29.484	31.391	21.847	29.484	31.391	21.847
2	3.831	3.831	3.148	8.254	8.540	6.425	25.539	25.539	20.987	55.023	56.930	42.834
3	1.116	1.095	2.741	9.370	9.635	9.166	7.439	7.301	18.271	62.462	64.231	61.105

* The measure of the adequacy of the Kaiser–Meyer–Olkin (KMO) draw equals 0.894, which is greater than 0.5; Bartlett's sphericity test is valid (variables are statistically significantly related to each other); chi2 equals 2478.376, and $p = 0.000$; # KMO = 0.892; Bartlett's sphericity test is valid; chi2equals 1307.295, and $p = 0.000$; ^ KMO = 0.852; Bartlett's sphericity test is valid; chi2 equals 1210.543, and $p = 0.000$; tot.—total number of respondents; y—persons who answered "yes"; n—people who answered "no".

In the case of the respondents who believe that the employer's image created on the basis of modern communication channels is more beneficial than the employer's image built by means of classical marketing communication channels, the first of the factors separated has a slightly higher eigenvalue (4.709) and explains a greater part (31.393%) of the total variability of the system analysed. The factor consists of five variables reflecting the four activities which entered the first factor for the total number of the respondents, accompanied by an activity regarding the establishing and consolidating of contacts with academic communities. It can therefore be said that this factor is characterized by a broader subject range as compared to the first factor identified for the total number of the respondents, also including activities related to building the relational capital.

The internal structure of the first factor is entirely different for the respondents who stated that the image shaped by modern communication channels is not better than the image shaped by classical communication channels. This factor includes four variables referring to activities consisting in conducting promotional campaigns both by means of classical media and modern media, such as social media. It is worth adding that the eigenvalue of this factor is much lower. It also explains the lowest part of the total variability of the phenomenon studied.

As for the second factor identified for the total number of the respondents, it includes four variables analogous to the first factor identified for the respondents who deny the possibility of creating a more favourable external image of the employer thanks to modern communication channels. The second factor, distinguished for the respondents indicating that such an image is better, has a similar structure. On the other hand, the second factor identified in the case of the respondents who have a different opinion is created by two variables concerning the building of knowledge capital within the organization, and in its the environment. The third factor includes one variable for each of the groups analysed, which is the organization of free work experience. This is the same variable for the total of the respondents and for the respondents who claim that the external image of an employer is better if it is built using modern communication channels. However, in the case of people representing a different opinion in this matter, the variable forming the third factor is active participation in job fairs, which is a variable of a completely different nature. It is worth adding that four variables did not enter into any factor for each group of respondents under analysis.

The factors recognized can be identified with the respondent segments characterized by the same attitude towards actions undertaken by employers. It is clearly visible that the factors are analogous to the total number of the respondents and to the respondents believing that the image shaped using modern communication channels is better, whereas they are entirely different in the case of the respondents who have the opposite opinion in this respect. Therefore, it can be assumed that the attitude towards the image-based effectiveness of using modern communication channels is a feature that differentiates opinions on activities which affect the shaping of a positive external image of an employer.

Table 3. The results of factor analysis of activities undertaken by enterprises depending on respondents' opinions on whether the employer's image created on the basis of modern marketing communication channels is more beneficial than the employer's image created on the basis of classical marketing communication channels.

Variable	Factor								
	1			2			3		
	tot.	y	n	tot.	y	n	tot.	y	n
offering internship programmes	**0.816**	**0.809**	0.054	0.132	0.151	**0.755**	0.077	0.029	0.362
organizing training	**0.798**	**0.796**	0.078	0.134	0.092	0.650	0.121	0.048	0.512
organizing paid work experience	**0.753**	**0.704**	−0.080	0.056	0.038	0.434	−0.373	−0.413	0.586
activities carried out at universities and high schools aimed at increasing the knowledge about a company as an employer	**0.730**	**0.764**	0.160	0.191	0.222	**0.753**	0.181	0.048	0.209
active participation in job fairs	0.675	0.675	0.100	0.166	0.251	0.131	0.365	0.276	**0.814**
maintaining contacts and cooperation with university environments through student representatives of a given enterprise	0.666	**0.779**	0.212	0.281	0.244	0.419	0.216	0.141	0.488
supporting activities undertaken by student organizations	0.603	0.616	0.182	0.403	0.441	0.264	0.004	0.015	0.703
offering employment to people aged 50+	0.395	0.404	0.236	0.385	0.364	0.000	−0.331	−0.437	0.633
organizing promotional campaigns on television, showing an enterprise as an attractive employer	0.091	0.103	**0.831**	**0.845**	**0.846**	0.071	0.100	0.157	0.199
organizing promotional campaigns on social media (Facebook, Instagram), showing an enterprise as an attractive employer	0.188	0.180	**0.824**	**0.826**	**0.842**	0.209	0.057	−0.014	0.164
organizing promotional campaigns on the radio, showing an enterprise as an attractive employer	0.067	0.086	**0.805**	**0.824**	**0.832**	0.075	0.069	0.032	0.184
organizing promotional campaigns on social media of a professional nature (LinkedIn, GoldenLine), showing an enterprise as an attractive employer	0.212	0.267	**0.831**	**0.822**	**0.806**	0.170	0.012	−0.062	0.178
organizing competitions for potential employees	0.367	0.396	0.411	0.540	0.519	0.061	−0.058	−0.093	0.559
participation in competitions/rankings for the best employer	0.519	0.549	0.434	0.528	0.489	0.244	0.070	0.094	0.603
organizing free work experience	0.322	0.394	0.170	0.126	0.152	0.589	**0.780**	**0.765**	−0.041

tot.—total number of respondents; y—persons who answered "yes"; n—people who answered "no".

4.3. Respondents' Differentiation Depending on Their Opinions on whether the Employer's Image Created on the Basis of Modern Marketing Communication Channels Is More Beneficial Than the Employer's Image Created on the Basis of Classical Marketing Communication Channels

In order to verify this assumption, the Kruskal–Wallis test was carried out. It turns out that a statistically significant difference can be distinguished in the case of nine out of fifteen activities analysed, for which the significance level is less than the 0.05 limit (Table 4). These include, among other things, activities that took up key positions in the hierarchy and were included in the first factor for the total number of the respondents and for the respondents who believe that the external image created through modern communication channels is better. Thus, it can be concluded that the hypothesis H1 turned out to be valid for respondents only for the nine actions mentioned.

Table 4. Significance of differences between responses regarding actions undertaken by enterprises depending on respondents' opinions on whether the employer's image created on the basis of modern marketing communication channels is more beneficial than the employer's image created on the basis of classical marketing communication channels.

Variable	Employer's External Image Created through Modern Channels Is Better Than Employer's Image Created through Classical Marketing Communication Channels	KW Test Value	Level of Significance 'p'
active participation in job fairs	no	152.32	**0.006**
	yes	178.68	
activities carried out at universities and high schools aimed at increasing the knowledge about a company as an employer	no	155.52	**0.040**
	yes	175.48	
offering internship programmes	no	148.25	**0.000**
	yes	182.75	
organizing paid work experience	no	147.18	**0.000**
	yes	183.82	
organizing free work experience	no	167.05	0.758
	yes	163.95	
organizing training	no	154.15	**0.018**
	yes	176.85	
maintaining contacts and cooperation with university environments through student representatives of a given enterprise	no	155.64	**0.039**
	yes	175.36	
participation in competitions/rankings for the best employer	no	154.31	**0.024**
	yes	176.69	
supporting activities undertaken by student organizations	no	158.26	0.139
	yes	172.74	
organizing competitions for potential employees	no	154.41	**0.026**
	yes	176.59	
organizing promotional campaigns on television, showing an enterprise as an attractive employer	no	157.35	0.102
	yes	173.65	
organizing promotional campaigns on the radio, showing an enterprise as an attractive employer	no	164.62	0.861
	yes	166.38	
organizing promotional campaigns on social media (Facebook, Instagram), showing an enterprise as an attractive employer	no	151.86	**0.006**
	yes	179.14	
organizing promotional campaigns on social media of a professional nature (LinkedIn, GoldenLine), showing an enterprise as an attractive employer	no	157.43	0.104
	yes	173.57	
offering employment to people aged 50+	no	163.38	0.672
	yes	167.62	

However, in the case of activities involving promotional campaigns, statistically significant differentiation was found only for campaigns on social media of the Facebook type. This is confirmed by the conclusion drawn from the results of the factor analysis. Activities of this type carried out in classical media were accompanied by promotional activities carried out in modern media. They were included in the same factor, regardless of whether it was about the total number of the respondents or for any of the other groups of the respondents analysed.

In the next stage of the analysis, dependencies between the respondents' opinions on the impact of activities undertaken by enterprises on shaping their positive external image as an employer and the respondents' opinions on whether the employer's image created on the basis of modern marketing communication channels is more beneficial than the employer's image created on the basis of classical marketing communication channels. As it results from Table 5, only five variables analysed have a significance level of less than 0.05. Therefore, only for these five variables can one talk about statistically significant dependencies between the activities undertaken by employers and the respondents' opinions saying that the external image of an employer created on the basis of modern channels is better than the external image of an employer created thanks to classical marketing communication channels. The five variables do not include any of the two variables that reflect the organization of promotional campaigns in social media. Thus, it can be said that in the case of the respondents, research hypotheses H2 and H3 are not valid. The identified relationships are characterized by low strength, as evidenced by the value of V-Cramer coefficient for individual variables, which is lower than 0.3. It is worth noting that these variables were not included in any factor identified in the factor analysis.

Table 5. Dependencies between respondents' opinions on the impact of activities undertaken by enterprises on shaping their positive external image as an employer, and respondents' opinions on whether the employer's image created on the basis of modern marketing communication channels is more beneficial than the employer's image created on the basis of classical marketing communication channels.

Variable	Chi-Square Test Value	V Cramer Coefficient Value	Level of Significance '*p*'
active participation in job fairs	11.938	0.190	**0.018**
activities carried out at universities and high schools aimed at increasing the knowledge about a company as an employer	5.351	0.127	0.253
offering internship programmes	13.939	0.206	**0.007**
organizing paid work experience	15.429	0.216	**0.004**
organizing free work experience	0.907	0.052	0.924
organizing training	8.398	0.160	0.078
maintaining contacts and cooperation with university environments through student representatives of a given enterprise	5.573	0.130	0.233
participation in competitions/rankings for the best employer	12.690	0.196	**0.013**
supporting activities undertaken by student organizations	5.518	0.129	0.238
organizing competitions for potential employees	10.419	0.178	**0.034**

Table 5. *Cont.*

Variable	Chi-Square Test Value	V Cramer Coefficient Value	Level of Significance 'p'
organizing promotional campaigns on television, showing an enterprise as an attractive employer	3.672	0.105	0.452
organizing promotional campaigns on the radio, showing an enterprise as an attractive employer	4.941	0.122	0.293
organizing promotional campaigns on social media (Facebook, Instagram), showing an enterprise as an attractive employer	9.090	0.166	0.059
organizing promotional campaigns on social media of a professional nature (LinkedIn, GoldenLine), showing an enterprise as an attractive employer	5.604	0.130	0.231
offering employment to people aged 50+	1.761	0.073	0.780

At the end of the analysis, it was checked whether respondent's age is a feature differentiating respondents' opinions about activities undertaken by enterprises which affect the shaping of a positive external image of an employer. The results of the Kruskal–Wallis test indicate that statistically significant differences occurred in the case of eight activities, which is indicated by the value of the significance level of less than 0.05 (Table 6), yet none of the two activities consisting in organizing promotional campaigns on social media. It means that for respondents the statement presented in the research hypothesis H4 is not true in the case these both activities.

Table 6. The significance of differences between responses concerning activities undertaken by enterprises depending on the age of respondents.

Variable	Age	KW Test Value	Level of Significance 'p'
active participation in job fairs	18–25	168.30	**0.091**
	26–35	173.24	
	36–45	132.75	
activities carried out at universities and high schools aimed at increasing the knowledge about a company as an employer	18–25	168.04	**0.02**
	26–35	183.69	
	36–45	124.95	
offering internship programmes	18–25	170.01	**0.001**
	26–35	179.24	
	36–45	111.45	
organizing paid work experience	18–25	168.32	0.057
	26–35	176.09	
	36–45	129.82	
organizing free work experience	18–25	169.28	0.253
	26–35	153.10	
	36–45	143.33	

Table 6. *Cont.*

Variable	Age	KW Test Value	Level of Significance 'p'
organizing training	18–25	168.63	**0.014**
	26–35	180.59	
	36–45	122.67	
maintaining contacts and cooperation with university environments through student representatives of a given enterprise	18–25	168.73	**0.000**
	26–35	195.45	
	36–45	107.35	
participation in competitions/rankings for the best employer	18–25	164.01	**0.009**
	26–35	208.19	
	36–45	137.72	
supporting activities undertaken by student organizations	18–25	166.12	0.071
	26–35	189.50	
	36–45	136.67	
organizing competitions for potential employees	18–25	161.84	**0.021**
	26–35	209.57	
	36–45	155.98	
organizing promotional campaigns on television, showing an enterprise as an attractive employer	18–25	163.57	0.444
	26–35	186.00	
	36–45	163.15	
organizing promotional campaigns on the radio, showing an enterprise as an attractive employer	18–25	162.84	**0.042**
	26–35	205.05	
	36–45	151.30	
organizing promotional campaigns on social media (Facebook, Instagram), showing an enterprise as an attractive employer	18–25	166.15	0.154
	26–35	185.40	
	36–45	140.43	
organizing promotional campaigns on social media of a professional nature (LinkedIn, GoldenLine), showing an enterprise as an attractive employer	18–25	165.86	0.425
	26–35	179.36	
	36–45	148.87	
offering employment to people aged 50+	18–25	159.09	0.008
	26–35	212.78	
	36–45	177.73	

5. Discussion

Based on the research conducted, it can be concluded that the results obtained constitute a significant contribution to theory, especially in the area of building an employer's image by using modern forms of marketing communication with potential employees. The aspects of communication through social media between an enterprise and the participants of the external labour market have so far been analysed in relation to the search for employers and the self-presentation of candidates. These issues were examined, among others, by Chiang and Suen [105], who focused exclusively on one of the specialist media platforms, which is LinkedIn, and Adams [106], who compared the effectiveness of LinkedIn with the effectiveness of Facebook in the process of finding an employer and employees. Adams pointed out that the relative importance of LinkedIn is greater in recruitment activities. Other authors compared in general terms the application of professional and non-professional social media in the process of not only recruitment but also selection [101,107], demonstrating that stronger reactions of candidates occurred in the case of using of professional media. Nikolaou [108] also drew attention to the differences between professional and non-professional social media, but he focused, similarly to Hallam [109] and Yokoyama [110], on the scope of information presented in the recruitment process, without examining the aspects of image. Other authors, including Stopfer and SGosling [111], studied social media, both professional and non-professional, with reference primarily to recruitment and selection activities, indicating their key importance in this respect.

On the other hand, the research results presented in this article indicate that in the case of image-related activities, the importance of social media of the Facebook type was comparatively greater (with a higher value of the average rating). In addition, image-related activities undertaken with the help of both types of social media were present in the same factor both for the total number of the respondents and for the other two groups of respondents, which shows that the respondents who noticed the usefulness of taking actions in this area via Facebook also saw the usefulness of analogous actions taken via LinkedIn.

The results of the research presented in this article indicate that the age of the respondents was not a feature that statistically significantly differentiated the responses both in the case of professional and non-professional social media. However, other authors [101] did not find such differentiation in the case of using professional media in recruitment and selection activities, yet they identified differentiation in relation to the application of non-professional social media in the recruitment and selection process, which is a clear difference compared with the application of social media for image-related purposes.

Moreover, research on the attitudes and emotions of potential employees in the role of candidates for work has been conducted so far, but these studies were mainly limited to the perception of the recruitment process itself [112,113] and the tools and methods used in the process (the so-called 'Chan and Schmitt model') [114]). The perception of an employer reflected in the employer's external image has not been taken into account. In contrast to the studies mentioned above, in which differences in the perception of the recruitment process itself were identified, depending on the tools used during the recruitment process, this article did not recognize statistically significant dependencies between the use of professional social media and the external image of an employer. Such dependencies were not identified for non-professional social media, either, although in their case one can speak of statistically significant differentiation of responses among the respondents who believe that the employer's image created on the basis of modern channels is more beneficial than the employer's image created through classical marketing communication channels and the responses among the respondents with a different opinion in this matter. It is worth emphasizing that in the case of energy companies, only a few studies on employer can be mentioned. It should be emphasized, however, that they concerned the other aspects than shaping external employer's image by using social media. For example, Heilmann, Saarenketo and Liikkanen [29] studied the motives and practices of energy companies in the scope of employer branding.

Among the few studies referring to the perception of an enterprise using social media in the recruitment process, research conducted by Priyadarshini, Kumar and Jha [115] can be mentioned. The research showed that the use of social media leads to the perception of an enterprise in a positive light, yet the use of professional and non-professional media was not compared in these studies, treating social media as one group of forms of communication with candidates. In addition, that research was limited to analysing the statements of representatives of this group of participants of the external labour market, that is, the candidates. It had the nature of qualitative research, which made it impossible to conduct statistical analysis.

Therefore, it can be concluded that in the literature on the subject, as regards the relationships between potential employees and the employer, considerations regarding social media are carried out primarily in relation to the recruitment and selection process, and, for example, the image-related aspects presented in this article are omitted. Among the benefits obtained through the use of social media, image-related issues are often neglected, limited to the issues related to better and faster recognition of candidates [116], etc. Although some authors notice the fact that using social media poses new challenges for employers, such as the formation of reputation [110,117] or branding [118] which is close to the issue of image, they do not examine those aspects in detail in the context of image.

6. Conclusions

The purpose of the research conducted was to determine the importance of using modern marketing communication channels in the process of shaping the external image of an enterprise as an employer. The analysis covered fifteen activities undertaken by enterprises in the area of building their image as an employer, taking into account the views of the respondents on whether an employer's image created on the basis of modern marketing communication channels (e.g., Internet, mobile applications) is more beneficial than an employer's image based on classical marketing communication channels (e.g., press, television, radio). The results of the research indicate that organizing promotional campaigns on social media took a lower position, especially in the case of professional media. A relatively higher rating was obtained in the case of non-professional media, although it was only 9th position.

Among the total number of respondents, a segment of people who consider that undertaking communication activities in traditional and modern media influences the shaping of a positive external image of an enterprise as an employer was distinguished. This segment was also identified among the other two groups of the respondents. However, no segment was identified for people who would attribute an image-creating role only to modern media and to the use of modern media in promotional campaigns.

Statistically significant differentiation with regards to opinions on whether an employer's image created through modern media is better was identified for non-professional media but neither in the case of professional media nor non-professional media age of respondents was a differentiating feature. Thus, as the main conclusion it can be said that hypothesis H1 turned out to be valid for non-professional media like Facebook, and hypothesis H4 turned out to be invalid for both types of social media.

The next important conclusions concern the assumed dependencies. Statistically significant dependencies were not identified either for professional media or for non-professional media between respondents' opinions on the impact of activities undertaken by enterprises to shape their positive external image as an employer and respondents' opinions on whether an employer's image created on the basis of modern marketing communication channels (e.g., Internet, mobile applications) is more beneficial than an employer's image created on the basis of classical marketing communication channels (e.g., press, television, radio). It means that for activities connected with organizing promotional campaigns in social media in the case of the respondents, research hypotheses H2 and H3 are not valid.

7. Implications, Limitations and Directions of Future Research

The results of the research conducted and the conclusions drawn on their basis have a cognitive and application value, allowing the cognitive research gap identified to be reduced. The managerial implications include, for example, the fact that the use of classical and modern compositions of social media is a better solution in building a positive image of an enterprise as an employer, rather than relying solely on one or the other group of media, especially in the case of segments of respondents expressing such expectations. From the point of view of shaping a positive external image of an employer, activities leading to increasing knowledge resources among both current and potential employees by an enterprise are of key importance. Internship, training and work experience (especially paid), as well as the presentation of an individual offer in secondary and higher schools, obtained average ratings of the highest values. Conclusions drawn from the research have particularly big implications for executives and managers of energy companies too. They allow to prepare and conduct appropriate activities in the employer's image field increasing competitiveness of energy companies on the labour market as well as leading to sustainable development. It is very important because energy enterprises should recruit and retain the best specialists, which can be achieved by applying modern solutions in the marketing communications process, including communication via social media.

The results of the research have also social implications, indicating the attitude of potential employees to selected activities undertaken by enterprises in the field of personnel policy and the image-related importance of those activities.

Obviously, the research conducted has some limitations, which include its subjective scope (the research included only adults under 45 years of age), its objective scope (only selected aspects of using social media in shaping the external image of the enterprise as an employer were studied) and its geographical scope (the research included only Polish representatives of young potential employees). In future research, the authors will strive to eliminate those limitations by expanding the subjective, objective and geographical scopes of the research. It is worth emphasizing that although the limitation listed on the first position is an intentionally made assumption (respondents represented the so-called mobile age), in the future, authors want to conduct studies among representatives of other age groups. This will allow, among others, a comparative analysis to be made between representatives of different generations of potential employees and between representatives of various cultural circles illustrating the system of values from various countries. Moreover, it may also be interesting to undertake research on the impact of the COVID-19 pandemic on employer's image in the context of using social media for its creation.

Author Contributions: Literature review—A.I.B. and G.W.; gathering the primary data—G.W.; other stages of the preparation and writing of this article—A.I.B. All authors have read and agreed to the published version of the manuscript.

Funding: This research received no external funding.

Conflicts of Interest: The authors declare no conflict of interest.

References

1. Szablewski, A.; Martin, M. *Determinants of Economic Competitiveness*; Lodz University of Technology Press: Łódź, Poland, 2009.
2. Rybaczewska, M. Employer Image of the Company from the Perspective of Its Customers. *Int. J. Contemp. Manag.* **2017**, *16*, 61–81. [CrossRef]
3. Saha, N.; Gregar, A. Human Resource Management: As a Source of Sustained Competitive Advantage of the Firms. *IPEDR* **2012**, *46*, 7–14. [CrossRef]
4. Raguž, I.V.; Zekan, Z.B.; Peronja, I. Knowledge as a source of competitive advantage in knowledge based companies. *Manag. Organ.* **2017**, *3*, 533–544.
5. Hamadamin, H.H.; Atan, T. The Impact of Strategic Human Resource Management Practices on Competitive Advantage Sustainability: The Mediation of Human Capital Development and Employee Commitment. *Sustainability* **2019**, *11*, 5782. [CrossRef]
6. Burmann, C.; Schaefer, K.; Maloney, P. Industry image: Its impact on the brand image of potential employees. *J. Brand Manag.* **2007**, *15*, 157–176. [CrossRef]
7. Mangold, W.G.; Miles, S.J. The employee brand: Is yours an all-star? *Bus. Horiz.* **2007**, *50*, 423–433. [CrossRef]
8. Alnıaçık, E.; Alnıaçık, Ü.; Erat, S.; Akçin, K. Attracting Talented Employees to the Company: Do We Need Different Employer Branding Strategies in Different Cultures? *Procedia Soc. Behav. Sci.* **2014**, *150*, 336–344. [CrossRef]
9. Dabirian, A.; Kietzmann, J.; Diba, H. A great place to work!? Understanding crowdsourced employer branding. *Bus. Horiz.* **2017**, *60*, 197–205. [CrossRef]
10. Theurer, C.P.; Tumasjan, A.; Welpe, I.M.; Lievens, F. Employer Branding: A Brand Equity-based Literature Review and Research Agenda. *Int. J. Manag. Rev.* **2018**, *20*, 155–179. [CrossRef]
11. Kim, K.H.; Jeon, B.J.; Jung, H.S.; Lu, W.; Jones, J. Effective employment brand equity through sustainable competitive advantage, marketing strategy, and corporate image. *J. Bus. Res.* **2012**, *65*, 1207–1211. [CrossRef]

12. Edwards, M.R.; Edwards, T. Employee responses to changing aspects of the employer brand following a multinational acquisition: A longitudinal study. *Hum. Resour. Manag.* **2013**, *52*, 27–54. [CrossRef]
13. Backhaus, K.; Tikoo, S. Conceptualizing and researching employer branding. *Career Dev. Int.* **2004**, *9*, 501–517. [CrossRef]
14. Sullivan, J. The 8 Elements of a Successful Employment Brand. 2004. Available online: https://www.ere.net/the-8-elements-of-a-successful-employment-brand/ (accessed on 15 March 2021).
15. Ahmad, N.A.; Daud, S. Engaging People with Employer Branding. *Procedia Econ. Financ.* **2016**, *35*, 690–698. [CrossRef]
16. Mölk, A.; Auer, M. Designing brands and managing organizational politics: A qualitative case study of employer brand creation. *Eur. Manag. J.* **2018**, *36*, 485–496. [CrossRef]
17. Sestino, A.; Prete, M.I.; Piper, L.; Guido, G. Internet of Things and Big Data as enablers for business digitalization strategies. *Technovation* **2020**, *98*, 102173. [CrossRef]
18. Watson, G.F.; Worm, S.; Palmatier, R.W.; Ganesan, S. The Evolution of Marketing Channels: Trends and Research Directions. *J. Retail.* **2015**, *91*, 546–568. [CrossRef]
19. Iankova, S.; Davies, I.; Archer-Brown, C.; Marder, B.; Yau, A. A comparison of social media marketing between B2B, B2C and mixed business models. *Ind. Mark. Manag.* **2019**, *81*, 169–179. [CrossRef]
20. Schivinski, B.; Dabrowski, D. The effect of social media communication on consumer perceptions of brands. *J. Mark. Commun.* **2016**, *22*, 189–214. [CrossRef]
21. Szwajca, D. The Role of Social Media in Corporate Reputation Management—The Results of the Polish Enterprises. *Found. Manag.* **2017**, *9*, 161–174. [CrossRef]
22. Aldousari, A.A.; Robertson, A.; Yajid, M.S.A.; Ahmed, Z.U. Impact of employer branding on organization's performance. *J. Transnatl. Manag.* **2017**, *22*, 153–170. [CrossRef]
23. Alnıaçık, E.; Alnıaçık, Ü. Identifying Dimensions of Attractiveness in Employer Branding: Effects of Age, Gender, and Current Employment Status. *Procedia Soc. Behav. Sci.* **2012**, *58*, 1336–1343. [CrossRef]
24. Chunping, Y.; Xi, L. The Study on Employer Brand Strategy in Private Enterprises from the Perspective of Human Resource Management. *Energy Procedia* **2011**, *5*, 2087–2091. [CrossRef]
25. Reis, G.G.; Braga, B.M. Employer attractiveness from a generation perspective: Implications for employer branding. *Rev. Adm.* **2016**, *51*, 103–116. [CrossRef]
26. Cretu, A.E.; Brodie, R.J. The influence of brand image and company reputation where manufacturers market to small firms: A customer value perspective. *Ind. Mark. Manag.* **2007**, *36*, 230–240. [CrossRef]
27. Puncheva-Michelotti, P.; Hudson, S.; Jin, G. Employer branding and CSR communication in online recruitment advertising. *Bus. Horiz.* **2018**, *61*, 643–651. [CrossRef]
28. Wołodźko, K.; Woźniak, J. The use by large polish organizations of information about CSR activities in e-recruitment. *Econ. Sociol.* **2017**, *10*, 47–60. [CrossRef] [PubMed]
29. Heilmann, P.; Saarenketo, S.; Liikkanen, K. Employer branding in power industry. *Int. J. Energy Sect. Manag.* **2013**, *7*, 283–302. [CrossRef]
30. Saeed, R.; Naeem, B.; Bilal, M.; Naz, U. Integrated marketing communication : A review paper. *Interdiscip. J. Contemp. Res. Bus.* **2013**, *5*, 124–133.
31. Corniani, M. Digital Marketing Communication. *Symphonya. Emerg. Issues Manag.* **2012**, *2*, 41–61. [CrossRef]
32. Sathish, L. A Study About Marketing Communications Effectiveness in The Business-To-Business Markets. *Int. Conf. Inter. Discip. Res. Eng. Technol.* **2015**, *1*, 99–104.
33. Marketing Communications. Available online: https://www.ama.org/topics/marcom/ (accessed on 20 June 2021).
34. Smith, T.M.; Gopalakrishna, S.; Chatterjee, R. A Three-Stage Model of Integrated Marketing Communications at the Marketing-Sales Interface. *J. Mark. Res.* **2006**, *43*, 564–579. [CrossRef]
35. Batra, R.; Keller, K.L. Integrating Marketing Communications: New Findings, New Lessons, and New Ideas. *J. Mark.* **2016**, *80*, 122–145. [CrossRef]
36. Owen, R.; Humphrey, P. The Structure of Online Marketing Communication Channels. *J. Manag. Mark. Res.* **2010**, *2*, 54–62.
37. Han, S.H.; Yoon, S.W.; Chae, C. Building social capital and learning relationships through knowledge sharing: A social network approach of management students' cases. *J. Knowl. Manag.* **2020**, *24*, 921–939. [CrossRef]
38. Cao, Y. Research on Integrated Marketing Communication in the Mobile Internet Environment. In Proceedings of the 2015 International Conference on Economics, Social Science, Arts, Education and Management Engineering, Wuhan, China, 12–13 December 2015. [CrossRef]
39. Key, T.M.; Czaplewski, A.J. Upstream social marketing strategy: An integrated marketing communications approach. *Bus. Horiz.* **2017**, *60*, 325–333. [CrossRef]
40. Leiter, M.P.; Harvie, P. Conditions for staff acceptance of organizational change: Burnout as a mediating construct. *Anxiety Stress Coping* **1998**, *11*, 1–25. [CrossRef]
41. Dziallas, M.; Blind, K. Innovation indicators throughout the innovation process: An extensive literature analysis. *Technovation* **2019**, *80*, 3–29. [CrossRef]
42. Image. Available online: https://www.ldoceonline.com/dictionary/image (accessed on 5 June 2021).
43. Gioia, D.A.; Schultz, M.; Corley, K.G. Organisational identity, image, and adaptive instability. *Acad. Manag. Rev.* **2000**, *25*, 63–81. [CrossRef]

44. Išoraitė, M. Brand Image Theoretical Aspects. *Integr. J. Bus. Econ.* **2018**, *2*, 116–122. [CrossRef]
45. Malik, M.E.; Naeem, B.; Munawar, M. Brand Image: Past, Present and Future. *J. Basic Appl. Sci. Res.* **2012**, *2*, 13069–13075.
46. Alvesson, M. Organization: From Substance to Image? *Organ. Stud.* **1990**, *11*, 373–394. [CrossRef]
47. Hatch, M.J.; Schultz, M. The Dynamics of Organizational Identity. *Hum. Relat.* **2002**, *55*, 989–1018. [CrossRef]
48. Dowling, G.R. Managing your corporate images. *Ind. Mark. Manag.* **1986**, *15*, 109–115. [CrossRef]
49. Gray, E.R.; Balmer, J.M.T. Managing Corporate Image and Corporate Reputation. *Long Range Plan.* **1998**, *31*, 695–702. [CrossRef]
50. Adeniji, A.A.; Osibanjo, A.O.; Abiodun, A.J.; Oni-Ojo, E.E. Corporate Image: A Strategy for Enhancing Customer Loyalty and Profitability. *J. South. Afr. Bus. Res.* **2015**. [CrossRef]
51. Wood, E. Corporate Identity. In *The Public Relations Handbook*; Theaker, A., Ed.; Routledge: London, UK, 2004; pp. 74–92.
52. Fombrun, C.J.; Van Riel, C.B.M. The Reputational Landscape. *Corp. Reput. Rev.* **1996**. [CrossRef]
53. Ferrell, O.C.; Hartline, M.D. *Marketing Strategy*; Cengage Learning: Boston, MA, USA, 2011.
54. Baruk, A.I. Contentment of employees vs their prosumeric activity in the scope of recommending an employer. *J. Bus. Ind. Mark.* **2017**, *32*, 742–751. [CrossRef]
55. Minchington, B. *Your Employer Brand: Attract, Engage, Retain*; Collective Learning Australia: Adelaide, Australia, 2006.
56. Jenner, S.J.; Taylor, S. Employer branding—fad or the future of HR? In *Research Insight: Employer Branding: The Latest Fad or the Future for HR?* Humpage, S., Ed.; Chartered Institute of Personnel and Development: London, UK, 2008.
57. Martin, G. *Employer Branding—Time for Some Long and 'Hard' Reflections*; Chartered Institute of Personnel and Development: London, UK, 2007; pp. 18–23.
58. Danaher, P.J.; Rossiter, J.R. Comparing perceptions of marketing communication channels. *Eur. J. Mark.* **2011**, *45*, 6–42. [CrossRef]
59. Tiago, M.T.P.M.B.; Veríssimo, J.M.C. Digital marketing and social media: Why bother? *Bus. Horiz.* **2014**, *57*, 703–708. [CrossRef]
60. Skiera, B.; Rangaswamy, A.; Malthouse, E.C.; Hennig-Thurau, T.; Gensler, S.; Lobschat, L.; Friege, C. The Impact of New Media on Customer Relationships. *J. Serv. Res.* **2010**, *13*, 311–330. [CrossRef]
61. Pratt, A.C. New media, the new economy and new spaces. *Geoforum* **2000**, *31*, 425–436. [CrossRef]
62. Manovich, L. What Is New Media? How Media Became New. Available online: https://www.mheducation.co.uk/openup/chapters/0335217109.pdf (accessed on 4 June 2021).
63. Lister, M.; Dovey, J.; Giddings, S.; Grant, I.; Kelly, K. *New Media: A critical Introduction*; Routledge: London, UK, 2008.
64. Winer, R.S. New Communications Approaches in Marketing: Issues and Research Directions. *J. Interact. Mark.* **2009**, *23*, 108–117. [CrossRef]
65. Witczak, O. Nowe media w budowaniu marki i wizerunku przedsiębiorstwa. *Studia Ekon.* **2013**, *140*, 80–97.
66. Rai, S. Engaging Young Employees (Gen Y) in a Social Media Dominated World—Review and Retrospection. *Procedia Soc. Behav. Sci.* **2012**, *37*, 257–266. [CrossRef]
67. Miles, S.J.; Mangold, W.G. Employee voice: Untapped resource or social media time bomb? *Bus. Horiz.* **2014**, *57*, 401–411. [CrossRef]
68. Felix, R.; Rauschnabel, P.A.; Hinsch, C. Elements of strategic social media marketing: A holistic framework. *J. Bus. Res.* **2017**, *70*, 118–126. [CrossRef]
69. Haryani, S.; Motwani, B. Discriminant model for online viral marketing influencing consumers behavioural intention. *Pac. Sci. Rev. Humanit. Soc. Sci.* **2015**, *1*, 49–56. [CrossRef]
70. Kannan, P.K.; Li, H. Digital marketing: A framework, review and research agenda. *Int. J. Res. Mark.* **2017**, *34*, 22–45. [CrossRef]
71. Stephen, A.T. The role of digital and social media marketing in consumer behaviour. *Curr. Opin. Psychol.* **2016**, *10*, 17–21. [CrossRef]
72. Wang, Z.; Kim, H.G. Can Social Media Marketing Improve Customer Relationship Capabilities and Firm Performance? Dynamic Capability Perspective. *J. Interact. Mark.* **2017**, *39*, 15–26. [CrossRef]
73. Yadav, M.; Joshi, Y.; Rahman, Z. Mobile Social Media: The New Hybrid Element of Digital Marketing Communications. *Procedia Soc. Behav. Sci.* **2015**, *189*, 335–343. [CrossRef]
74. Zhu, Y.Q.; Chen, H.G. Social media and human need satisfaction: Implications for social media marketing. *Bus. Horiz.* **2015**, *58*, 335–345. [CrossRef]
75. Bizzi, L. The hidden problem of Facebook and social media at work: What if employees start searching for other jobs? *Bus. Horiz.* **2018**, *61*, 23–33. [CrossRef]
76. Pitt, C.S.; Botha, E.; Ferreira, J.J.; Kietzmann, J. Employee brand engagement on social media: Managing optimism and commonality. *Bus. Horiz.* **2018**, *61*, 635–642. [CrossRef]
77. Kaplan, A.M.; Haenlein, M. Users of the world! The challenges and opportunities of Social Media. *Bus.Horiz.* **2010**, *53*, 59–68. [CrossRef]
78. Constantinides, E.; Fountain, S. Web 2.0: Conceptual foundations and marketing issues. *J. Direct Data Digit. Mark. Pract.* **2008**, *9*, 231–244. [CrossRef]
79. Huang, J.; Baptista, J.; Galliers, R.D. Reconceptualizing rhetorical practices in organizations: The impact of social media on internal communications. *Inf. Manag.* **2013**, *50*, 112–124. [CrossRef]
80. Kapoor, K.K.; Tamilmani, K.; Rana, N.P.; Patil, P.; Dwivedi, Y.K.; Nerur, S. Advances in Social Media Research: Past, Present and Future. *Inf. Syst.Front.* **2018**, *20*, 531–558. [CrossRef]

81. Oestreicher-Singer, G.; Zalmanson, L. Content or community? A digital business strategy for content providers in the social age. *MISQ* **2013**, *37*, 591–616. [CrossRef]
82. Kietzmann, J.H.; Hermkens, K.; McCarthy, I.P.; Silvestre, B.S. Social media? Get serious! Understanding the functional building blocks of social media. *Bus. Horiz.* **2011**, *54*, 241–251. [CrossRef]
83. Tavleen, K. Role of Social Media in Building Image of an Organization as a Great Place to Work. *Proc. ASBBS* **2013**, *20*, 546–553.
84. Jucaitytė, I.; Maščinskienė, J. Peculiarities of Social Media Integration into Marketing Communication. *Procedia Soc. Behav. Sci.* **2014**, *156*, 490–495. [CrossRef]
85. Dwivedi, Y.K.; Ismagilova, E.; Rana, N.P.; Raman, R. Social Media Adoption, Usage And Impact In Business-To-Business (B2B) Context: A State-Of-The-Art Literature Review. *Inf. Syst. Frontiers.* **2021**. [CrossRef]
86. Eren, E.; Vardarlıer, P. Social Media's Role in Developing an Employees Sense of Belonging in the Work Place as an HRM Strategy. *Procedia Soc. Behav. Sci.* **2013**, *99*, 852–860. [CrossRef]
87. Solis, B.; Breakenridge, D.K. *Putting the Public Back in Public Relations*; FT Press: Upper Saddle River, NJ, USA, 2009.
88. Cornelissen, J.P. *Corporate Communication: A Guide to Theory and Practice*; Sage Publications: London, UK, 2014.
89. Brecht, F.; Koroleva, K.; Guenther, O. Increasing global reach: Using social network sites for employer branding. In *Wirtschaftsinformatik Proceedings, Proceedings of the 10th International Conference on Wirtschaftsinformatik, Zurich, Switzerland, 18 February 2011*; University of Zurich: Zurich, Switzerland, 2011.
90. O'Connor, K.W.; Schmidt, G.B.; Drouin, M. Helping workers understand and follow social media policies. *Bus. Horiz.* **2016**, *59*, 205–211. [CrossRef]
91. Dutta, S.; Fraser, M. When Job Seekers Invade Facebook. The McKinsey Quarterly. 2009. Available online: https://www.mckinsey.com/industries/high-tech/our-insights/when-job-seekers-invade-facebook (accessed on 21 June 2021).
92. Holland, J. The role of mobile marketing communications in media strategy. *Innov. Mark.* **2010**, *6*, 18–25.
93. Kaplan, A.M. If you love something, let it go mobile: Mobile marketing and mobile social media 4 × 4. *Bus. Horiz.* **2012**, *55*, 129–139. [CrossRef]
94. Watson, C.; McCarthy, J.; Rowley, J. Consumer attitudes towards mobile marketing in the smart phone era. *Int. J. Inf. Manag.* **2013**, *33*, 840–849. [CrossRef]
95. Harrigan, P.; Miles, M.P.; Fang, Y.; Roy, S.K. The role of social media in the engagement and information processes of social CRM. *Int. J. Inf. Manag.* **2020**, *54*, 102151. [CrossRef]
96. Haber, I.E.; Toth, M.; Hajdu, R.; Haber, K.; Pinter, G. Exploring Public Opinions on Renewable Energy by Using Conventional Methods and Social Media Analysis. *Energies* **2021**, *14*, 3089. [CrossRef]
97. Vasilev, Y.; Cherepovitsyn, A.; Tsvetkova, A.; Komendantova, N. Promoting Public Awareness of Carbon Capture and Storage Technologies in the Russian Federation: A System of Educational Activities. *Energies* **2021**, *14*, 1408. [CrossRef]
98. Martin, S. The rise of Facebook-based communication in the energy sector. *Z. Energ.* **2016**, *40*, 89–96. [CrossRef]
99. Kumar, J.C.R.; Majid, M.A. Renewable energy for sustainable development in India: Current status, future prospects, challenges, employment, and investment opportunities. *Energy Sustain. Soc.* **2020**, *10*, 2. [CrossRef]
100. Joshi, A.; Kale, S.; Chandel, S.; Pal, D.K. Likert Scale: Explored and Explained. *Br. J. Appl. Sci. Technol.* **2015**, *7*, 396–403. [CrossRef]
101. Aguado, D.; Rico, R.; Rubio, V.J.; Fernández, L. Applicant reactions to social network web use in personnel selection and assessment. *J. Work. Organ. Psychol.* **2016**, *32*, 183–190. [CrossRef]
102. Ostertagova, E.; Ostertag, O.; Kovač, J. Methodology and Application of the Kruskal-Wallis Test. *Appl. Mech. Mater.* **2014**, *611*, 115–120. [CrossRef]
103. Sztemberg-Lewandowska, M. *Analiza Czynnikowa w Badaniach Marketingowych*; Wydawnictwo Uniwersytetu Ekonomicznego we Wrocławiu: Wrocław, Poland, 2008.
104. Abdi, H.; Williams, L.J. Principal component analysis. *Wiley Interdiscip. Rev. Comput. Stat.* **2010**, *2*, 433–459. [CrossRef]
105. Chiang, J.K.H.; Suen, H.Y. Self-presentation and hiring recommendations in online communities: Lessons from LinkedIn. *Comput. Hum. Behav.* **2015**, *48*, 516–524. [CrossRef]
106. Adams, S. New survey: LinkedIn more dominant than ever among job seekers and recruiters, but Facebook poised to gain. *Forbes* **2013**, *10*, 28–29.
107. Black, S.L.; Johnson, A.F.; Takach, S.E.; Stone, D.L. Factors affecting applicant's reactions to the collection of data in social network websites. *Acad. Manag.* **2018**, *2012*, 11742. [CrossRef]
108. Nikolaou, I. Social networking web sites in job search and employee recruitment. *Int. J. Sel. Assess.* **2014**, *22*, 179–189. [CrossRef]
109. Hallam, J. *The Social Media Manifesto: A Guide to Using Social Technology to Build a Successful Business*; Palgrave Macmillan: Basingstoke, UK, 2013.
110. Yokoyama, M.H. How social network sites (SNS) have changed the employer -employee relationship and what are the xt challenges for human resource (HR)? *REGE* **2016**, *23*, 2–9. [CrossRef]
111. Stopfer, J.M.; Gosling, S.D. Online social networks in the work context. In *The Psychology of Digital Media at Work*; Derks, D., Bakker, A., Eds.; Psychology Press: London, UK, 2013; pp. 39–59.
112. Anderson, N.; Salgado, J.F.; Hülsheger, U.R. Applicant reactions in selection: Comprehensive meta-analysis into reaction generalization versus situational specificity. *Int. J. Sel. Assess.* **2010**, *18*, 291–304. [CrossRef]
113. McLarty, B.D.; Whitman, D.S. A Dispositional Approach to Applicant Reactions: Examining Core Self-Evaluations, Behavioral Intentions, and Fairness Perceptions. *J. Bus. Psychol.* **2016**, *31*, 141–153. [CrossRef]

114. Chan, D.; Schmitt, N. An Agenda for Future Research on Applicant Reactions to Selection Procedures: A Construct-Oriented Approach. *Int. J. Sel. Assess.* **2004**, *12*, 9–23. [CrossRef]
115. Priyadarshini, C.; Kumar, Y.; Jha, R.R. Employer Attractiveness Through Social Media: A Phenomenological Study. *Qual. Rep.* **2017**, *22*, 969–983. [CrossRef]
116. Li, C. *Open Leadership: How Social Technology Can Transform the Way You Lead*; Jossey-Bass: San Francisco, CA, USA, 2010.
117. McAfee, A. *Enterprise 2.0: New Collaborative Tools for your Organization's Toughest Challenges*; Harvard Business Press: Boston, Poland, 2009.
118. Grzesiuk, K.; Wawer, M. Employer Branding Through Social Media: The Case of Largest Polish Companies. Section: Contemporary Business Management Challenges and Opportunities. In Proceedings of the 10th International Scientific Conference Business and Management 2018, Vilnius, Lithuania, 3–4 May 2018; pp. 381–390.

energies

Article

Segmentation of Food Consumers Based on Their Sustainable Attitude

Michał Gazdecki [1,*], Elżbieta Goryńska-Goldmann [1], Marietta Kiss [2] and Zoltán Szakály [2]

1 Department of Economics and Economy Policy in Agribusiness, Poznań University of Life Sciences (PULS), Wojska Polskiego Str. 28, 60-637 Poznań, Poland; gorynska-goldmann@up.poznan.pl
2 Faculty of Economics and Business, Institute of Marketing and Commerce, University of Debrecen (UD), 4032 Debrecen, Hungary; kiss.marietta@econ.unideb.hu (M.K.); szakaly.zoltan@econ.unideb.hu (Z.S.)
* Correspondence: michal.gazdecki@up.poznan.pl

Abstract: The proposed study aims to segment consumers based on a sustainable approach to the consumption of food. The shift in consumers' attitudes towards more balanced food consumption can be one of the sustainability drivers for entire food chains and may result in more sustained energy usage in the whole food chain and implementation of farm to fork strategy to the practice. We considered consumers' attitudes as a multidimensional construct. Under this assumption, we asked respondents a series of questions related to the cognitive, behavioral, and affective components of an attitude. Data were collected from a market survey run among 433 consumers. We identified three consumer segments. The "Doers" segment exhibits sustainable behavior to a greater extent than the others. At the same time, they have less knowledge about the concept of food sustainability while the affective dimension was developed on an average level. The "Conscious" segment had well-developed cognitive and affective dimensions (which might indicate their openness to the information about sustainability positive feelings), however, it was not reflected in their behavior. Finally, the "Reluctant" segment, did not show a sustainable attitude towards food consumption in any of the analyzed dimensions. Answering the question of how common sustainable attitudes are may help in determining the market potential and in developing product and promotion strategies.

Keywords: sustainable consumption; responsible consumption; consumer segmentation; sustainable attitude; food products; consumers behavior

Citation: Gazdecki, M.; Goryńska-Goldmann, E.; Kiss, M.; Szakály, Z. Segmentation of Food Consumers Based on Their Sustainable Attitude. *Energies* **2021**, *14*, 3179. https://doi.org/10.3390/en14113179

Academic Editor: Dimitrios A. Georgakellos

Received: 24 April 2021
Accepted: 24 May 2021
Published: 29 May 2021

1. Introduction

It is hard to disagree that to combat climate changes the current consumption patterns must be altered globally [1–9]. The process of transformation of consumption plays an important role in facing environmental challenges, both globally and locally, and the sense of responsibility for the society, future generations, and the planet [10–20]. The Sustainable Development Goal (SDG) 12 Ensure sustainable consumption and production patterns defined by the European Commission have become a focal point of action until 2030 (Agenda 2030) [21]. This is reflected in the published strategy Farm to Fork [22,23] and the newest edition of the report "The State of Food Security and Nutrition in the World 2020" [24], where it was underscored that consumption patterns conforming to sustainable food consumption (SFC) will be playing a key role in achieving sustainable development, climate goals and satisfying the needs of the ever-growing population. Such processes are crucial nowadays as food markets are assessed as unbalanced. Production of raw materials and food processing requires a high energy input, creating negative side effects, such as greenhouse gas emission, food losses, and environmental burden.

The socio-economic transformation brought about qualitative and quantitative changes in food consumption. Other than by reducing the quantity of used goods and services, SFC can be also achieved by creating more appropriate consumption patterns, all with respect to the basic life needs and aspirations to improve the lives of current and future generations. Any changes to the structure of food consumption require the support of public policy, due to the presence of the lock-in effect [25,26]. Consumption patterns that make food consumption more sustainable should be popularized, and consumers should be perceived as actors playing the main role in creating a transition to a more sustainable food system [20,27–33]. Partha et al. [6] have correctly noted that the success of transition will depend on how individuals and households are convinced to change their consumption patterns and how the state cares, during the transformation of the model, about any vulnerable groups (i.e., those unaware or not motivated to alter their consumption) and indifferent groups [34].

Given the reasonable use of goods and services [26,35], the possibility of elimination of numerous, negative social and environmental effects and the preservation of natural resources for future generations, the sustainable food consumption model is desired [36], and it is necessary to implement strategies aimed at achieving it. This justifies the need to introduce systematic studies on consumers aiming to identify their ability to alter behaviors and a context that fosters such a change [37–41]. The understanding of phenomena present in the area of consumption forms the basis to design processes present in food supply chains, thanks to which they will be able to evolve in the direction of more sustainable processes [14,37,42–45].

Changes in consumption stem from the changing preferences and behaviors of consumers. Vermeir et al. [18] emphasize that although food preferences may be difficult to change, they may still develop in short periods, due to the dynamic transformations happening in the social, cultural, and economic environment. The analysis of behaviors of consumers, while considering the economy of sustainable development, calls for the analysis of signs of behaviors and their causes [3,4,46]. SFC studies, whose scope has been constantly expanding, show that the consumers' interest in SFC is lower than expected [43,47,48], while the scale of signs is not yet large [49,50]. The latest analyses show that despite the popularity of the idea of sustainable consumption and production, Europeans have not yet learned the concept of sustainable development (SD) well and cannot always identify it, with a marginally small number of the surveyed pointing that it is connected with nutrition [51,52]. The share of organic food in EU's market is estimated at 4%, and only 0.3% in Poland (with over 7% in Denmark and 4.4% in Germany) [53,54]. Still, the importance of food's health and environmental aspects are reflected in the growing trends, such as conscious, smart, critical, ethical, green, responsible, ecological, fair, shared, individual consumption, prosumption, deconsumption, food sharing, and freeganism [8,25,48,55–60]. "The Reflection Paper towards a Sustainable Europe by 2030" [61] shows that around 43 million people in the EU still cannot afford to eat a good quality meal every second day.

Further transformations of attitudes and behaviors of food consumers are required for SFC to grow, with such attitudes and behaviors possible to evaluate and identify based on segmentation [47,62]. There are plenty of literary resources discussing the issues of segmentation, with most of them often leading to distinguishing sub-groups, in turn allowing the identification of behaviors and motivations, along with an in-depth analysis of the character of attitudes [37,51]. Although many studies focus on the selection of sustainable food and consumption behaviors, the research efforts on consumer segmentation from the point of view of SFC are still limited [36,41].

Consumer segmentation is performed based on various criteria. Some of the criteria characterize the consumer as an individual and some as a group of consumers. For example, the first group may include consumption patterns, purchasing behaviors, motivation, the manner of perceiving a product, content perception, the level of satisfaction of needs, etc. [47,63,64]. The use of segmentation criteria describing groups of consumers has developed along with the world's globalization and expansion of planning horizons of target markets by international corporations with macro-criteria (such as economic prosperity, political and cultural system, infrastructure), which started to be used for international market segmentation purposes [65,66] segmentation of groups of potential customers, from abroad.

The identification of consumer segments based on balancing their attitudes towards food consumption is of fundamental importance for the further development of SFC, especially in the practical aspect. Any efforts aiming to alter unsustainable attitudes and behaviors of consumers should be focused on learning and understanding ways to influence consumers to change their behaviors towards more sustainable food, positioning sustainable food, creating effective strategies and information campaigns [7,26,46,56,62,67,68]. The knowledge of consumer profiles allows the creation of a more efficient information and education policy, maintained both by the government and NGOs.

The purpose of this paper is to segment consumers based on a sustainable approach to the consumption of food. The prospects for a more sustainable consumption depend on the ability to improve the innovativeness level of consumers, enterprises, and science and of government and public institutions. Undoubtedly, the creation of a favorable environment that promotes development of consumption sustainability requires a closer cooperation between the academic environment, industry, government, and social organizations. In the context of the topic addressed in this paper, it is important to know consumer attitudes towards sustainable food consumption. The consumer segmentation presented in this paper narrows that gap.

2. Review of Studies on Consumer Segmentation with Regard to SFC

The review of studies on consumer segmentation in relation to the concept of SFC has been prepared as the result of searching databases of scientific publications. The search covered the articles made accessible from 1990 until March 2020 in Web of Science and Scopus. Publications were selected if the searched phrases appeared in one of the following elements: thematic sections, keywords, titles, and abstracts. The following phrases were selected for searching: "food", "consumption", "segmentation", "segment", followed by "sustainable consumption" and "cluster analysis".

The search of the Scopus base yielded 24 records, while the Web of Science database returned 13 records. When assessing the usefulness of publications using Moher's et al. [69] method, twenty-seven papers published after 2010 were identified (Figure 1).

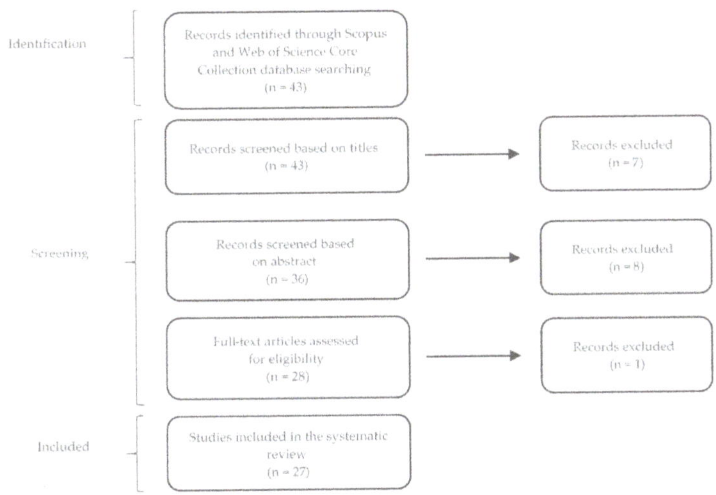

Figure 1. Selection of literary sources.

Dominant subject areas (categories) of the publications were agricultural and biological sciences, environmental sciences, social science, economics, econometrics and finance, business, green sustainable science technology, and environmental engineering. Analyzing the findings by country, it was found that a significant part of them were from Europe, especially Italy, Germany, and the Netherlands. The studies within the analyzed area have been performed, among others, by: Wageningen University and Research Centre (4 publications); Parthenope University of Naples (2); Gent University (2); Aarthus University (2); University of Gottingen (2). The articles have been analyzed and evaluated in terms of usefulness by two, independent scientific institutions (UPP and UD). To be included in the project, an article had to be prepared based on original data. The details of every publication have been identified by the first researcher, then verified by the second researcher and presented in the collective list in Table 1. Any publications that failed to satisfy these criteria have been excluded from analysis.

Table 1. Overview of the 27 selected papers.

No.	Author	Ref. No.	Regional Scope	No. of Respondents	Investigated Factors	No. of Segments	Types of Segments
1	Vanhonacker, F., Van Loo, E.J., Gellynck, X., Verbeke, W.	[68]	Belgium	$n = 221$	- Socio-demographic - Economic - Behavioral - Environmental sustainability	5	Conscious; Active; Unwilling; Ignorant; Uncertain.
2	Verain, M.C.D., Dagevos, H., Antonides, G.	[7]	Netherlands	$n = 942$	- Socio-demographic - Economic - Behavioral - Psychosocial	4	Unsustainers; Curtailers; Product-oriented consumers; Sustainers.
3	Bronnmann, J., Asche, F.	[70]	Germany	$n = 485$	- Socio-demographic - Economic - Behavioral - Environmental sustainability	2	Model 1; Model 2.
4	Thøgersen, J.	[71]	10 European countries, covering the five regions North, South, East, West and Central Europe	n ≈ 335 in each country (total ≈ 3350)	- Socio-demographic - Lifestyle	5	FRL segment 1 ("Everyday food providers"); FRL segment 2 ("Food ignoramuses"); FRL segment 3 ("Enthusiastic food consumers"); FRL segment 4 ("Uninvolved food consumers"); FRL segment 5 ("Traditional family oriented food consumers").
5	Lavelle, M.J., Rau, H., Fahy, F.	[72]	Ireland (Northern Ireland and the Republic of Ireland; County Galway, Derry/Londonderry, and Dublin)	$n = 1500$	- Socio-demographic - Economic - Behavioral - Environmental sustainability	4	Dark; Greens; Browns; Yellows; Light-Greens.
6	Gerini, F., Alfnes, F., Schjøll, A.	[47]	Norwegian	$n = 948$	- Socio-demographic - Economic - Behavioral	3	Segment purchasing the most organic food; Occasionally purchase organic products; Segment attempts to avoid organic products.
7	Verain, M.C.D., Sijtsema, S.J., Dagevos, H., Antonides, G.	[73]	Netherlands	$n = 829$	- Socio-demographic - Economic - Behavioral - Behavioral	3	Cluster 1 ("Pro-selves"); Cluster 2 ("Average consumers"); Cluster 3 ("Conscious consumers").
8	Stranieri, S., Baldi, L.	[74]	Italy (Lombardy Region)	$n = 351$	- Socio-demographic - Economic - Behavioral - Environmental sustainability	3	Favorable; Sceptical; Mistrust.
9	Van Huy, L., Chi, M.T.T., Lobo, A., Nguyen, N., Long, P.H.	[75]	Vietnam	$n = 203$	- Socio-demographic - Economic - Psychographic - Lifestyle	3	Conservatives; Trendsetters; Unengaged.

Table 1. *Cont.*

No.	Author	Ref. No.	Regional Scope	No. of Respondents	Investigated Factors	No. of Segments	Types of Segments
10	Mózner Z. V.	[76]	Hungary	$n = 975$	- Socio-demographical - Economic - Environmental sustainability	6	Cluster 1 ("Meat- and vegetable-based meal consumers"); Cluster 2 ("Milk and meat consumers"); Cluster 3 ("Average consumers"); Cluster 4 ("Fruit, vegetable and dairy product consumers"); Cluster 5 ("Bread and bakery product consumers"); Cluster 6 ("No milk and dairy product consumers").
11	Annunziata, A., Mariani, A.	[62]	Italy (Campania Region)	$n = 200$	- Socio-demographic - Economic - Behavioral	3	Green; Egocentric; Sustainability oriented.
12	Aprile, M.C., Mariani, A.	[77]	Nepal (South of Italy	$n = 400$	- Socio-demographic - Economic - Environmental sustainability	4	Orientated to sustainability labels, Kind-hearted; Power seekers; Environmentalists.
13	Prokeinová, B. R., Paluchová, J.	[78]	Slovakia	$n = 318$	- Socio-demographic - Economic - Behavioral - Environmental sustainability	4	Consumers—non-students in terms of consumer behavior when purchasing food; Consumers—students in terms of consumer behavior when purchasing food; Consumers—non-students in terms of consumption; Consumers—students in terms of consumption.
14	La Lama, G.C. M-D., Estévez-Moreno, L.X., Villarroel, M., María, G.A., Sepúlveda, W.S.	[79]	Mexican	$n = 843$	- Demographic - Environmental sustainability	3	Skeptical; Concerned; Ethical.
15	Hölker, S., von Meyer-Höfer, M., Spiller, A.	[80]	Germany	$n = 1049$	- Socio-demographic - Economic - Behavior - Values	5	Cluster A; Cluster B; Cluster C; Cluster D; Cluster E.
16	Palmieri, N., Forleo, M.B.	[26]	Italy	$n = 257$	- Socio-demographic information - Behavioral	7	Cluster 1 ("Scared & closed"); Cluster 2 ("Scared & open"); Cluster 3 ("Neophiliac & sensitive"); Cluster 4 ("Neutral"); Cluster 5 ("Phobic & careful"); Cluster 6 ("Not phobic & open"); Cluster 7 ("Fearless & unconcerned").
17	Sogari G., Pucci T., Aquilani B., Zanni L.	[81]	Italy	$n = 2597$	- Demographic - Economic - Environmental sustainability - Behavioral	2	Millennials; Non-millennials.
18	Janßen D., Langen N.	[82]	Germany	$n = 787$	- Behavioral: - Environmental sustainability	3	Price-sensitives; Label choosers; Price-conscious label discriminators.
19	Yildirim, S., Candan, B.	[83]	Turkey	$n = 453$	- Demographic - Economic - Values - Personal value	2	Cluster 1 (" Self—developers—Social environmentalist"); Cluster 2 ("Self-challengers—Functional environmentalist").
20	Hasanzade, V., Osburg, V.-S., Toporowski, W.	[84]	Germany	$n = 249$	- Demographic - Behavioral - Affective factors (i.e., assessing)	3	Ethically motivated consumers; Price oriented consumers; Price-quality oriented consumers.

Table 1. *Cont.*

No.	Author	Ref. No.	Regional Scope	No. of Respondents	Investigated Factors	No. of Segments	Types of Segments
21	Risius, A., Hamm, U., Janssen, M.	[85]	Germany	n = 459	- Socio-demographic - Economic - Behavioral:	5	Average consumers (Class 1); COO oriented consumers (Class 2); Sustainability oriented consumers (Class 3); Premium-price oriented consumers (Class 4); Low-price oriented consumers (Class 5).
22	Wang, O., Somogyi, S.	[86]	China	n = 643	- Socio-demographic - Economic - Psychological - Behavioral	2	Sustainable-selfish-pioneer; Sustainable-selfish-conservative.
23	Jakubowska, D., Radzyminska, M.	[87]	Poland, Czech Republic	n = 631 (323—Czech Republic; 308 – Poland)	- Values - Environmental sustainability	2	Cluster 1; Cluster 2.
24	Ghvanidze, S., Velikova, N., Dodd, T.H., Oldewage-Theron W.	[88]	United States, United Kingdom, Germany	n = 1048	- Socio-demographic - Economic - Behavioral	4	Cluster 1 ("Apathetic Consumers"); Cluster 2 ("Health-Conscious Diners"); Cluster 3 ("Holistic Perfectionists"); Cluster 4 ("Ethical Advocates").
25	Su, C.-H., Tsai, C.-H., Chen, M.-H., Lv, W.Q. U.S.	[89]	United States	n = 812	- Sociodemographic - Economic - Environmental sustainability (i.e., environmental awareness)	3	Gen Z consumers with high environmental consciousness (sustainable activists); Gen Z consumers moderate ecological awareness (sustainable believers); Gen Z consumers with low environmental consciousness (sustainable moderates).
26	Hrubá, R.	[90]	Czech Republic	n = 331	- Sociodemographic - Lifestyle	2	Cluster 1; Cluster 2; Cluster 3; Cluster 4.
27	Krystallis, A.	[91]	Greece	n = 506	- Socio-demographic - Economic - Affective factors (i.e., assessment) - Behavioral	4	Cluster 1 ("Eco-friendly, intensive farming supports"); Cluster 2 ("Indifferent"); Cluster 3 ("Sustainable farming supports"); Cluster 4 ("Ambiguous").

The following categories of variables were specified after the analysis of the papers (Table 1): environmental sustainability, socio-demographic, psychographic, economic, behavioral, affective factors, lifestyle, values. Food consumption is a complex issue and calls for a broader perspective [41,92]. In order to explain the consumers' behaviors, the researchers used in their studies wide ranges of variables connected to the lifestyle [71,75], personal traits [7,86], values [80,90,93], preferences, purchasing behaviors [7,26,47,62,70,78,81,84–86], behaviors related to food consumption, the consumers' attitude to sustainable food consumption [62,75,78,91], environmental issues [68,70,72–74,76,77,79,81,82,88–91] and behaviors that would limit sustainable choices [62,70,72,73,75,76,78,79,81,88,89].

The authors of the analyzed works employed various statistical methods to identify consumer segments. Factor analysis and cluster analysis are among the most used ones [26,62,68,71–76,79,80,87–90,93]. The other methods included modelling (e.g., logit model [70]; latent class modelling (LCM) [85]; conceptual model [81]; mixed logit model [47], conjoint analysis model [62,84,91]), the use of selected techniques with regard to data mining [78] and multivariate analysis [77]. The following part of the paper presents individual categories of variables identified in the analyzed papers.

2.1. Socio-Demographics

Socio-demographics have been used in most of the cited works. This type of variable is one of the commonly used variables in consumer behavior studies [7,73,81]; they are easy to measure and strongly determine consumer's behaviors, which in turn makes them work well as a basis of segmentation.

Gender of consumers was considered in 21 papers (Table 2). For instance, Gen Z segments of the U.S. sustainable food market differ with respect to gender. In addition, Verain et al. [7] stated that "unsustainers" and "product-oriented" consumers were more often male than female compared with "curtailers". Analyzing two sub-samples characterized by different age: "millennials" (18–35 years) and "non-millennials" (36–88 years) revealed that gender is significant and positive only in the category of "non-millennials", while the income level is significant and positive for "millennials". Sogari et al. [81] who ran a study on the vine market noted that "if in the past wine was seen as mainly a male beverage, in recent years more and more females are becoming wine aficionados". Significant gender differences across segments were shown by Su et al. [89], who analyzed consumer groups based on environmental consciousness.

Table 2. Socio-demographic factors used in the analyzed papers.

Socio-Demographic Factors	Numbers Refer to the Studies List in Table 1
Age group	1, 2, 3, 5, 6, 7, 8, 9, 10, 11, 12, 13, 14, 15, 16, 17, 19, 20, 21, 24, 26, 27
Gender/Sex	1, 2, 3, 5, 6, 7, 9, 10, 11, 12, 13, 14, 15, 16, 17, 19, 20, 21, 24, 25, 27
Educational level	1, 2, 3, 5, 6, 7, 8, 9, 10, 11, 12, 13, 14, 15, 16, 17, 19, 22, 24, 25, 27
Household size/Mean household size	3, 5, 9, 10, 11, 21
Living environment (rural or urban)/Domicile/Residence/Type of settlement	1, 8, 10, 12, 13
Marital status/Partnership/Married	3, 11, 22, 27
Occupation	19, 22, 27
Social status	12, 13
Origin/Ethnicity	14, 26
Family member/Kids/Children/	3, 9, 11
Family composition (number of household members)/Household composition	2, 7
Residential region/Area of residence	25, 27
Country class	4, 24,
Leisure time activity	10
Housing tenure	5

Age of consumers was taken into consideration in 22 papers; in 7 out of the 22 papers noted, significant age differences depended on the consumer segment [7,68,70,73–75,81]. For example, Vanhonacker et al. [68] concluded that the "Unwilling" segment were the youngest on average and significantly younger than the "Active" and the "Uncertain" (the oldest segment). Further, the "Conscious" were significantly younger than the "Uncertain". The "Unwilling" segment was the opposite of the "Active" segment and combined a high personal footprint with a low personal relevance.

Education of consumers was considered in 21 of the analyzed papers. Seven articles showed that education significantly diversifies the consumers' behaviors and attitudes [68,70,72,74,75,79,88]. Vanhonacker et al. [68] found that the large majority of the "Conscious", "Active" and "Unwilling" consumers were higher educated, while a more balanced distribution in education level was found among the "Ignorant" and the "Uncertain". Ghvanidze et al. [88] stated that the "Apathetic" segment had high levels of education and income.

Other variables used in the discussed papers, but on a smaller scale, were, among others: household size [62,70,72,75,76,85], living environment/residence (or type of settlement) [68,74,76–78], family composition/household composition [7,73], and country class [71]. Some of the least used variables in this category included social status, origin/ethnicity, living environment (rural or urban), occupation, family composition, household size, family member, housing tenure, partnership, kids, residential region, country class and leisure time activity.

2.2. Environmental Sustainability

11 out of the 27 articles used environmental sustainability variables (Table 3) [68,70,72,74,76–79,81,82,87,90]. Various, detailed variables were identified among these, such as footprint-related, certificates, pro-environmental habitual activity, environmental awareness and related to the product and/or production process.

Table 3. Environmental sustainability factors used in the analyzed papers.

Environmental Sustainability Categories	Environmental Sustainability Items	Numbers Refer to the Studies List in Table 1
Footprint related	Self-evaluation of personal footprint	1
	Personal relevance of the ecological footprint	1
	Ecological footprint of food consumption (ecological footprint (gha) = quantity consumed per year per person (kg) × ecological footprint intensity (gha/kg))	10
Certificates	Sustainable certification	3
	Sustainability labels on products	18
	Attitude to using sustainability labels: (1) Propensity to read labels before consuming food products, (2) Degree of knowledge regarding sustainability labels.	12
Pro-environmental habitual activity	Bought reusable products instead of disposable ones, Reduced energy use, Reduced water use, Shopped or paid a bill online, Avoided products with excess packaging, Repaired items instead of purchasing new ones	5
Sustainability attitude	Attitude about animal welfare	14
	Environmental attitudes and values: (1) Believe the condition of the natural environment affects the quality of produced food, (2) Environment pollution is a consequence of today's people's lifestyle.	23
Environmental awareness	Environmental involvement, Environmental values for purchasing environmentally friendly products, Importance of perception of sustainable food, Environmental protection, Product attributes, Food choices associated with healthy eating habits.	25
Related to the product and/or production process	Shelf life of food products (e.g., fresh-cut salad with an extended shelf-life date)	8
	Main constructs have been extrapolated: (1) Product/Process, (2) Supply chain, (3) Energy types and their use, (4) Soil/Landscape preservation	17

Two of the discussed publications focused on the selection and consumption of food and the related impact on the environment measured by the "ecological footprint" [68,76]. Vanhonacker et al. [68] show that numerous consumers fail to see the impact of animal production on the natural environment. They also point to the presence of alternative behaviors in relation to conventional meat consumption, for which eating habits and cultural patterns must be adapted. However, the readiness to pay higher prices is significantly lower than the readiness to consume. Mózner [76], however, has noted that consumers who ate more fruit, vegetables, and milk products did not have a smaller ecological footprint in terms of the entire food consumption.

Three publications examined the issue of food certification [70,77,82]. The used certification systems differ in terms of information value, methods of assigning and conducting monitoring activities. The impact of the origin certificate on consumers has been confirmed by Bronnmann and Asche [70], who showed that consumers were more willing to pay a higher price and were more eager to make purchases, all in relation to wild fish. Similar results confirming the readiness to pay a higher price have been also obtained by Aprile and Mariani [77], Janßen and Langen [82], and La Lama et al. [79].

The consumers' focus on environmental protection issues have been also considered in the studies by Jakubowska and Radzymińska [87] and by La Lama et al. [79]. Consumers can be divided into ones expressing strong pro-environment attitudes (and health-related values) and ones that do not take environmental aspects into account when selecting food [87]. Similar results have been obtained by Su et al. [89] who stated that the sustainable food market can be segmented according to environmental awareness. Sogari et al. [81] have identified a connection between the selection of a product and the consumer's engagement in the purchasing process and their environmental protection awareness.

2.3. Behavioral Factors

Behavioral changes have been used in 19 out of 27 articles [7,26,47,62,68,70,72,74,76,78,80–82,84–86,88,91]. An overview of examples of the factors that have been included in the segmentation studies of the cited authors can be found in Table 4. The behavioral factors consisted mostly of buying behavior and general food choice motives/attribute importance by food category sustainability, general/life attitude, consumption habits, occasional behavior, ethical issues, consumer preferences for product information related to environmental issues.

Table 4. Behavioral factors used in the analyzed papers.

Behavioral Categories	Behavioral Items	Numbers Refer to the Studies List in Table 1
Buying behavior	Consumption frequency	8, 17
	Responsibility for food purchases within household	1
	Purchaser or not main food purchaser in the household	3
	Purchase frequency of organic food products	6
	Purchase intentions for sustainable shellfish species	22
	Food shopping frequency	8
	Consumers' purchase frequency	17
	Average price for product	17
	Use social media to gather information on product before buying	17
	Consumer preferences and their attitudes towards food choice	16
	Consumer perceptions and their attitudes towards edible seaweed	16
	Willingness to pay measures for the respective attributes	18
General food choice motives/attribute importance by food category sustainability	Price	2, 8, 20, 21
	Healthiness	2, 11
	Taste	2
	Trust	8
	Country of origin	21
	Sustainability label	21
	Safety of product	8
	Claim	21
	Freshness	8
	No-buy option	21
	Environmentally friendly packaging	11
	Locally produced to support local farmers	11
	Produced in an uncontaminated environment	11
	Obtained in an environmentally friendly way	11
General/life attitude	Animal welfare rights	11, 20
	Healthy life attitude	8
	Food involvement	2
	Life values	7
	Health-conscious lifestyles	24

Table 4. *Cont.*

Behavioral Categories	Behavioral Items	Numbers Refer to the Studies List in Table 1
Ethical issues	Diet, agreement regarding the dietary guideline	7, 15
	Produced in full respect of human rights, without exploiting women and children	11
	Ethical concern about production processes	24
Consumption habit	Stage in the transition towards healthier towards more sustainable eating	7
	Sustainable consumption	13
	Consumption frequency variables	27
	Healthy dietary patterns	24
Occasional behavior	(1) Purchased energy efficient appliance, (2) Installed insulation, (3) Switched to renewable energy supplier, (4) Purchased an energy efficient car	5
Consumer preferences for product information related to environmental issues	Concerns about the environment	20
	Low carbon emission	11
	Obtained in an environmentally friendly way	11
	Produced in an uncontaminated environment	11
	Environmentally conscious behavior	24

A few of the issues used for the purposes of establishing consumer segmentation are their behaviors related to purchasing ecological food products and their willingness to pay the price. For example, considering the willingness to pay measures for the respective attributes, it has been noted that the buyers of ecological food were more inclined to pay a premium for ecological products [47,62]. Vanhonacker et al. [68] and Palmieri and Forleo [26] have confirmed that consumers were more inclined to buy products perceived as more sustainable. Consumers exhibiting a more sustainable approach are also more willing to buy innovative products.

An important aspect is also the identification of factors behind the selection of food, with health and quality attributes playing an important role [62]. Risius et al. [85], who segmented fish consumers, emphasized the major importance of the country of origin. Lots of attention was also paid to connecting consumer values with purchasing behaviors. An example could be the attention paid by humans to the natural environment or the well-being of animals [93]. Verain et al. [7] found that food involvement and personal norms with regard to healthy and sustainable food positively affect sustainable food choices. Annunziata and Mariani [62] emphasized that the importance of ethical values (i.e., that food was produced with complete observance of human rights or with no abuse to women and children) and environmental values (e.g., that food was produced in an uncontaminated environment, in an eco-friendly manner and with support for local farmers).

The behaviors are related to the level of the consumers' awareness about sustainability, as confirmed by Prokeinová and Paluchová [78] with regard to the younger consumers who choose environmentally friendly and socially acceptable products more willingly than their parents.

2.4. *Psychographic Factors*

Psychographic variables have been used as a basis for segmentation in three of the analyzed papers [7,75,86]. Verain et al. [7] have confirmed that social and personal norms, ability, subjective knowledge, and food involvement significantly differentiate meat consumers' behaviors. For example, "product-oriented consumers differ from each other in that the product-oriented attach more importance to social norms and have a higher ability to judge sustainably produced food, and subjective knowledge on sustainable products". Van Huy et al. [75] have focused, among other aspects, on the attention to healthy food, love of cooking, convenience, and love of local and organic food. Wang and Somogyi [86] have examined the level of acceptability of crustaceans from sustainable production among Chinese consumers. They have found that personal standards significantly influence the purchase intentions of consumers.

2.5. Economic Factors

A wide range of economic variables, such as net household income per month, financial status of household, and employment status, were used as segmentation criteria or profiling variables. For example, Verain et al. [7] have identified four consumer segments that differ significantly regarding economic factors, e.g., segments named "Curtailers" have lower incomes compared to the "Sustainers" segment. The study by Van Huy et al. [75] has highlighted a nexus between the identified segments and the level of income of the consumers. Yildirim and Candan [93] have concluded that most green product buyers were at a high level of income. These results show that a more prominent presence of sustainable attitudes can be expected from the wealthier consumers.

The employment status has been included in three articles, and only in one study, significant differences across segments have been found by Lavelle et al. [72], who stated that respondents differ in their uptake of occasional and habitual pro-environmental behaviors. Considerable differences exist between the two behavior groups with regard to key socio-demographic variables, such as employment status and income, as well as residential location and housing tenure.

2.6. Affective Factors

Two of the presented articles considered the affective factors. In one of them, Krystallis [91] has taken into account consumers' assessments related to environmental factors, the perception of processing and well-being of animals in the case of processed meat. On the other hand, Hasanzade et al. [84] have analyzed the connection between ethical product attributes (e.g., ethical criteria of animal welfare, environmental protection, and labor and human rights) with the selected elements of behaviors.

2.7. Lifestyle and Values Factors

As shown by Aydin and Ünal [93], the lifestyle of consumers is related to the sustainability of consumption behaviors. In the publications of Thøgersen [71] and Van Huy et al., [75] the food-related lifestyle (FRL) model developed by Grunert et al. [94] has been used. For example, Thøgersen [71] investigates how the country of residence and FRL interact in shaping (un)sustainable food consumption patterns. The analysis has revealed that the outcome variables vary significantly across FRL segments. Further, after controlling for FRL, the direct effect of country class is highly significant for meatless suppers and marginally significant for buying organic food but non-significant for eating beef and food innovativeness. Van Huy et al. [75] have stated that the FRL model enables better understanding of how consumers employ food and its culinary aspects to achieve certain values in their lives. The acceptance of a specific lifestyle by consumers plays an important role in creating a "green economy" [95].

Consumer value-related approach represents a study by Hölker et al. [80] which developed consumer segmentation based on the human–animal relationship. The segmentation has been prepared according to specific values, such as animal rights, utilitarianism, new contractarian approach, abolitionism, original anthropocentrism, and anthropocentrism with indirect duties. At the same time, Yildirim and Candan [83] have found that green product buyers could be segmented into subgroups according to their personal values and consumption values. Significant differences regarding personal values, especially related to environmental matters, have been confirmed by Jakubowska and Radzymińska [87] as well.

The presented content proves that there are studies on consumer attitudes and behaviors in relation to SFC. Most of the researchers have focused on capturing individual factors or groups of factors allowing the identification of symptoms of sustainable behaviors of consumers. The researchers focused on environmental aspects (food choices and practices beneficial to the environment). It is also understood that other factors can also impact attitudes, which are basic determinants of human behavior [96]. Some of the studies have specified the attitude of consumers towards food categories (e.g., organic products), while some other towards products (e.g., fish, seaweed, lettuce, wine). The authors of all studies considered the behavioral component and many studies also accounted for the attitude's affective component [26,71–75,79,86–89]. It is worth noting that accounting for consumer attitudes in three dimensions was an infrequently taken approach, which forms a theoretical basis for the considerations in our article. A similar perspective has also been applied by, for instance, Jakubowska and Radzymińska [87], La Lama et al. [79], Palmieri and Forleo [26], Thøgersen [71], Van Huy et al. [75], Vanhonacker et al. [68].

3. Methodological Approach

The article is based on results of the authors' own surveys conducted by means of direct interviews. The study aims to identify sustainable consumer behaviors with regard to food consumption.

The general methodological approach is presented in Figure 2. A plan for collecting the data was developed, with the assumption that the goal would be to distinguish consumer segments by the level of sustainability of food consumption. The plan covered: developing a method of selecting consumers for the study and organizing the study itself, namely training interviewers and determining the time frame of the study. Afterward, a questionnaire was developed and put to test, ultimately serving to collect the research material. Once the data had been collected, the raw material was subjected to formal and substantive evaluation. The prepared material was then subjected to clustering by means of a non-hierarchical clustering algorithm. The distinguished segments were then described.

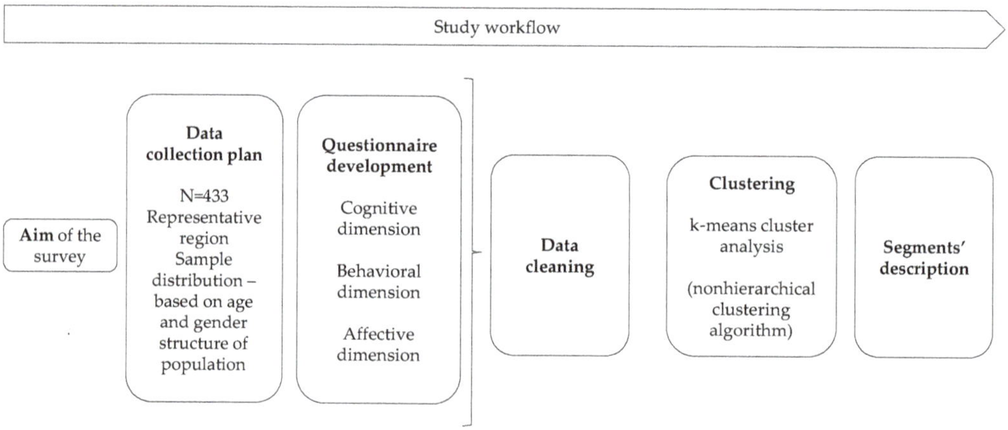

Figure 2. Methodological approach of the research.

3.1. Selection of Respondents

The respondents selected for the study are adults, over 18 years of age, living in the region of Wielkopolska, Poland. The region was selected because: (1) the inhabitants of economically developed regions usually have higher education levels and higher revenues, etc., which means this concept is more widespread in such regions; (2) new directions in food consumption and nutrition usually spread among larger urban agglomerations, inhabitants of regions developed socially and economically, for whom they are certain, sought for a model of food consumption; (3) Wielkopolska is one of the best economically developed regions in Poland (in terms of gross GDP and growth, foreign capital involvement concentration and investment expenses, unemployment rate, revenues, education, and human capital potential) [97].

The quota sampling method was used for the selection of respondents for the study. Public statistics data served as a basis to determine the sample's structure that reflected the structure of Wielkopolska inhabitants in terms of age and gender. The sample was 433 individuals. The social and demographic characteristics of the survey's participants is presented in Table 5.

Table 5. Respondents sample structure.

Characteristic	Item	Percent
Gender	Female	54.04
	Male	45.96
Age (years)	18–24	16.40
	25–34	16.40
	35–44	19.40
	45–54	15.47
	55–64	16.40
	65 and more	15.94
Education	Elementary and junior high	5.31
	Basic vocational	17.78
	Secondary	24.48
	Post-secondary/senior high	18.48
	Higher	33.26
Number of household's members	1 person	9.93
	2 persons	29.33
	3 persons	24.25
	4 persons	21.27
	5 persons and more	14.78
Self-assessment of household budget	We live in serious poverty—we have not enough to meet daily needs	0.91
	We live frugally—we have to manage our finances very carefully	13.86
	We live like average people—we have enough for daily needs, but must save for major purchases	44.57
	We live well—we can afford a lot, without having to save for specific purposes	36.03
	We live very well—we can afford certain luxuries	4.62

3.2. Questionnaire Development and Data Collection

Coming from the concept of the approach proposed by Breckler [98], which assumes three-dimensionality of attitudes towards SC, the questionnaire used in the survey has been divided into three parts. Each part contained questions related to the cognitive, behavioral, and affective components, respectively. Questions characterizing food consumption sustainability were asked within each area.

The first version of the questionnaire was subjected to pilot tests. The test consisted of 30 personal interviews with individuals who represented the target group. The purpose of the test was to check the questionnaire:

- in the formal aspect, namely whether the instructions and the formatting of the document are clear and facilitate the efficient work of the interviewers,
- in the content aspect, i.e., identification of questions for which there is little variation in answers, making them impossible to use as a basis for segmentation or profiling variables.

After collecting remarks from interviewers who did pilot tests and following the analysis of respondents' answers, the necessary corrections were introduced, mainly related to the formulation of questions. The final questionnaire, used in the study, is the result of the introduction of these changes.

The face-to-face interviews for the study were performed between November and December 2019.

3.3. *Data Cleaning*

The collected material was subjected to post hoc data cleaning approach. The procedure of data cleaning assumes they are verified by the researchers (Exploratory Data Analysis) [99] and consisted of checking outliers in order to identify any errors that would occur during uploading or digitizing the data; checking the presence of any missing values and replacing them with a median, if found; and checking for any duplicated records in the database and removing such duplicates, if found. The cleaned database was subjected to statistical analysis.

3.4. *Data Analysis-Clustering*

We used a k-means cluster analysis (non-hierarchical clustering algorithm), which groups objects on a set of user-selected characteristics. The resulting clusters should exhibit high internal (within-cluster) homogeneity and high external (between-cluster) heterogeneity. Thus, if the classification is successful, objects within clusters will be close together when plotted geometrically and different clusters will be far apart. Cluster analysis was chosen because of the data reduction procedure, which is done objectively by reducing the information from an entire population or sample to information about specific groups, as a large number of observations are meaningless unless classified into manageable groups [100].

Based on selected questions within each of the SC dimensions (cognitive, behavioral, and affective), 3 indices were created, by transforming all variables into dichotomic ones and aggregating sums of answers. There are two types of variables:

- dichotomic (1/0)—no transformation was done;
- ordinal (scales 1–3, 1–5 or 1–7)—top box answers were recoded into 1.

Because of the different number of variables used for each dimension, it was necessary to make sure we worked with roughly equal amplitudes within each index. Cluster analysis is sensitive to different scales, as it uses distance measures, therefore by using a mean and standard deviation values of each section, we standardized them with z-score transformation ($z = (x - \mu)/\sigma$) [101], to eliminate the impact of larger standard deviation.

Description of Dimensions

As the three-dimensional model of attitudes was employed, each dimension was reflected by the number of questions in the questionnaire (Table 6). For the cognitive dimension, 6 questions (22 variables in the data) were used to measure and to differentiate respondents by their awareness of the sustainable consumption concept. The behavioral dimension covers 4 questions, representing 28 variables in the data. In the affective dimension, we used 2 questions, based on which 17 variables in the data were created.

Table 6. Segmentation variables used for the study.

Dimension	Question	Variables	Score
Cognitive	(1) Have you heard about the following trends—consumer phenomena?	- Deconsumption - Responsible consumption - Collaborative consumption - Eco-consumption - Prosumption - Freeganism - Smart shopping - Cocooning - Globalization - Over-consumption - Regionalization - Ethno-centrism - Tradition	1 point if yes
	(2) Which of the listed trends are interesting to you? Please indicate the ones with which you identify to the highest degree (multiples choice possible)	- Responsible consumption - Collaborative consumption - Eco-consumption - Prosumption - Freeganism	1 point if yes
	(3) Are you familiar with the concept of sustainable consumption?	- Yes - No	1 point if yes
	(4) Do you look for information about food, consumption, and food market?	- Yes - No	1 point if yes
	(5) How often do you acquire information (e.g., by reading, watching TV programs, listening to the radio) about food, consumption, and food market?	- Daily - 4–5 times a week - 2–3 times a week - Once a week - 2–3 a month - Once a month - Less frequent than once a month	1 point if: Daily 4–5 times a week 2–3 times a week
	(6) During the last year, did you get any information about sustainable food consumption?	- Yes - No	1 point if yes

Table 6. *Cont.*

Dimension	Question	Variables	Score
	(7) Have there been any changes in nutrition in your household in the last 5 years?	- Yes - No	1 point if yes
	(8) Have there been any changes in food sourcing in your household in the last 5 years?	- Yes - No	1 point if yes
Behavioral	(9) Please indicate your attitude to the examples of behaviors listed below, on a numerical scale ranging from 1 to 5, where 1 means "I strongly disagree" and 5 "I strongly agree".	- I segregate waste, if possible - I buy food products in recyclable and biodegradable packaging - I go shopping with my reusable bag - I try to store, process, or pass on to others any food that has not been fully consumed - I buy food products from local/domestic producers, despite the fact that they may be more expensive - I am willing to spend more time buying food that I want - I save money and limit waste production by buying food in bigger packaging - I use low-processed food when preparing meals at home - I avoid highly-processed food - I avoid wasting food in my household - I support local producers by buying their products - I treat meals as a chance to start relations with others - I care about the nutrition of my family members	1 point if 5 or 4 on the scale
	(10) In the past (3–5 years ago) was it: less important = 1, equally important = 2, more important = 3, for you?	- I segregate waste, if possible - I buy food products in recyclable and biodegradable packaging - I go shopping with my reusable bag - I try to store, process, or pass on to others any food that has not been fully consumed - I buy food products from local/domestic producers, despite the fact that they may be more expensive - I am willing to spend more time buying food that I want - I save money and limit waste production by buying food in bigger packaging - I use low-processed food when preparing meals at home - I avoid highly-processed food - I avoid wasting food in my household - I support local producers by buying their products I treat meals as a chance to start relations with others - I care about the nutrition of my family members	1 point if 3 on the scale

Table 6. *Cont.*

Dimension	Question	Variables	Score
	(11) Which of the actions listed below have you heard about?	- "Po stronie natury"/English: „On the nature side by Żywiec Zdrój" (Żywiec Zdrój) - "Zadbaj o środowisko"/English: „Take care about the environment" (Carlsberg Polska) - "Danone dla środowiska"/English: Danone for the environment (Danone) - "Tworzenie wspólnej wartości"/„Creating the value together" (Nestle) - "Życie w sposób zrównoważony"/„Sustanaible life" (Unilever) - "Kto nie przyniesie—odpada"/Who will not bring—will be dismissed (Coca-Cola HBC Polska, Tesco Polska and Carlsberg Polska)	1 point if yes
Affective	(12) The following is a list of various statements and each one of them has potential benefits and inconveniences for you. Consider every approach and decide if it is ultimately BENEFICIAL or DISADVANTAGEOUS for you.	- I buy local products, but spend more time shopping - I support local economy by consuming local products, but pay more for them - I buy food in smaller portions, but use more packaging - By not shopping "to stockpile for later" I lose the benefit of promotional prices for certain food products - I improve the quality of consumed food, but at the same time I reduce the amount of consumed food - Avoiding excessive, disadvantageous consumption requires self-control and self-discipline - I am ready to pay more for higher quality products - Being a sustainable consumer means that I have to be engaged in the life of a local community - Responsible shopping means that I have to be more committed and spend time on preparing shopping lists - Exchanging, lending and other forms of sustaining communitarianism mean that I have to spend time and be committed - Buying any new products or services means that I have to spend time and effort on research	1 point if "beneficial"

3.5. Clustering

When running the analysis, the number of clusters was specified as 2, 3 and 4, to compare them and find the best solution. As the 2-cluster solution gave the result of data division into negative values of cognitive, behavioral, and evaluative indices for one cluster and positive for another one, that does not provide an insightful interpretation. As we needed to choose a better solution from two options, i.e., 3-clusters and 4-clusters, we run a validation procedure and assessed differences of cognitive, behavioral, affective indexes between clusters. Considering that in the 4-cluster solution there are more pairs with an insignificant difference, we decided to choose a 3-cluster solution for further interpretation.

The relative centers of each cluster on a standardized scale, are presented in Table 7, and reflect differences between the clusters. The first cluster has the highest values of behavioral aspect, which means that this group represents the most sustainable pattern of behavior. The second one is the only one with positive values for cognitive aspects, so respondents with the highest awareness level. The third group includes respondents with the lowest values for each aspect.

Table 7. Final cluster centers.

	Cluster		
	1	**2**	**3**
Zscore (cognitive)	−0.42567	1.1596	−0.55525
Zscore (behavioral)	0.81631	0.14434	−0.76766
Zscore (affective)	0.14786	0.68968	−0.6525

The ANOVA analysis allows determination of the importance of each index with the relative weight. For the 3 indices, F values are large and all of them are significant ($p < 0.001$), indicating that they have a significant impact on the results for the 3-cluster solution (Table 8).

Table 8. ANOVA analysis for 3 clusters.

	Cluster		Error		F	Sig.
	Mean Square	**df**	**Mean Square**	**df**		
Zscore (cognitive)	125.531	2	0.421	430	298.328	0.000
Zscore (behavioral)	95.835	2	0.559	430	171.468	0.000
Zscore (evaluative)	68.157	2	0.688	430	99.118	0.000

To validate the analysis, we created cluster membership and compare, whether mean values of each cluster are significantly different within each index. All 3 clusters are significantly different, so we can reject the null hypothesis, that the group means are all equal in all indices:

- Affective—$F(2.430) = 99.118$, $p < 0.001$;
- Behavioral—$F(2.430) = 171.468$, $p < 0.001$;
- Cognitive—$F(2.430) = 298.328$, $p < 0.001$.

Nevertheless, we have results showing that not all group means are equal, so we should check whether pair comparison shows a significant difference.

To sum up, there are 3 clusters that describe and differentiate all observations:

- Cluster 1—low cognitive, high behavioral, high affective;
- Cluster 2—high cognitive, low behavioral, high affective;
- Cluster 3—low cognitive, low behavioral, low affective.

4. Results

As a description of the results, we will first present a synthetic description of segments, followed by the characteristics of their representatives within three attitude dimensions. We identified three segments (clusters) and based on the dominant characteristics named them as: cluster 1—"Doers", cluster 2—"Conscious" and cluster 3—"Reluctant" their characteristics can be presented on the basis of the data in Table 9.

Table 9. Consumer segments.

	Feature Level	Doers	Cluster Conscious	Reluctant	The Whole Sample
Attitude dimension	Cognitive, Zscore	−0.4256 Low	1.1596 High	−0.5552 Low	.
	Behavioral, Zscore	0.81631 High	0.14434 Moderate	−0.7676 Low	.
	Affective, Zscore	0.14786 Moderate	0.68968 High	−0.6525 Low	.
		Percent (%) or mean value			
Gender	Woman	56	61	47	54
	Man	44	39	53	46
Age	18–24 years	14	24	13	16
	25–34 years	11	27	13	16
	35–44 years	16	21	21	19
	45–54 years	17	18	13	15
	55–64 years	22	5	21	16
	65 and more years	20	6	20	16
	Mean age	48.10	36.47	47.03	44.19
Education	Other	6	2	9	6
	Vocational school	19	5	27	18
	Secondary school	27	20	26	24
	Post-secondary school	24	15	17	18
	University	24	58	21	33
Number of people in a household	1–2 people	39	35	42	39
	3–4 people	46	50	43	46
	5 and more people	15	15	15	15
Village/town/city	Village	42	28	24	31
	Town below 10,000 inhabitants	5	5	3	4
	Town from 10,000 to 49,999 inhabitants	14	11	10	12
	Town from 50,000 to 99,999 inhabitants	7	7	11	9
	Town 100,000 inhabitants and more	31	48	51	44
Material conditions	Bad	0	1	10	4
	Neither good, nor bad	38	33	47	40
	Good	61	66	41	55
	DK/ND	1	0	2	1
Self-assessment of household budget	We live in serious poverty—we have not enough to meet daily needs	0	0	1	0
	We live modestly—we have to manage our finances very carefully	13	4	23	14
	We live like average people—we have enough for daily needs, but must save for major purchases	47	41	45	45
	We live well—we can afford a lot, without having to save for specific purposes	36	45	29	36
	We live very well—we can afford certain luxuries	2	10	2	5
	DK/ND	1	0	1	1

Table 9. *Cont.*

	Feature Level	Doers	Cluster Conscious	Reluctant	The Whole Sample
Total family income	PLN 1001-2000	13	8	18	14
	PLN 2001-3000	22	11	24	20
	PLN 3001-4000	26	23	28	26
	PLN 4001-5000	21	27	16	21
	PLN 5001-6000	9	15	5	9
	PLN 6001 and more	9	15	9	11

Cluster 1—segment "Doers"

This segment included 31% of food consumers. These are action-oriented people, whose consumption-oriented behaviors show a higher number of sustainable behaviors than in the case of other people, despite the fact that their knowledge about sustainable consumption is low and their opinions about this concept are moderately positive. This means that the reasons behind their (sustainable) behaviors may come from other areas, e.g., the socialization or upbringing process, material situation, living environment, etc. This segment shows a significantly higher percentage of people living in rural areas. The members of this segment have diversified incomes, whose distribution is similar to the one in the examined sample. These people are slightly older than the average for the examined group.

Cluster 2—segment "Conscious"

This segment included 30% of food consumers. The individuals in this segment have a high value of indexes within the cognitive and affective dimensions of attitudes, but a noticeably lower intensity of sustainable behaviors. These consumers have a better understanding of and a more positive attitude towards the sustainability concept but have not yet introduced a higher number of sustainable behaviors. However, their higher awareness may be considered a good foundation to grow into a fully sustainable attitude. This segment has a higher representation of women. It is also a segment with a major share of young people and the highest share of individuals with higher education in the identified segments (which may determine the higher level of awareness). The group's higher level of income and satisfaction from material status should be also underscored.

Cluster 3—segment "Reluctant"

This segment included 39% of food consumers. This group consists of individuals with negative attitudes towards the idea of sustainable consumption, manifested both in the low awareness of concepts, negative opinions of them, and low importance of sustainable behaviors. The group is nearly equally represented by men and women, with age distribution similar to the distribution in the examined group. The representatives of this segment may be encountered in towns of various sizes. The profiling data show that this segment is more often populated by individuals with lower education, whose material situation is poor (they declare lower satisfaction with their incomes and that their incomes are lower).

Segment characteristics in the cognitive sphere

Significant differences can be spotted among segments in the cognitive sphere. The individuals in the "Conscious" segment had a high awareness of the terms used for describing sustainable consumption behaviors. In total, 87% of its representatives have met with the term "sustainable consumption". In the case of "Doers" and "Reluctant" segments, this was 36% and 19%, respectively. We can also observe a high activity of the "Conscious" segment individuals in the search for information about food and a high amount of the retained information. The "Reluctant" segment individuals declare looking for information about food in the least degree (17% of the segment's representatives). Additionally, they pay little attention to information about consumption sustainability.

Segment characteristics in the behavioral sphere

The readiness to introduce changes in nutrition is mainly declared by individuals in the "Doers" and "Conscious" segments (67% and 70%, respectively)—however, these changes are of a different nature. The representatives of the "Doers" segment refer to the sustainability concept to a higher degree (e.g., increasing the share of fruit and vegetables in their diets, preparing meals by cooking them themselves). They also more frequently declared limiting undesirable behaviors such as eating fried meals and meals with a high-fat content, consuming sugar, salt, and stimulants. The "Doers" representatives limited meat consumption to the highest degree. The members of the "Conscious" segment would introduce similar changes, but they were declared with lower frequency. This segment dominates only in terms of introducing healthy and organic products to the diet. The nutritional changes were the least common in the "Reluctant" segment, declared by around 1/3 of the segment's participants. Moreover, they were related to the sustainable consumption principles to a lower degree than in the remaining segments.

The individuals in the "Does" and "Conscious" segments more often declared changes related to sourcing food for their households. The use of large retail facilities (such as supermarkets and hypermarkets) grew among the "Doers". This could be related to the fact that half of the members of this segment live outside towns, which is where structural changes to retail have been happening in the recent years, leading to changes in purchasing models. Different behaviors are represented by the representatives of the "Conscious" and "Reluctant" segments, both very similar to each other in terms of the place of living. The "Conscious" segment was more eager to use smaller, specialized shops and marketplaces. On the other hand, the "Reluctant" segment preferred large-size retailer and increased the importance of online shopping for food.

The "Does" and "Conscious" segments are changing their behaviors related to sourcing food in the direction convergent with sustainability principles, but they are doing it differently. For example, the "Doers" more frequently declare limiting wastage. For the "Conscious" segment, however, making food products autonomously and purchasing low-processed products are more typical.

Segment characteristics in the affective sphere

Certain similarities emerge between the "Doers" and the "Conscious" segments, and a major difference in relation to the "Reluctant" segment. The two first segments have a positive attitude towards sustainable consumption principles and appreciate the impact of such behaviors on the local communities, environment, and local economy. Still, the individuals in the "Conscious" segment are firmer in this regard than the "Doers". For example, they have a higher propensity to pay more for organic products or spend extra time on sustainable behaviors, e.g., searching for information on nutrition principles. The individuals in the "Reluctant" segment, on the other hand, assess various signs of sustainable consumption as disadvantageous for them.

Contrary to our expectations, not all consumers are interested in food and sustainable consumption. The "Doers" are interested in food for various reasons, which fail to cover the issue of sustainable consumption. There may be an impression that this topic is disregarded by such individuals. The "Conscious" segment consisted of consumers actively looking for information about food and nutrition, sensitive to the issues of sustainable consumption.

5. Discussion

The results we obtained expand the current consumer segmentation models with regard to attitudes towards sustainability. Similar to several studies conducted in other countries [7,47,62,73–75,79,82,84,89], we proposed three segments of consumers.

Consumers have varying awareness on consumption sustainability (behavioral aspect), buying and eating behaviors reinforcing responsibility towards the planet and future generations (behavioral aspect), the perception of promotional messages related to the balancing of food product choices, and the subjective evaluation of benefits for consumers resulting from sustainable behaviors (affective aspect). The study is one of numerous papers about profiling food consumers based on a three-element attitude dimension in relation to SFC.

5.1. Cognitive Dimension

Our studies have revealed that the information about environmental, economic, and social consequences of excessive consumption are factors that foster the popularization of sustainable attitudes. The force of various media should be used to reduce adverse behaviors [16,75,102] by building SC awareness and affecting consumption behaviors, as supported by e.g., Hasanzade et al. [84], Prokeinová and Paluchová [78], Sogri et al. [81], Wang and Somogyi [86], Verain et al. [73]. A better understanding of the characteristics of consumers via segmentation facilitates the preparation of a more effective communication strategy. Public institutions, food producers, and commerce should take steps leading to a better understanding of food and nutrition, potential benefits for the environment, responsibility for the planet and future generations. The existence of a connection between the type of consumers and susceptibility to messages communicated, for example, via social media has been indicated by Sogari et al. [81] claiming that "the greater the importance the consumer places on the product/process dimension of environmental sustainability, the higher the self-selection in market segments". The researchers also pointed to the need to increase the possibilities of communicating the activities of enterprises in environmental protection. Hasanzade et al. [84] have added to the literary references the differentiation of consumer segments due to behaviors resulting from the way they react to messages, additionally showing the need to carefully select information about products (e.g., the product's ethical character). However, not all studies confirm the need to differentiate communication efforts. According to Verain et al. [73] there exist messages that can have a universal character and reach all consumer groups.

5.2. Behavioral Dimension

Labelling products as environmentally friendly and promoting "new food" may be helpful in reinforcing sustainable consumer choices [74,85,103]. The results of our studies show that marking food products as healthy and organic may be important for the "Conscious" segment. This segment is similar to the "Trendsetters" segment that appeared in the study by Van Huy et al. [75].

Nevertheless, our study fails to provide a detailed insight into the issue of a sustainable diet, which is discussed, for example, by Verain et al. [7,73] who have confirmed that consumer segments are differentiated by the approach to health and eating healthy, organic food products. The interest in healthy eating has been also noted by Van Huy et al. [75], who noted that the interest in healthy nutrition is positively related to organic, locally sourced food.

Verain et al. [73] have proved that individuals oriented toward sustainable growth have been limiting the consumption of meat after learning about the benefits this approach provides both to their health and the environment. This is also confirmed by our study since the inclination to eat meat has been used to differentiate the segments. Meat consumption has been limited to the highest degree by the "Doers", closely trailed by the "Conscious" segment.

The results of our study show a relation between the place of residence of consumers and their sustainable attitudes. It seems that people living in rural areas have "natural" sustainable attitudes (mainly in the behavioral dimension), which may be related to the specific characteristic of the rural environment as a place for living and working. La Lama et al. [79] also point to the role played by the place of residence of consumers, additionally pointing to other factors such as income, lifestyle, access to frequently updated information, and telecommunication technologies.

Vermeir et al. [18] have stated that "many consumers express environmental concern but do not consistently act on it. That is, consumer attitudes toward environmental sustainability are mainly positive, but there is a notable gap between favorable disposition and actual purchase of sustainable food products, i.e., the attitude-behavior gap". Similar conclusions can be formulated on the basis of our segmentation—the "Conscious" segment shows a well-developed cognitive and affective dimension of the attitude, with a less developed behavioral dimension.

5.3. *Affective Dimension*

The affective dimension proves the consumers' positive approach towards the concept and principles of sustainable consumption and their commitment to the idea. The results we have obtained show that this is fostered by the consumers' higher material status, although this is not the only condition. The "Conscious" segment is quicker to see health benefits coming from SC than the "Doers". This segment shows similarities to consumers of cluster 3 in the research conducted by Verain et al. [73], who pay close attention to SC attributes. In the segmentation done by La Lama et al. [79] one of the segments is "Skeptical". It consists of consumers with an egocentric approach and little contact with modern food production practices, who are reluctant to pay higher prices for welfare-friendly products. The segment is similar to our "Reluctant" segment, showing a negative attitude towards the sustainability concept. Groups of consumers with a negative attitude towards SC have a significant impact on the development of the market of products covered by this concept. The research conducted by Gerini et al. [47] has shown that consumers may exhibit positive attitudes towards sustainable products, while being reluctant to pay higher prices for them. If most consumers are not willing to pay more for, e.g., organic products, whose production is more costly, such products will not be able to capture a higher share of the market. The growth of the market for sustainable products and services should be supported by good access to information [27,47,104,105]. The factor supporting the growth of this market may include references to consumer ethnocentrism and localism, as discussed, for example, by Van Huy et al. [75] on the basis of studies conducted among consumers in Vietnam, and to personal responsibility [47].

6. Conclusions

Conclusions in two areas may be formulated on the basis of our work. The first conclusion applies to the review of papers on food consumer segmentation based on the sustainability concept. The second one covers the segmentation and characterization of such segments in the region of Wielkopolska, Poland.

It may be concluded, on the basis of the literature review, that eight categories of variables are used for the purposes of segmentation of consumers with regard to food consumption. These were: environmental sustainability, socio-demographic, psychographic, economic, behavioral, affective factors, lifestyle, and consumers' values. The factors related to the environment and consumption behaviors should be considered the dominating category. The multi-faceted nature of consumer attitudes leads researchers to use a wide range of variables related to lifestyle, values, preferences, and consumers' attitude towards sustainable food consumption and environmental issues. Given the significant diversity of the used factors, it may be also observed that individual papers most usually put emphasis on one of the dimensions of the attitude. The theoretical foundation of the approach to segmentation proposed by us is the concept of a three-element attitude structure, thanks to which factors that may be indicators of sustainability can be approached comprehensively. The results we obtained prove that such an approach is efficient in the segmentation of consumers.

The consumers' segmentation model proposed in that paper contributes to the knowledge about consumers' sustainable behavior and might be used for further research development and as well as by practitioners and consumer policymakers. Identified segments represent a different potential for adoption of sustainable behaviors what implies the necessity of implementing various methods of promoting the idea of sustainability among them.

Two of the identified segments ("Doers" and "Conscious") represent a certain escalation of sustainable attitudes, and the characteristics of these segments show that they may grow in the future. The representatives of the third segment ("Reluctant") are negatively oriented towards the sustainability concept. Taking into account the distribution of the size of individual segments, it may be assumed that around 60% of consumers make up a group that may positively modify their attitudes towards sustainable consumption. Consequently, this group is the target of activities that may be taken to support the development of sustainable attitudes. A universal set of activities promoting sustainable food consumption attitudes should cover activities in two main areas. The first one would be to build consumer awareness in relation to the concepts of sustainability and sustainable consumption, along with the derived benefits. This area should mainly impact the "Doers"—thanks to the reinforcement of the consumers' knowledge it will be possible to achieve the effect of further intensification of behaviors. The second area would be activities incentivizing to take sustainable activities, on the basis of the already developed consumer awareness.

Understanding and recognizing consumers' attitudes and behaviors is useful for industrial practitioners and decision-makers making efforts to transition into more sustainable food systems. Information and communication strategies should be built upon full knowledge about food consumers from a given region, taking into consideration the three-dimensional nature of their attitudes. By adapting the content of messages to the profiles of specific consumer segments, emphasis should be applied to informing about benefits coming from the consumption of sustainable food, in order to motivate the sensitive segments and to raise the consumers' awareness about the benefits stemming from the pursuit of the sustainable food consumption model.

Limitations and Recommendations for Further Research

A conclusion can be drawn from our literary research that the food consumption segmentation proposition presented in the article is one of very few propositions that apply to the concept of making food consumption attitudes sustainable and is based on a three-element attitude concept. Maintaining such an approach in the future will make result comparison possible, leading to a better understanding of the consumers.

Our study is limited by a relatively low sample size and its regional reach, which implicates the need to continues similar studies, but in pan-regional and international scope. It is also worth expanding the future studies with qualitative elements, for a better understanding of mechanisms of shaping sustainable attitudes.

The knowledge of signs of sustainable consumption and factors that determine it is still emerging and needs intensive studies. It would be beneficial to direct such future studies at methods of combating any factors that hinder the popularization of sustainable attitudes. Examples of valuable directions would be the impact of retail structure on making food consumption behaviors sustainable, or the characteristics of one's living environment as a factor determining the sustainability of attitudes. Additionally, depending on the type of food considered by consumers, their attitudes can be differentiated.

Author Contributions: Conceptualization M.G., E.G.-G., M.K. and Z.S.; methodology M.G., E.G.-G., M.K. and Z.S.; software M.G. and E.G.-G.; validation M.G. and E.G.-G.; formal analysis M.G. and E.G.-G.; investigation M.G. and E.G.-G.; resources M.G., E.G.-G., M.K. and Z.S.; data curation M.G. and E.G.-G.; writing—original draft preparation M.G. and E.G.-G.; writing—review and editing M.G., E.G.-G., M.K. and Z.S.; visualization M.G. and E.G.-G.; supervision M.G. and E.G.-G.; project administration M.G. and E.G.-G.; funding acquisition, M.G. and E.G.-G. All authors have read and agreed to the published version of the manuscript."

Funding: This research received no external funding.

Data Availability Statement: Not applicable.

Acknowledgments: Not applicable.

Conflicts of Interest: The authors declare no conflict of interest.

References

1. European Commission. *Proposal for a Concil Decision on the Conclusion on Behalf of the European Union of the Paris Agreement Adopted under the United Nations Framework Convention on Climate Change$—Publications Office of the EU. COM/2016/0395 Final—2016/0184 (NLE)*; European Commission: Brussels, Belgium, 2016.
2. Ivanova, M.; Baste, I.; Lee, B.; Belliethathan, S.; Gelil, I.A.; Gupta, J.; Haas, P.M.; Habtezion, Z.; Halpaap, A.; Mo-hamed-Katerere, J.C.; et al. *Global Environmental Outlook 5, United NationsEnvironment Programme: Chapter 17, GlobalResponses*; University of Massachusetts Boston: Boston, MA, USA, 2012; Volume 36.
3. OECD. Promoting Sustainable Consumption. In *Good Practices in OECD Countries*; OECD: Paris, France, 2008.
4. Sala, S.; Anton, A.; McLaren, S.J.; Notarnicola, B.; Saouter, E.; Sonesson, U. In quest of reducing the environmental impacts of food production and consumption. *J. Clean. Prod.* **2017**, *140*, 387–398. [CrossRef]
5. Scholl, G.; Rubik, F.; Kalimo, H.; Biedenkopf, K.; Söebech, Ó. Policies to promote sustainable consumption: Innovative approaches in Europe. *Nat. Resour. Forum* **2010**, *34*, 39–50. [CrossRef]
6. Partha, D.; Ulph, A.; Southerton, D. *Sustainable Consumption: Multi-Disciplinary Perspectives*, 1st ed.; Sir Partha, D., Ulph, A., Southerton, D., Eds.; Oxford University Press: New York, NY, USA, 2014; ISBN 978-0-19-967935-5.
7. Verain, M.C.; Dagevos, H.; Antonides, G. Sustainable food consumption. Product choice or curtailment? *Appetite* **2015**, *91*, 375–384. [CrossRef]
8. Vermeir, I.; Verbeke, W. Sustainable Food Consumption: Exploring the Consumer "Attitude—Behavioral Intention" Gap. *J. Agric. Environ. Ethics* **2006**, *19*, 169–194. [CrossRef]
9. Almond, R.E.A.; Grooten, M.; Petersen, T. *Living Planet Report 2020—Bending the Curve of Biodiversity Loss*; World Wildlife Fund (WWF): Gland, Switzerland, 2020.
10. Chesson, J. Sustainable Development: Connecting Practice with Theory. *J. Environ. Assess. Policy Manag.* **2013**, *15*, 1350002. [CrossRef]
11. Comim, F.; Tsutsumi, R.; Varea, A. Choosing sustainable consumption: A capability perspective on indicators. *J. Int. Dev.* **2007**, *19*, 493–509. [CrossRef]
12. Dubas-Jakóbczyk, K.; Kocot, E.; Czerw, A.; Juszczyk, G.; Karwowska, P.; Menne, B. Health as an Investment in Poland in the Context of the Roadmap to Implement the 2030 Agenda for Sustainable Development and Health 2020. Available online: https://www.euro.who.int/en/countries/poland/publications/health-as-an-investment-in-poland-in-the-context-of-the-roadmap-to-implement-the-2030-agenda-for-sustainable-development-and-health-2020-2018 (accessed on 19 November 2020).
13. European Commission. Future of the Common Agricultural Policy. 2020. Available online: https://ec.europa.eu/info/food-farming-fisheries/key-policies/common-agricultural-policy/future-cap_en (accessed on 19 January 2021).
14. FAO. *Online Food Loss and Waste Database. Dataset of Food Loss and Waste Esti Mates from Grey Literature, National and Sectoral Reports*; FAO: Rome, Italy, 2020.
15. Guzmán-Rincón, A.; Carrillo-Barbosa, R.; Martín-Caro, L.; Álamo, E.; Rodríguez-Cánovas, B. Sustainable Consumption Behaviour in Colombia: An Exploratory Analysis. *Sustainabilty* **2021**, *13*, 802. [CrossRef]
16. Lazzarini, G.A.; Zimmermann, J.; Visschers, V.H.; Siegrist, M. Does environmental friendliness equal healthiness? Swiss consumers' perception of protein products. *Appetite* **2016**, *105*, 663–673. [CrossRef] [PubMed]
17. Oslo Roundtable Oslo Roundtable on Sustainable Production and Consumption. Available online: https://enb.iisd.org/consume/oslo004.html (accessed on 19 November 2020).
18. Vermeir, I.; Weijters, B.; de Houwer, J.; Geuens, M.; Slabbinck, H.; Spruyt, A.; van Kerckhove, A.; van Lippevelde, W.; de Steur, H.; Verbeke, W. Environmentally Sustainable Food Consumption: A Review and Research Agenda from a Goal-Directed Perspective. *Front. Psychol.* **2020**, *11*, 1603. [CrossRef] [PubMed]
19. Wang, H. Analysis on the Changes in Consumer Behavior and Marketing Countermeasure. *SHS Web Conf.* **2015**, *17*, 1007. [CrossRef]
20. Bjelle, E.L.; Wiebe, K.S.; Többen, J.; Tisserant, A.; Ivanova, D.; Vita, G.; Wood, R. Future changes in consumption: The income effect on greenhouse gas emissions. *Energy Econ.* **2021**, *95*, 105114. [CrossRef]
21. Agenda 2030 Sustainable Consumption and Production/Department of Economic and Social Affairs. Available online: https://sdgs.un.org/topics/sustainable-consumption-and-production (accessed on 12 February 2021)
22. European Commission. *Communication from the Commission to the European Parliament, the Council, the European Economic and Social Committee and the Committee of the Regions. Closing the Loop—An EU Action Plan for the Circular Economy*. European Commission COM (2015) 614 final. Available online: https://eur-lex.europa.eu/legal-content/EN/TXT/DOC/?uri=CELEX:52015DC0614&from=EN (accessed on 20 April 2021).
23. European Commission. *Communication from The Commission to The European Parliament, The Council, The European Economic and Social Committee And The Committee Of The Regions A Farm to Fork Strategy for a Fair, Healthy and Environ-Mentally-Friendly Food System*; European Union: Brussels, Belgium, 2020; p. 23.
24. FAO, IFAD, UNICEF, WFP & WHO. *The State of Food Security and Nutrition in the World 2020*; FAO: Rome, Italy, 2020.
25. Lorek, S.; Spangenberg, J.H. Sustainable consumption within a sustainable economy—Beyond green growth and green economies. *J. Clean. Prod.* **2014**, *63*, 33–44. [CrossRef]

26. Palmieri, N.; Forleo, M.B. The potential of edible seaweed within the western diet. A segmentation of Italian consumers. *Int. J. Gastron. Food Sci.* **2020**, *20*, 100202. [CrossRef]
27. Gazdecki, M.; Goryńska-Goldmann, E. *Relationships on Food Markets—Consumers' Perspective*, 1st ed.; Poznan University of Life Science: Poznan, Poland, 2018; ISBN 978-83-7160-922-0.
28. Łuczka, W.; Smoluk-Sikorska, J. *Sustainable Consumption—Between Theory and Practice*; Vytautas Magnus University: Kaunas, Lithuania, 2017; pp. 1161–1162.
29. Sargant, E. *Sustainable Food Consumption*; Environmental Policy; Wageningen Academic Publishers: Wageningen, The Netherlands, 2014; Volume 11, ISBN 978-90-8686-263-4.
30. Vittersø, G.; Tangeland, T. The role of consumers in transitions towards sustainable food consumption. The case of organic food in Norway. *J. Clean. Prod.* **2015**, *92*, 91–99. [CrossRef]
31. Potter, H.K.; Röös, E. Multi-criteria evaluation of plant-based foods—Use of environmental footprint and LCA data for consumer guidance. *J. Clean. Prod.* **2021**, *280*, 124721. [CrossRef]
32. EIT Food. *The Future of Food. Conference Report*; EIT Food: Leuven Belgium, 2020; p. 26.
33. Puigdueta, I.; Aguilera, E.; Cruz, J.L.; Iglesias, A.; Sanz-Cobena, A. Urban agriculture may change food consumption towards low carbon diets. *Glob. Food Secur.* **2021**, *28*, 100507. [CrossRef]
34. Peano, C.; Merlino, V.M.; Sottile, F.; Borra, D.; Massaglia, S. Sustainability for Food Consumers: Which Perception? *Sustainability* **2019**, *11*, 5955. [CrossRef]
35. Aschemann-Witzel, J.; Ares, G.; Thøgersen, J.; Monteleone, E. A sense of sustainability? How sensory consumer science can contribute to sustainable development of the food sector. *Trends Food Sci. Technol.* **2019**, *90*, 180–186. [CrossRef]
36. Kamenidou, I.C.; Mamalis, S.A.; Pavlidis, S.; Bara, A.E.-Z.G. Segmenting the Generation Z Cohort University Students Based on Sustainable Food Consumption Behavior: A Preliminary Study. *Sustainability* **2019**, *11*, 837. [CrossRef]
37. Henchion, M.; McCarthy, M.; Resconi, V.C.; Troy, D. Meat consumption: Trends and quality matters. *Meat Sci.* **2014**, *98*, 561–568. [CrossRef]
38. Higgs, S. Social norms and their influence on eating behaviours. *Appetite* **2015**, *86*, 38–44. [CrossRef]
39. Mullee, A.; Vermeire, L.; Vanaelst, B.; Mullie, P.; Deriemaeker, P.; Leenaert, T.; de Henauw, S.; Dunne, A.; Gunter, M.J.; Clarys, P.; et al. Vegetarianism and meat consumption: A comparison of attitudes and beliefs between vegetarian, semi-vegetarian, and omnivorous subjects in Belgium. *Appetite* **2017**, *114*, 299–305. [CrossRef]
40. Rizkalla, N.; Erhan, T.P. Sustainable Consumption Behaviour in The Context of Millennials in Indonesia—Can Environmental Concern, Self-efficacy, Guilt and Subjective Knowledge Make a Difference? *Manag. Sustain. Bus. Manag. Solutions Emerg. Econ.* **2020**, *25*, 43. [CrossRef]
41. Verain, M.C.; Bartels, J.; Dagevos, H.; Sijtsema, S.J.; Onwezen, M.C.; Antonides, G. Segments of sustainable food consumers: A literature review. *Int. J. Consum. Stud.* **2012**, *36*, 123–132. [CrossRef]
42. Maciejewski, G.; Mokrysz, S. Wróblewski, Łukasz Segmentation of Coffee Consumers Using Sustainable Values: Cluster Analysis on the Polish Coffee Market. *Sustainability* **2019**, *11*, 613. [CrossRef]
43. Muñoz-Sánchez, V.-M.; Pérez-Flores, A.-M. The Connections between Ecological Values and Organic Food: Bibliometric Analysis and Systematic Review at the Start of the 21st Century. *Sustainability* **2021**, *13*, 3616. [CrossRef]
44. Majewski, E.; Komerska, A.; Kwiatkowski, J.; Malak-Rawlikowska, A.; Wąs, A.; Sulewski, P.; Gołaś, M.; Pogodzińska, K.; Lecoeur, J.-L.; Tocco, B.; et al. Are Short Food Supply Chains More Environmentally Sustainable than Long Chains? A Life Cycle Assessment (LCA) of the Eco-Efficiency of Food Chains in Selected EU Countries. *Energies* **2020**, *13*, 4853. [CrossRef]
45. Bajan, B.; Mrówczyńska-Kamińska, A.; Poczta, W. Economic Energy Efficiency of Food Production Systems. *Energies* **2020**, *13*, 5826. [CrossRef]
46. Vanhonacker, F.; Verbeke, W. Buying higher welfare poultry products? Profiling Flemish consumers who do and do not. *Poult. Sci.* **2009**, *88*, 2702–2711. [CrossRef] [PubMed]
47. Gerini, F.; Alfnes, F.; Schjøll, A. Organic- and Animal Welfare-labelled Eggs: Competing for the Same Consumers? *J. Agric. Econ.* **2016**, *67*, 471–490. [CrossRef]
48. Furno, M.; del Giudice, T.; Cicia, G. Organic consumers' profile beyond the mean. *Org. Agric.* **2021**, 1–13. [CrossRef]
49. Łuczka, W. Zrównoważona konsumpcja i uwarunkowania jej rozwoju. *Handel Wewnętrzny* **2016**, *6*, 136–145.
50. Scalvedi, M.L.; Saba, A. Exploring local and organic food consumption in a holistic sustainability view. *Br. Food J.* **2018**, *120*, 749–762. [CrossRef]
51. Rejman, K.; Kaczorowska, J.; Halicka, E.; Laskowski, W. Do Europeans consider sustainability when making food choices? A survey of Polish city-dwellers. *Public Health Nutr.* **2019**, *22*, 1330–1339. [CrossRef] [PubMed]
52. Śmiglak-Krajewska, M.; Wojciechowska-Solis, J. Consumption Preferences of Pulses in the Diet of Polish People: Motives and Barriers to Replace Animal Protein with Vegetable Protein. *Nutrients* **2021**, *13*, 454. [CrossRef]
53. BÖWL Die Bio-Branche 2018. Zahlen. Daten. Fakten. Available online: https://www.boelw.de/service/mediathek/broschuere/die-bio-branche-2018/ (accessed on 12 February 2021).
54. Nestorowicz, R. Asymetria wiedzy a rozwój rynku żywności ekologicznej w Polsce. *Handel Wewnętrzny* **2018**, *5*, 212–224.
55. Balderjahn, I.; Peyer, M.; Paulssen, M. Consciousness for fair consumption: Conceptualization, scale development and empirical validation. *Int. J. Consum. Stud.* **2013**, *37*, 546–555. [CrossRef]
56. Bryła, P. Organic food consumption in Poland: Motives and barriers. *Appetite* **2016**, *105*, 737–746. [CrossRef]

57. Carrington, M.; Chatzidakis, A.; Goworek, H.; Shaw, D. Consumption Ethics: A Review and Analysis of Future Directions for Interdisciplinary Research. *J. Bus. Ethic* **2021**, *168*, 215–238. [CrossRef]
58. Goryńska-Goldmann, E.; Adamczyk, G.; Gazdecki, M. Awareness of Sustainable Consumption and Its Implications for the Selection of Food Products. *J. Agribus. Rural. Dev.* **2016**, *10*, 301–308. [CrossRef]
59. Luchs, M.G.; Phipps, M.; Hill, T. Exploring consumer responsibility for sustainable consumption. *J. Mark. Manag.* **2015**, *31*, 1449–1471. [CrossRef]
60. Łuczka, W.; Kalinowski, S. Barriers to the Development of Organic Farming: A Polish Case Study. *Agriculture* **2020**, *10*, 536. [CrossRef]
61. European Commission. *Reflection Paper towards a Sustainable Europe by 2030*; European Commission: Bruxelles, Belgium, 2019.
62. Annunziata, A.; Mariani, A.; Angela, M. Consumer Perception of Sustainability Attributes in Organic and Local Food. *Recent Patents Food, Nutr. Agric.* **2018**, *9*, 87–96. [CrossRef] [PubMed]
63. Nie, C.; Zepeda, L. Lifestyle segmentation of US food shoppers to examine organic and local food consumption. *Appetite* **2011**, *57*, 28–37. [CrossRef] [PubMed]
64. Niedermeier, A.; Emberger-Klein, A.; Menrad, K. Which factors distinguish the different consumer segments of green fast-moving consumer goods in Germany? *Bus. Strat. Environ.* **2021**, *30*, 1–6. [CrossRef]
65. Grunert, K.G. International segmentation in the food domain: Issues and approaches. *Food Res. Int.* **2019**, *115*, 311–318. [CrossRef]
66. Steenkamp, J.-B.E.; Ter Hofstede, F. International market segmentation: Issues and perspectives. *Int. J. Res. Mark.* **2002**, *19*, 185–213. [CrossRef]
67. De Jonge, J.; van Trijp, H.C.M. Meeting Heterogeneity in Consumer Demand for Animal Welfare: A Reflection on Existing Knowledge and Implications for the Meat Sector. *J. Agric. Environ. Ethic-* **2012**, *26*, 629–661. [CrossRef]
68. Vanhonacker, F.; van Loo, E.J.; Gellynck, X.; Verbeke, W. Flemish consumer attitudes towards more sustainable food choices. *Appetite* **2013**, *62*, 7–16. [CrossRef]
69. Moher, D.; Liberati, A.; Tetzlaff, J.; Altman, D.G.; The PRISMA Group. Preferred reporting items for systematic reviews and meta-analyses: The PRISMA statement. *PLoS Med.* **2009**, *6*, e1000097. [CrossRef]
70. Bronnmann, J.; Asche, F. Sustainable Seafood from Aquaculture and Wild Fisheries: Insights from a Discrete Choice Experiment in Germany. *Ecol. Econ.* **2017**, *142*, 113–119. [CrossRef]
71. Thøgersen, J. Sustainable food consumption in the nexus between national context and private lifestyle: A multi-level study. *Food Qual. Prefer.* **2017**, *55*, 16–25. [CrossRef]
72. Lavelle, M.J.; Rau, H.; Fahy, F. Different shades of green? Unpacking habitual and occasional pro-environmental behavior. *Glob. Environ. Chang.* **2015**, *35*, 368–378. [CrossRef]
73. Verain, M.C.D.; Sijtsema, S.J.; Dagevos, H.; Antonides, G. Attribute Segmentation and Communication Effects on Healthy and Sustainable Consumer Diet Intentions. *Sustainabilty* **2017**, *9*, 743. [CrossRef]
74. Stranieri, S.; Baldi, L. Shelf Life Date Extension of Fresh-Cut Salad: A Consumer Perspective. *J. Food Prod. Mark.* **2017**, *23*, 939–954. [CrossRef]
75. Van Huy, L.; Chi, M.T.T.; Lobo, A.; Nguyen, N.; Long, P.H. Effective Segmentation of Organic Food Consumers in Vietnam Using Food-Related Lifestyles. *Sustainability* **2019**, *11*, 1237. [CrossRef]
76. Mózner, Z.V. Sustainability and consumption structure: Environmental impacts of food consumption clusters. A case study for Hungary. *Int. J. Consum. Stud.* **2014**, *38*, 529–539. [CrossRef]
77. Aprile, M.C.; Mariani, A. Young People's Propensity to Use Sustainability Labels on Food Products: A Case Study in the South of Italy. *Qual. Access Success* **2015**, *16*, 75–79.
78. Prokeinová, R.B.; Paluchová, J. Identification of the Patterns Behavior Consumptions by Using Chosen Tools of Data Mining—Association Rules. AGRIS -Line Pap. *Econ. Inform.* **2014**, *6*, 1–10.
79. La Lama, G.C.M.-D.; Estévez-Moreno, L.X.; Villarroel, M.; Rayas-Amor, A.A.; María, G.A.; Sepúlveda, W.S. Consumer Attitudes Toward Animal Welfare-Friendly Products and Willingness to Pay: Exploration of Mexican Market Segments. *J. Appl. Anim. Welf. Sci.* **2018**, *22*, 13–25. [CrossRef]
80. Hölker, S.; Von Meyer-Höfer, M.; Spiller, A. Animal Ethics and Eating Animals: Consumer Segmentation Based on Domain-Specific Values. *Sustainability* **2019**, *11*, 3907. [CrossRef]
81. Sogari, G.; Pucci, T.; Aquilani, B.; Zanni, L. Millennial Generation and Environmental Sustainability: The Role of Social Media in the Consumer Purchasing Behavior for Wine. *Sustainability* **2017**, *9*, 1911. [CrossRef]
82. Janßen, D.; Langen, N. The bunch of sustainability labels—Do consumers differentiate? *J. Clean. Prod.* **2017**, *143*, 1233–1245. [CrossRef]
83. Yıldırım, S.; Candan, B. Segmentation of Green Product Buyers Based on Their Personal Values and Consumption Values. *Environ. Values* **2015**, *24*, 641–661. [CrossRef]
84. Hasanzade, V.; Osburg, V.-S.; Toporowski, W. Selecting decision-relevant ethical product attributes for grocery shopping. *Manag. Decis.* **2018**, *56*, 591–609. [CrossRef]
85. Risius, A.; Hamm, U.; Janssen, M. Target Groups for Fish from Aquaculture Consumer Segmentation Based on Sustainability Attributes and Country of Origin. *Aquaculture* **2019**, *499*, 341–347. [CrossRef]
86. Wang, O.; Somogyi, S. Consumer adoption of sustainable shellfish in China: Effects of psychological factors and segmentation. *J. Clean. Prod.* **2019**, *206*, 966–975. [CrossRef]

87. Jakubowska, D.; Radzymińska, M. Health and environmental attitudes and values in food choices: A comparative study for Poland and Czech Republic. *Oeconomia Copernic.* **2019**, *10*, 433–452. [CrossRef]
88. Ghvanidze, S.; Velikova, N.; Dodd, T.H.; Oldewage-Theron, W. Are Sustainable Consumers Health Conscious? A Segmentation Study of Wine Consumers. *J. Food Prod. Mark.* **2019**, *25*, 690–711. [CrossRef]
89. Su, C.-H.; Tsai, C.-H.; Chen, M.-H.; Lv, W.Q. U.S. Sustainable Food Market Generation Z Consumer Segments. *Sustainability* **2019**, *11*, 1–14. [CrossRef]
90. Hruba, R. Lifestyle Segmentation of Czech Food Shoppers: How Sustainability and Corporate Social Responsibility Correspond to Consumers´ Lifestyles/International Business Information Management Association (IBIMA). *Sustain. Econ. Dev. Appl. Innov. Manag.* Available online: https://ibima.org/accepted-paper/lifestyle-segmentation-of-the-czech-food-shoppers-how-sustainability-and-corporate-social-responsibility-correspond-to-consumers-lifestyles (accessed on 20 April 2021).
91. Krystallis, A. Production. In *Markets, Business and Sustainability*; Bentham Science Publishers: Sharjah, UAE, 2015; p. 229. ISBN 978-1-68108-026-0.
92. Hrubá, R. Lifestyle segmentation of Czech food shoppers. *Èkonomia* **2019**, *25*, 53–59. [CrossRef]
93. Aydın, H.; Ünal, S. A Study on the Effects of the Consumer Lifestyles on Sustainable Consumption. *Inquiry* **2016**, *1*. [CrossRef]
94. Grunert, K.G.; Brunsø, K.; Bisp, S. *Food-Related Life Style: Development of a Cross-Culturally Valid Instrument for Market Surveillance*; Aarhus University, MAPP Centre: Aarhus, Denmark, 1993.
95. Binder, M.; Blankenberg, A.-K. Green lifestyles and subjective well-being: More about self-image than actual behavior? *J. Econ. Behav. Organ.* **2017**, *137*, 304–323. [CrossRef]
96. Fishbein, M.; Ajzen, I. *Predicting and Changing Behavior: The Reasoned Action Approach*; Taylor & Francis: London, UK, 2011; ISBN 978-1-136-87473-4.
97. Rosner, A.; Stanny, M. *Socio-Economic Development of Rural Areas in Poland*; The European Fund for the Development of Polish Villages Foundation (EFRWP): Institute of Rural and Agricultural Development, Polish Academy of Sciences (IRWIR PAN): Warsaw, Poland, 2017; ISBN 978-83-941050-6-8.
98. Breckler, S.J. Empirical validation of affect, behavior, and cognition as distinct components of attitude. *J. Pers. Soc. Psychol.* **1984**, *47*, 1191–1205. [CrossRef]
99. Dasu, T.; Johnson, T. Exploratory Data Mining and Data Cleaning: An Overview. In *Exploratory Data Mining and Data Cleaning*; Wiley: Hoboken, NJ, USA, 2003; pp. 1–16; ISBN 978-0-471-44835-8.
100. Hair, J.F.; Black, W.C.; Babin, B.J.; Anderson, R.E. *Multivariate Data Analysis: Pearson New International Edition*, 7th ed.; Pearson: London, UK, 2013; ISBN 978-1-292-02190-4.
101. Kreyszig, E. *Advanced Engineering Mathematics*, 10th ed.; Wiley: Hoboken, NJ, USA, 2018; p. 444.
102. Goryńska-Goldmann, E. Disadvantageous Behavior for Achieving Sustainable Food Consumption According to Consumer Opinions. *Rocz. Ann.* **2019**, *3*. [CrossRef]
103. Miglak-Krajewska, M.; Wojciechowska-Solis, J.; Viti, D. Consumers' Purchasing Intentions on the Legume Market as Evidence of Sustainable Behaviour. *Agriculture* **2020**, *10*, 424. [CrossRef]
104. Thøgersen, J.; Jørgensen, A.-K.; Sandager, S. Consumer Decision Making Regarding a "Green" Everyday Product. *Psychol. Mark.* **2012**, *29*, 187–197. [CrossRef]
105. Marek-Andrzejewska, E.M.; Wielicka-Regulska, A. Targeting Youths' Intentions to Avoid Food Waste: Segmenting for Better Policymaking. *Agriculture* **2021**, *11*, 284. [CrossRef]

energies

Review

Microorganisms as New Sources of Energy

Jasminka Talapko *,†, Domagoj Talapko †, Anita Matić and Ivana Škrlec *

Faculty of Dental Medicine and Health, Josip Juraj Strossmayer University of Osijek, HR-31000 Osijek, Croatia
* Correspondence: jtalapko@fdmz.hr (J.T.); iskrlec@fdmz.hr (I.Š.)
† These authors contributed equally to this work.

Abstract: The use of fossil energy sources has a negative impact on the economic and socio-political stability of specific regions and countries, causing environmental changes due to the emission of greenhouse gases. Moreover, the stocks of mineral energy are limited, causing the demand for new types and forms of energy. Biomass is a renewable energy source and represents an alternative to fossil energy sources. Microorganisms produce energy from the substrate and biomass, i.e., from substances in the microenvironment, to maintain their metabolism and life. However, specialized microorganisms also produce specific metabolites under almost abiotic circumstances that often do not have the immediate task of sustaining their own lives. This paper presents the action of biogenic and biogenic–thermogenic microorganisms, which produce methane, alcohols, lipids, triglycerides, and hydrogen, thus often creating renewable energy from waste biomass. Furthermore, some microorganisms acquire new or improved properties through genetic interventions for producing significant amounts of energy. In this way, they clean the environment and can consume greenhouse gases. Particularly suitable are blue-green algae or cyanobacteria but also some otherwise pathogenic microorganisms (*E. coli*, *Klebsiella*, and others), as well as many other specialized microorganisms that show an incredible ability to adapt. Microorganisms can change the current paradigm, energy–environment, and open up countless opportunities for producing new energy sources, especially hydrogen, which is an ideal energy source for all systems (biological, physical, technological). Developing such energy production technologies can significantly change the already achieved critical level of greenhouse gases that significantly affect the climate.

Keywords: bioenergy; biomass waste; hydrogen; microorganisms; renewable energy sources

Citation: Talapko, J.; Talapko, D.; Matić, A.; Škrlec, I. Microorganisms as New Sources of Energy. *Energies* **2022**, *15*, 6365. https://doi.org/10.3390/en15176365

Academic Editor: Dimitrios A. Georgakellos

Received: 8 August 2022
Accepted: 29 August 2022
Published: 31 August 2022

1. Introduction

The basic feature of life is oxidoreduction, which creates energy from matter [1]. However, some microorganisms can embed solar energy in very complex mechanisms of production of low-energy compounds from so-called nature pollutants caused by natural pollutants created by the technology of processing oil, sugar cane, and natural oils (harmful technologies) [2]. In addition to photosynthesis, some microorganisms, such as cyanobacteria, can decompose water into the desired oxygen and even more desirable hydrogen, and some can directly produce hydrogen via anaerobic processes [3]. Some, in turn, can convert classic environmental pollutants into very highly potent energy compounds (methane, alcohol) [4]. Thus, the genotypic and phenotypic traits of many species of microorganisms can direct the production of energy products to more perfect and efficient technologies and environmental purifiers [5].

Current technologies of energy production (energy) are a big problem (technical, environmental, and financial), because in addition to environmental pollution, they require significant investment (initial research, adaptation of new technologies, remediation as the final stage of production) [6]. However, natural pollutants (in terms of quantities, environmental impact, and permanent need for disposal) significantly reduce the benefits of conventional energy from fossil fuels (oil, gas, coal) and represent a subsequent often unsolvable problem of the remediation of CO_2, NO_2, SO_2, and other oxides [7]. Technologies are being developed to use waste products (biorefinery) to produce renewable energy, as they permanently pollute the environment in the repeated energy production cycle [8]. Thus, microorganisms are undoubtedly crucial in developing waste purification and use strategies [9]. Bioenergy research is the center of scientific and technological research in the strategy of finding cost-effective biorefineries [10] as a way out of the current high level of air, water, and soil pollution to find photosynthetic and non-photosynthetic microorganisms that can produce clean energy (directly) or clean hydrogen [11].

An increasingly common research target of potential raw materials for biofuel production is microalgae obtained from adjusted wastewater. However, this may also significantly impact the environment, especially when compared with other renewable energy sources [12]. This can be particularly important when disposing of farm wastewater, representing an increasing environmental issue [13].

The present review aims to demonstrate the activity of biogenic and biogenic–thermogenic microorganisms that produce methane, alcohols, lipids, triglycerides, and hydrogen and contribute to creating renewable energy sources from waste biomass.

2. Microbial Technologies for Biofuel Production

The main reason for the increased interest in biomass as an energy source is the application contribution to the sustainable development paradigm. In addition, biomass sources are often present at the local level, and conversion into biofuel is possible with low initial costs [14]. Per the Renewable Energy Directive (EU Directive 2018/2001) [15], a common framework for the promotion of energy from renewable sources in the EU was established, setting a binding target for final gross consumption in the EU, with the total share of energy from renewable sources having to be 32% by 2030. This directive also promotes using non-food crops to produce biofuels and limits the number of biofuels and bioliquids produced from food or animal feed [15,16]. Methane is the so-called greenhouse gas produced indirectly by organic waste landfills (mainly in anaerobic processes) [17] and directly produced by all living beings (especially ruminants) [18]. There are two known sources of methane production: biological and non-biological. Non-biological methane is formed as a result of some geological processes. However, most methane (over 90%) is produced by the action of microorganisms and is a biological way (source) of methane production. This process of biological methane production is called methanogenesis, and microorganisms that carry out the same process are called methanogens [19,20]. Methanogens belong to the *Archaea* domain, which differs from bacteria because they do not possess peptidoglycan in the cell wall. Still, in *Methanosarcina*, it is a protein; in *Methanosarcina*, it is a heteropolysaccharide; and in *Methanobacterium* and *Methanobrevibacter*, it is replaced by pseudomurein [21]. The most crucial methanogen in the rumen is *Methanobacteriales ruminantium*, which contains pseudomurein in its cell wall. It needs formate, coenzyme M, hydrogen, and carbon dioxide for methane production [22]. For the process of methanogenesis, coenzymes F420, M, and HSHTP and lipids that methanogens have as cofactors are essential [23]. Cofactor F420 is necessary for the activity of hydrogenase as well as the formation of dehydrogenase enzymes, while coenzyme M acts as a terminal methyl carrier in the process of methanogenesis [24].

It is estimated that microorganisms annually produce and consume about one billion tons of methane [25]. However, the methane removal process can also occur in biological and non-biological ways. The most significant is the non-biological pathway in the Earth's atmosphere (specifically, the troposphere and stratosphere), where various chemical reactions under ultraviolet radiation decompose methane. In chemical reactions, the issue is associated with the breaking of the covalent bond in methane–carbon-hydrogen, which is one of the strongest bonds among all hydrocarbons [26]. Regardless, methane is used in a process called the catalytic steam reforming of methane, where methane is first converted into synthetic gas, i.e., into a mixture of hydrogen and carbon monoxide. Then, it serves as a raw material for producing hydrogen, methanol, and other chemicals, where the catalyst is nickel, and the reaction takes place at temperatures from 700 °C to 1100 °C [27].

Pyrolysis is the process of burning methane, in which formaldehyde (HCHO or H_2CO) is formed in the first step, to which the HCO radical is added, after which carbon monoxide (CO) is formed [28].

The photocatalytic oxidation of methane is similar to the natural atmospheric process that oxidizes CH_4 to CO_2 [29]. Ultraviolet light is used to split the oxygen molecule into two free radicals that react with methane to produce products such as CH_3OOH, CH_3OH, HCOOH, CO_2HOCH_2OOH, and water. In photocatalytic reactors, catalysts increase the formation of free radicals and thus the rate of the methane reaction [30].

$$CH_4 + O_2 \rightarrow CO + H_2 + H_2OCH_4 + H_2O \rightarrow 700\text{-}1100\ ^{\circ}C\ NiCO + 3H_2$$

Unlike non-biological methods, biological methods of methane decomposition are carried out by the action of microorganisms called methanotrophs, and the process is named methanotrophy [19] (Figure 1). Methanotrophs can use methane to produce methanol, and *Geobacter sulfurreducens* and *Shewanella oneidensis* can use the mechanism of specific electron transfer from the membrane's outer surface to visible surfaces. This phenomenon can be used in bioelectrochemical devices to produce biohydrogen [31]. In addition, methanotrophs have a significant role in reducing the production of large amounts of greenhouse gases via their formation below the surface of the Earth (below the ground) and the utilization of methane produced in the soil conversion of methane emissions into the atmosphere [32,33]. Methane oxidation begins with reducing oxygen to peroxide and then to methanol with the action of the monooxygenase enzyme (MMO) [34].

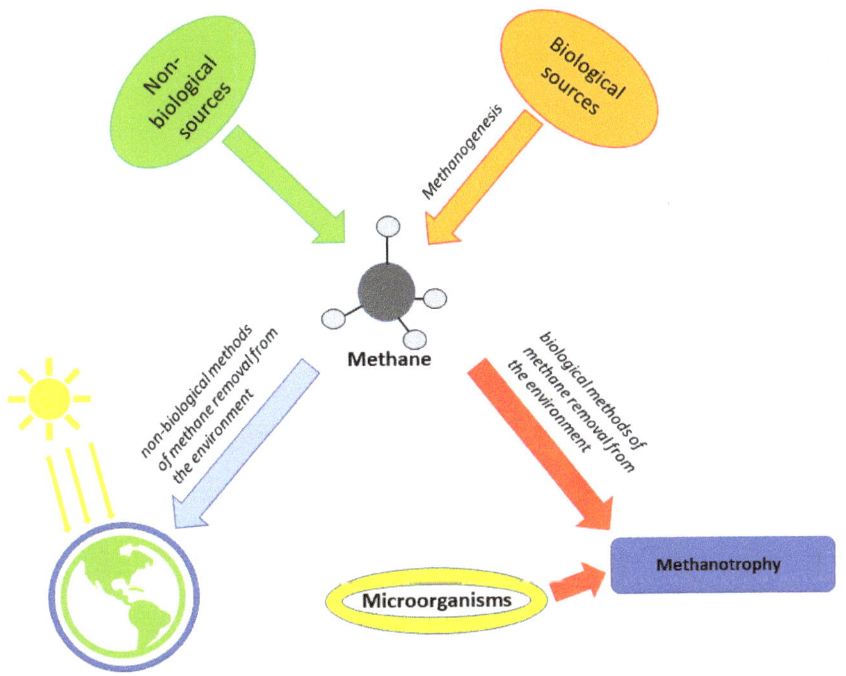

Figure 1. The process of formation and decomposition of methane via biological and non-biological means.

3. Production of Ethanol and Butanol

One of the most immediate and vital applications of biomass is the fermentation of biomass and the production of bioethanol, the most common renewable fuel today. Various microorganisms can be involved in the fermentation process for bioethanol production

(Table 1) [35]. Bioethanol is the leading liquid biofuel, with a global production of 29 billion gallons in 2019. The top producers are the United States and Brazil, which account for 84% of global production [36].

Table 1. Bioethanol yield from different microorganisms.

Microorganism		Substrate	%	Ref.
Bacteria	*Zymomonas mobilis*	Corn steep liquor	98%	[37]
Yeasts	*Saccharomyces cerevisiae*	Barley straw	82%	[38]
	S. cerevisiae	Coffee grounds	87.2%	[39]
	Kluyveromyces marxianus SUB-80-S	Poplar and eucalyptus biomass	50–72%	[40]
	K. marxianus IMB3	Wild millet	86%	[41]
	Candida shehatae	SX media (3% xylose and 0.67% YNB (nitrogen base without yeast) without amino acid)	71.6%	[42]
Mold	*Fusarium oxysporum*	Beer trope	60% of the theoretical yield	[43]

Yeasts can produce ethanol via the direct decarboxylation of pyruvate formed via biomass oxidation [44]. At the same time, bacteria (*E.coli*) with coenzyme A activate the acyl group during the decarboxylation of pyruvate and convert it to ethanol (reduction) (Figure 2) [45].

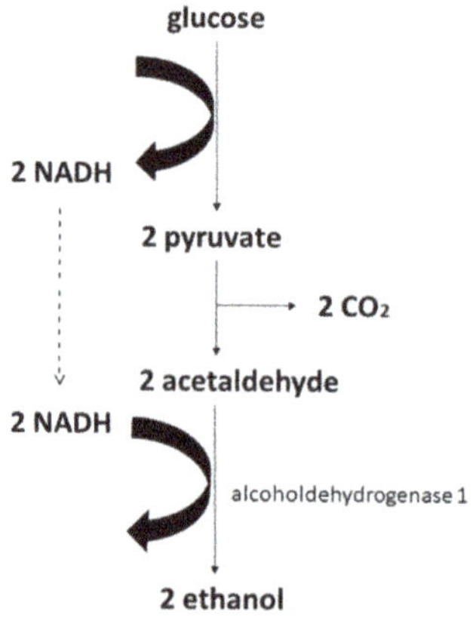

Figure 2. Ethanol formation from glucose in yeasts.

Ethanol production via direct decarboxylation (*Saccharomyces cerevisiae*) is more efficient than that of *E. coli*. Butanol can be commercially similarly produced from sugar–starch biomass [46] (Figure 3). In addition, it can be made from so-called polysaccharides; from acetone–butanol fermentation (anaerobic process); and from and acetone, CO_2, and hydrogen (*Clostridium acetobutylicum*) [47].

Some microorganisms are used in the gasification process. They can partially oxidize biomass by means of air or oxygen at about 800–1000 °C, whereby microalgae biomass is converted into a gaseous product—syngas—which means that microalgae are a suitable raw material for gasification [48,49]. Syngas is a mixture of hydrogen, CO, CO_2, methane, and nitrogen [50]. It is used as a turbine fuel but primarily as a feedstock for producing methanol, ethanol and synthetic hydrocarbons, butanol, methane, butyric, and acetic acid [51]. Microalgal bio-oils also contain metals (Fe, Mg, Ni, Zn), which can be removed via heat treatment [52]. Microbial oil can be obtained from microalgae, yeasts, and molds, and triglycerides (oleic, linoleic, and palmitic acid) can also be obtained [53]. These raw materials can be used to produce biodiesel. Some rare yeast and fungus species can yield various substrates (lignocellulosic biomass, industrial waste glycerol, whey fat, and molasses) [54]. Some microorganisms can grow in sewage sludge and wastewater [55].

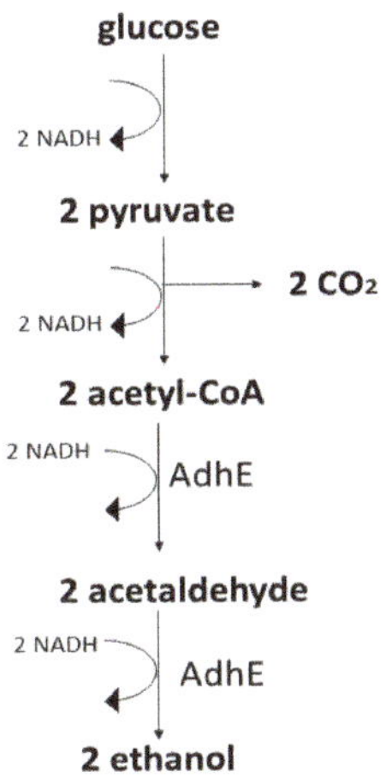

Figure 3. Ethanol formation from glucose in bacteria. AdhE—bifunctional CoA-dependent ethanol/aldehyde dehydrogenase.

Main Metabolic Pathways for Ethanol in the Most Prominent Microorganisms

The most common microorganisms used in ethanol production are yeast *Saccharomyces cerevisiae* and bacterium *Escherichia coli*. They possess specific metabolic pathways and different types of catalytic enzymes for producing biofuels [56]. For example, *Saccharomyces cerevisiae* produces ethanol via the direct decarboxylation of pyruvate, while *E. coli* activates the acyl group during the decarboxylation of CoA, which is then reduced to ethanol [57,58].

A more efficient route in the production of ethanol is considered to be a route without the use of CoA. This pathway can also be achieved in other microorganisms by means of genetic engineering techniques. However, this method represents a significant challenge because synthesizing an artificial metabolic pathway requires highly sophisticated tools to control mRNA and protein levels for the synthetic pathway to be functional [59].

Zymomonas mobilis is another well-studied strain with a known genome that produces ethanol. It is important to note that it has a significantly higher ethanol yield than *Saccharomyces cerevisiae*. Furthermore, ethanol yield considerably increases after genetic manipulation, i.e., after introducing genes encoding catabolic enzymes mannose and xylose, so the theoretical yield of ethanol within 72 h can reach 89.8% [60].

The costs of bioethanol production from lignocellulosic raw materials are high, so for commercial reasons, this production method is still not used [61,62]. However, the production of second-generation bioethanol is in the development phase. For this production, microbial strains that can produce ethanol from glucose and xylose, the main fermentable sugars, are necessary [63].

4. Biodiesel Production

Biodiesel is the first alternative fuel and, at the same time, the most widespread biofuel in Europe. It is obtained from oil and fat through the transesterification process and is very similar in composition to mineral diesel fuel [64]. Biodiesel production is an inherently complex system that requires optimization, keeping profitability and environmental sustainability in mind [65]. Recent studies suggest an unquestionable benefit for the environment. Economic profitability depends on feedstock sources and choice, technological process and production capacities, and transport to the consumer [66,67]. Sources of third-generation feedstock, microalgae, have an unquestionable advantage over other sources [68].

Regarding the need for increasing the amounts of energy (due to direct use in internal-combustion engines or the production of heat and electricity), guided by the imperative to reduce CO_2, modern science and technology are giving the first positive results [69]. These are the so-called biofuels produced as a product of microorganisms from biological materials and even from organic waste biomass [70]. In addition to the already considered bioenergy agents (methane, methanol, ethanol), we especially highlight the importance of biodiesel production. The European Union is at the forefront of applying such technologies in biodiesel consumption, which is about 105 billion liters—about 53% of the total world biodiesel consumption [71].

One of the most promising biodiesel production methods is the production of lipids, triglycerides, and other oil molecules from rapeseed, soybeans, and some other specialized plants, which incorporate this otherwise undesirable greenhouse gas into lipid molecules via photosynthesis [72]. At the same time, these plants have different types of fixatives (nitrofixatives, *Azotobacter* sp. and *Azospirillum* sp.; phosphofixatives, *Acinetobacter junii* and *Pseudomonas fluorescens*) in the soil, from which they benefit via the rhizome system [73]. Rhizome nodules fix nitrogen and phosphorus, conducive to plant growth and the formation of products (lipids, triglycerides) [74].

Biodiesel can then be produced directly from the vegetable oils of the above-mentioned plants [75] (Figure 4). Likewise, a biodiesel composition similar to vegetable oils can be obtained via the transesterification of *Rhodotorula glutinis* oil, and *Yarrowia lipolytica* can be used to produce microbial oil [76]. Genetic engineering can increase the tolerance of lipids or fatty acids in microorganisms, and some types of bacteria and fungi can tolerate higher amounts of accumulated triacylglycerol [77]. In terms of efficiency, bacteria show significantly more favorable properties than fungi (higher growth rate, easier maintenance, and the possibility of genetic adaptation) [78], because it is known that bacteria are subject to genetic improvement and possess the property of rapid growth, which can be used for the highly efficient production of microbial oil. Because of this, even *Escherichia coli*, under certain circumstances, can directly produce biodiesel in the form of fatty acid esters and can ferment biomass from renewable carbon sources (specialized or waste biomass) [79,80]. Thus, microbial oils are becoming a very likely source (raw material) in biodiesel production mainly due to the faster growth of microorganisms, easier maintenance, and the possibility of genetic adaptations [79,81].

The possibilities of direct electricity production by means of biochemical treatment systems are also being studied very intensively, and microbial fuel cell (MFC) devices convert chemical energy into electricity (without the Carnot cycle) from biodegradable raw materials and even from wastewater [82,83]. Furthermore, potentially electrogenic bacteria can be identified in the MFC device; in it, microorganisms feed on organic compounds, releasing electrons to the electrode, thus generating electricity [84]. In summary, it is clear that modern science and technology have efficient responses to increasing environmental pollution (especially greenhouse gases—CO_2 and methane) [85,86].

Main Metabolic Pathways for Biodiesel in the Most Prominent Microorganisms

Biodiesel belongs to the group of renewable energy sources and represents an ideal replacement for petroleum-based diesel fuels. It is produced using transesterifying fatty acid sources with short-chain alcohols, giving monoalkyl esters of long-chain fatty acids [87]. For this process to be realized, it is essential that microorganisms can produce fatty acids and short-chain alcohols that are available for transesterification and that they possess acyltransferases with more significant activity for short-chain alcohols. Genetic engineering significantly transforms microorganisms into forms that produce biodiesel with high efficiency [88].

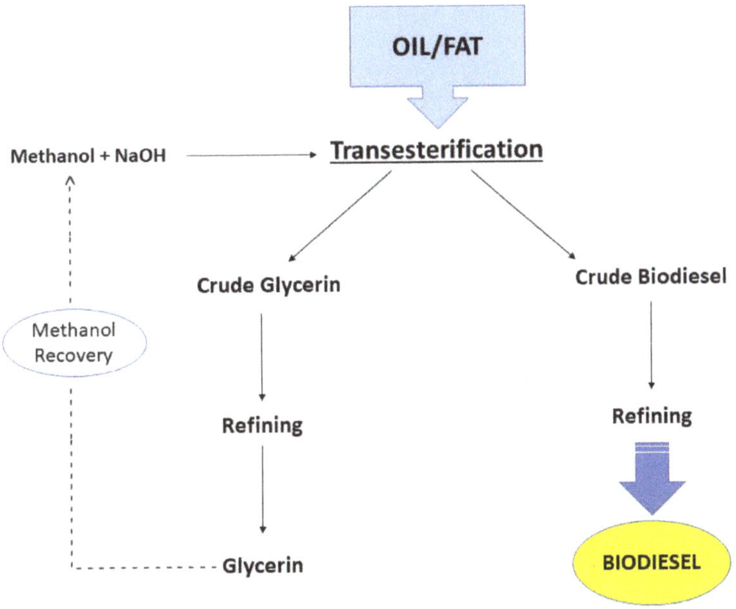

Figure 4. Biodiesel production pathway.

Microalgae attract particular attention as a raw material for biodiesel production (Figure 5). Namely, it is an economically profitable raw material for oil. They are characterized by easy cultivation, diverse metabolic activities, and a high content of fatty acids [89,90]. The results of the research study by Huang et al. suggest that the problems with fossil energy could be alleviated by the additional processing of microalgae residues after the lipid extraction process, using the pyrolysis process [91]. A moderately fast pyrolysis temperature (~700 °C) is essential for higher bio-oil production and a lower limit of pollutants [92].

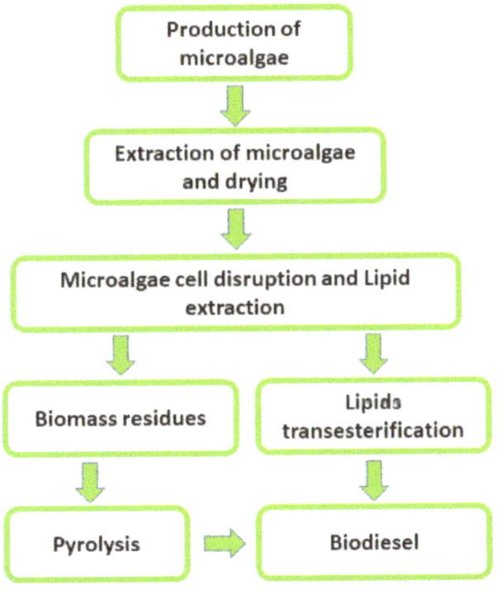

Figure 5. Pathways of processing microalgal lipids into biodiesel. Modified according to [93]. Reprinted with permission from ref. [93]. 2021 © Portal hrvatskih znanstvenih i stručnih časopisa—Hrčak.

Among the numerous autotrophic algae are *Botryococcus braunii, Chlorella vulgaris, Crypthecodinum cohnii, Dunaliella primolecta, Navicula pelliculosa, Scenedsmus acutus, Crypthecodinium cohnii, Monallanthus primolecta, Monallanthusocornussel olia,* and *Teallanthus chloridea sul*. The oil content in microalgae varies from 1 to 70% (Table 2).

Table 2. Lipid yields from different microalgae.

Microalgae	Substrate	Mass Proportion of Oil (%)	Ref.
Botryococcus braunii	dry matter	25–75	[94]
Chlorella sp.	dry matter	28–32	[94]
Crypthecodinium cohnii	dry matter	20	[94]
Nannochloropsis sp.	dry matter	31–68	[94]
Phaeodactylum tricornutum	dry matter	20–30	[94]
Schizochytrium sp.	dry matter	50–77	[94]
Cylindrotheca sp.	dry matter	16–37	[95]
Nitzschia sp.	dry matter	45–47	[95]

The cultivation conditions include the composition of the nutrient medium, pH, temperature, the efficiency of light delivery to the cells of microalgae, the intensity and wavelength of light, the speed and method of mixing the nutrient medium in the bioreactor, and the ratio of the concentration of dissolved oxygen and CO_2 in the nutrient medium [96,97]. Accordingly, the biotechnological production of microalgal lipids is determined by the physiological potential of the microalgae (Table 3), that is, the conditions and procedure of conducting the bioprocess in the bioreactor system [98]. Therefore, it is considered that microalgae are an inevitable trend in the development of future biodiesel, provided that for the industrial production of biomass, i.e., lipids of microalgae, the optimal conditions, and procedures for running the bioprocess in the bioreactor system are chosen so that an ecologically and economically sustainable bioprocess of the production of biomass, i.e., lipids of microalgae, is obtained for the production of biofuels (biodiesel) [99].

Numerous studies have established that residual biomass contains carbohydrates from which ethanol can be produced through fermentation. It is the specific share and yield during biodiesel production [100].

Table 3. Yields of lipids from different microorganisms.

Microorganism		Substrate	Mass Proportion of Oil (%)	Ref.
Bacteria	*Acinetobacter calcoaceticus*	dry matter	27–38	[95]
	Arthrobacter sp.	dry matter	>40	[95]
	Bacillus alcalophilus	dry matter	18–24	[95]
	Rhodococcus opacus	dry matter	24–24	[95]
Yeasts	*Candida curvata*	dry matter	58	[95]
	Cryptococcus albidus	dry matter	65	[95]
	Lipomyces starkeyi	dry matter	64	[95]
	Rhodotorula glutinis	dry matter	72	[95]
	Trichosporon oleaginosus	Lignocellulosic substrate	80	[101]
Molds	*Aspergillus oryzae*	dry matter	57	[95]
	Humicola lanuginosa	dry matter	75	[95]
	Mortierella isabellina	dry matter	86	[95]
	Mortierella vinacea	dry matter	66	[95]

5. Hydrogen

Biohydrogen represents an essential factor in solving global energy problems [102]. It is a substitute for primary fossil fuels and their derivatives. Its main advantage is that the product of its combustion with oxygen is water, and not CO and CO_2, which are greenhouse gases [103]. Therefore, it is expected to play a crucial role in the future energy infrastructure. Biohydrogen has a gross energy or heat value of 142 MJ/kg, which is significantly higher than those of natural gas or crude oil, whose values are 52 and 45 MJ/kg [102], while petrol has a value of 44 MJ/kg [103]. The global demand for hydrogen is predicted to increase from 70 million tons in 2019 to 120 million tons by 2024. Hydrogen development should also fulfill the seventh United Nations goal on affordable and clean energy [104].

Hydrogen is the first atom from which everything in the universe was created. The energy produced by fusion reactions (stars) was sufficient for forming all other elements, which created the conditions for evolution and the creation of life [105–108]. Traces of the life of the most primitive microorganisms (recognized today through the simplest microorganisms such as prions and viruses) used oxidation–reduction processes in which, under anaerobic conditions, hydride was oxidized to sulfides, nitrides, and phosphides and generated enough energy to start the process (which is still insufficiently explained) that could constitute life [109].

Modern life requires unimaginable amounts of energy [110,111], and fire is the simplest form of clean energy. In the same way, hydrogen is slowly and rapidly introduced into our daily lives [112]. Moreover, all mineral fuels provide energy by burning hydrogen (wood, coal, oil) [113], while nuclear processes, such as fusion, use hydrogen as fuel [114,115]. However, other products of hydrogen combustion (organic hydrocarbons, oxides) are today putting in question the continuation of life as we know it [116]. In addition, so-called greenhouse gases threaten civilization to such an extent, arousing the necessity for the creation of new mechanisms for increasing amounts of needed energy [117].

Microbial universality through life-saving adaptations has created natural reactors for producing biofuels and future fuels, i.e., hydrogen [118,119]. Moreover, they can produce it (extract it) from hydrocarbons, thus launching a more certain perspective for civilization [120]. The development and selection of microbial biorefineries are the result of the creation of syntrophic communities (a symbiotic form of joint metabolism) [121,122]. One example of syntrophy is methanogenic communities in which reducing equivalents, e.g., hydrogen and formate, are transferred among syntrophic partners [20,123]. In the coal seam, the anaerobic fermentation of organic matter includes hydrolysis, acidogenesis, acetogenesis, and methanogenesis [22,124]. In the first phase, bacteria hydrolyze macromolecules into simple sugars, amino acids, and fatty acids [125,126]. Then, acidogenic bacteria decompose them into propionic acid, butyric acid, and alcohol [127,128]. Microorganisms capable of acetogenesis then convert them to acetic acid, hydrogen, and CO_2, and ultimately methanogenic microorganisms can produce methane [129]. Thus, the production of carbon-based biogas significantly improves protection against the formation of unwanted gases (primarily sulfur). As noted, coal conversion into methane requires the synergistic action of three groups of microorganisms (syntrophic community) [130,131]. They are mainly from genera *Clostridium*, *Enterobacter*, *Klebsiella*, and *Citrobacter* [118,132]. Methanogenic bacteria, based on mcrA and the phylogeny of ribosomal genes, are classified into seven orders, among which *Methanopyrales*, *Methanococcales*, *Methanobacteriales*, *Methanomicrobiales*, and *Methanocellales* comprise hydrogenotrophic methanogens. At the same time, the *Methanomasiliiicoccales* guild is obligated to perform methylogenotropic respiration [20].

A more complex form of microorganisms can be considered as the factory of electrochemical devices for producing electricity and biohydrogen [133,134]. The mechanism of biohydrogen generation can start from wastewater and some other types of organic matter [135–137]. By creating an electro-biofilm, the mechanism of electron transfer to conductive surfaces is triggered [138,139]. These electrons can then be used to produce electricity and hydrogen [140], similar to the so-called electrochemical cells (BECs), molecular machines that transfer electrons from a microbial membrane [141,142]. Microalgae can further produce hydrogen via the reaction of the photolysis of water, i.e., by converting water into hydrogen ions and oxygen, after which they convert these hydrogen ions into hydrogen, all under anaerobic conditions [143,144]. Likewise, photosynthesis can produce hydrogen through two stages [145]. The first stage is created via photosynthesis, in which acid production is separated from hydrogen production. In the second stage, microalgae are denied access to sulfur and are forced to change their cellular metabolism for survival by forming starch, from which they produce hydrogen [146,147]. As a result, the amount of hydrogen produced gradually reduces, but this process does not create undesirable and harmful by-products [148]. Biological hydrogen can also be made via the fermentation of lignocellulosic raw materials and cotton-sludge hydrolates. It can be produced by bacteria isolated from higher organisms (such as fish and termites) [149]. These are predominantly *Enterobacter*, *Klebsiella*, *Clostridium*, and *Citrobacter* [150]. Several metabolic pathways exist for biohydrogen production, and anaerobic fermentation is the most efficient and rapid way to produce it [151,152] (Table 4).

Table 4. Methods for hydrogen production and their efficiency.

Feedstock	Production Method	Energy Efficiency (%)	Ref.
Biomass	Bioelectrolysis (microbial electrolysis)	70–80	[103]
	Biothermolysis (co-fermentation hydrolysis)	35–45	[103]
	Thermolysis (pyrolysis)	35–50	[153]
	Thermolysis (gasification)	35–50	[103]
	Thermolysis (partial oxidation)	60–75	[103]
	Biophotolysis (photofermentation)	<1	[153]
Microalgae	Biophotolysis (photofermentation)	<1	[154]
Microorganism	Biolysis (dark fermentation)	60–80	[155]

The production of hydrogen using distinct methods and using various feedstock implies different capital investment costs and the costs of hydrogen production itself, as shown in Table 5. The level of production-technology innovation, the accessibility of existing infrastructure, and the feedstock cost significantly impact this cost.

Table 5. Hydrogen production costs by different methods.

Process	Energy Source	Feedstock	Capital Cost (M USD)	Hydrogen Cost (USD/kg)	Ref.
Biomass pyrolysis	Generated steam	Biomass	53.4–3.1	1.25–2.20	[102]
Biomass gasification	Generated steam	Biomass	149.3–6.4	1.77–2.05	[102]
Direct biophotolysis	Solar	Water + algae	50 USD/m^2	2.13	[102]
Indirect biophotolysis	Solar	Water + algae	135 USD/m^2	1.42	[102]
Dark fermentation	-	Biomass	-	2.57	[102]
Photo-fermentation	Solar	Biomass	-	2.83	[102]

5.1. Main Metabolic Pathways for Hydrogen in the Most Prominent Microorganisms

For bacteria that participate in the production of biohydrogen, such as *Geobacter sulfurreducens* and *Shewanella oneidensis*, it is significant that they possess specific molecular mechanisms that facilitate the transfer of electrons from the outer membrane of the microorganism to visible surfaces, after which this feature can be used to produce biohydrogen and, accordingly, bioelectric energy [156]. Thus, such bioelectrochemical cells (BECs) represent an exceptional significance in the potential production of bioenergy from wastewater and organic biomass [157].

Microbial electrolysis cells (MECs) and microbial fuel cells (MFCs) are primarily used to produce biohydrogen and bioelectricity. Based on a biological perspective, both species function in a similar manner, and accordingly, common microorganisms can be used for bioenergy production. These microorganisms are called electrogenic or exoelectrogenic [158]. However, it is essential to note that the output energy from MECs and MFCs is insufficient for practical application and commercialization [159].

5.2. Hydrogen Production via Photofermentation with Photofermenting Bacteria

Biological methods for producing hydrogen are in increasing focus because they can use renewable raw materials such as the remains of plant biomass, organic waste, and sunlight [160]. There are two main ways in which microorganisms produce hydrogen:

photosynthesis and fermentation. The process of photosynthesis is dependent on light and includes direct biophotolysis, indirect biophotolysis, and photofermentation. On the other hand, dark photofermentation is essentially anaerobic fermentation, and this process is not dependent on light [161]. Microorganisms produce hydrogen at room temperature and pressure, significantly reducing the need for additional energy. Photosynthetic hydrogen is produced by microorganisms, such as photosynthetic bacteria, algae, and cyanobacteria. Fermentative hydrogen production is carried out by fermentative microorganisms, such as strict anaerobes, e.g., strains of *Clostridium* sp. thermophilic rumen bacteria and methanogenic facultative anaerobes or mixed cultures [162].

Based on the available data, the conclusion is that fermentative hydrogen production has more potential for practical application than photosynthetic hydrogen production. Hydrogen production via fermentation is currently more profitable in energy gain than photosynthesis [163]. This is supported by the facts that fermentation bacteria have fast growth and that oxygen does not affect the anaerobic process to a large extent; they do not need light; and they have a higher level of hydrogen production. It is also important to note that there is a large selection of substrates and that the methods (techniques) of setting up bioreactors are simple [164–166]. The production of hydrogen by means of dark fermentation can be carried out under different thermodynamic conditions so that it can be carried out under mesophilic, thermophilic, and hyperthermophilic conditions. However, the production degree is still more favorable at higher temperatures [167].

It is known that there is a significant difference between theoretical and practical energy yield, which can be seen from the following example:

Theoretical yield—12 moles of H2 can be produced from each mole of glucose.

Practical yield—a maximum of 3.8 moles of H2 can be produced from each mole of glucose.

Yields can be increased by combining two metabolic pathways and using compatible mixed bacterial cultures [79].

6. Cyanobacteria

Hydrogen from bacteria is also produced via photosynthesis, because bacteria do not consume the created hydrogen but rather retain it [168]. Namely, the process of photosynthesis of cyanobacteria begins with the fixation of solar energy due to water splitting, so this energy is stored by hydrogenase in sugars [169,170]. Electrons undergo a series of reactions produced by the ATP energy carrier and the reduction equivalents of NADPH, which are required for CO_2 fixation or sugar production [171]. These electrons and protons (energy) can be diverted to primarily produce hydrogen, that is, by linking the photosystem to hydrogenase. Modified cyanobacteria produce less sugar than unmodified ones at the expense of greater hydrogen production [169,172]. This fusion can function indefinitely. It is also transmitted during cell division, and since the oxygen created is blocking the processes, metabolism is transferred to anoxygenic photosynthesis. All these processes offer a new concept for the production of green hydrogen [173].

In addition to genetic modifications that create unique and desirable traits in some bacteria, some bacteria are also subject to natural phenomena that can still produce everything necessary for growth and reproduction under unfavorable conditions [174]. Extremophiles, which live where there is no liquid water and no solar energy, can use hydrogen from the air as a fuel to create water for their needs, specifically, by capturing hydrogen and oxygen from the air [175]. Thus, hydrogen drives chemosynthesis with enzyme RuBisCo, which otherwise uses sunlight to capture CO_2 [176].

Thus, about 400 species of evolutionarily adapted microorganisms live without free water and can use inorganic energy sources (hydrogen, CO) as sources that drive metabolism (chemosynthesis) [177,178]. Some marine bacteria work similarly. Seawater has sufficient amounts of hydrogen and CO, which can be sources of energy for bacteria from families *Rhodobacteraceae*, *Flavobacteriaceae*, and *Sfhingomonodaceae* [179,180]. A representative of this group of ultramicrobacteria, *Sfinopyxis alascensis*, grows mycotrophically on hydrogen by expressing NiFe hydrogenase [181,182].

Thus, one of the possible solutions for obtaining new renewable energy sources is hydrogen (H_2), which could be obtained via photosynthesis, which requires sunlight, water, and cyanobacteria. However, in doing so, it is necessary to consider the development of cost-effective production technologies [183].

7. Future Perspective

The biggest challenge in employing microorganisms to produce biofuels is producing a considerable amount of fuel more cheaply and efficiently than traditional fossil fuels. To replace petrol with bioethanol should be cheaper, which is very demanding to cover the necessary daily quantities. For example, in the USA, about 19 million barrels of petrol are consumed daily, and producing this amount on an industrial scale is challenging. Therefore, future biofuel productivity should be prioritized to increase microbial biofuel's acceptability [184]. Some of the most common advantages and disadvantages of the biofuel production process are shown in Table 6. Due to the politically increasingly unstable oil market, many countries are turning to renewable energy sources. Biofuels (bioethanol and biodiesel) are a sustainable energy source due to their high chemical similarity, carbon neutrality, and comparable energy content, and microorganisms are crucial to their synthesis. Depending on the feedstock's evolutionary hierarchy and the manufacturing technique, biofuels are divided into four generations. Biofuel production advances with each generation toward achieving sustainability and financial success in energy production. They are created to most effectively address the issues of the energy crisis, pollution, global warming, and waste management. Microorganisms used to be mere biomass decomposers, but because of advancements in biotechnology, gene editing, and synthetic biology, they now produce biofuel [185].

Table 6. Advantages and disadvantages of biofuel production.

Biofuel	Advantage	Disadvantage	Ref.
Bioethanol	Renewable sources; low cost; algae can rapidly absorb carbon dioxide, accumulate high concentrations of lipid and carbohydrates, be easily cultivated, and require less land than terrestrial plants	High costs of lignocellulosic feedstock; inputs of energy and water; requirements for large volume bioreactors and distillation columns; generation of large volumes of waste or low-value coproducts	[35]
Biodiesel	Renewable, sustainable, environmentally friendly, and biodegradable sources; low cost and high conversion rate; ideal replacement for petrol; reducing greenhouse gases; less harmful carbon emission; ecologically and economically sustainable bioprocess; use of existing engines without changes	High energy consumption; environmentally unfriendly processing including chemical catalysts, high cost, and limited supply of feedstocks; complex production processes; downstream technology; simultaneously produced waste; production is dependent on large quantities of water and oil	[79]
Biohydrogen	Renewable sources; cleanliness; low greenhouse gas emissions; biohydrogen has the advantage of being able to use a wide range of substrates to produce hydrogen; the first stage of the waste treatment and valorization process uses mild temperatures and does not need the external addition of metal catalysts; clear environmental benefits	Low performance; high capital cost investment; expensive materials; complex maintenance; variable energy loss; decreasing hydrogen production with the increase in the volume of the reactor; hydrogen storage; global-warming potential; land use; terrestrial- and freshwater-ecotoxicity potential; ecotoxicity potential; human-toxicity potential	[103,186,187]

The production processes for the second and third generations of biofuels are quite complex, which results in high energy costs. Additionally, the feedstock for the third generation has very complex requirements for structure, storage, and content. The expenses mentioned above explain the capital intensity of manufacturing second- and third-generation fuels, respectively, and the decision by most nations to choose first-generation biofuels [188]. The microbial lipids produced by microorganisms are the ideal feedstock for biodiesel synthesis due to their high production rate and independence from environmental conditions such as soil and climate. In the study by Wang et al., several ideas for generating biodiesel using microbes from inexpensive lignocellulosic biomass are addressed [79]. A country that intends to develop alternative fuels must have enough land to prevent a food shortage and enact stringent controls limiting the proportions of raw materials provided to the food and fuel markets. The ratios in which a blend of biodiesel with diesel and bioethanol with petrol can be created must also be governed by state standards. Developing second- and third-generation biofuel production, which uses significantly less land and is mostly not arable, despite having a higher capital during production, needs help and subsidies [188].

8. Conclusions

Numerous studies are being conducted based on the growing need to find new renewable energy sources that could replace gas and oil and reduce harmful effects to the ecosystem. As a result, scientists are increasingly turning to biofuels based on microorganisms. In doing so, it is necessary to use increasingly advanced genetic engineering technologies. The imperative of preserving and surviving civilization can be met—using enough species of microorganisms that we can find in our immediate environment. In this way, microorganisms are ready to lead us into new human–environmental relationships to be our companions in a more confident and secure future.

Author Contributions: Conceptualization, J.T. and D.T.; writing of the manuscript, J.T., D.T., A.M. and I.Š.; updating of the text, J.T. and I.Š.; literature searches, J.T., D.T., A.M. and I.Š.; figure drawings, A.M.; critical reviewing of the manuscript, J.T. and I.Š.; organization and editing of the manuscript, J.T. and I.Š. All authors have read and agreed to the published version of the manuscript.

Funding: This research study was funded by a grant from the Croatian Ministry of Science and Education and dedicated to multi-year institutional financing of scientific activity at Josip Juraj Strossmayer University of Osijek, Faculty of Dental Medicine and Health, Osijek, Croatia, grant number IP11-FDMZ-2021.

Institutional Review Board Statement: Not applicable.

Informed Consent Statement: Not applicable.

Data Availability Statement: Not applicable.

Conflicts of Interest: The authors declare no conflict of interest. The funders had no role in the design of the study; in the collection, analyses, or interpretation of data; in the writing of the manuscript; or in the decision to publish the results.

References

1. Gupta, A.; Gupta, R.; Singh, R.L. Microbes and Environment. In *Principles and Applications of Environmental Biotechnology for a Sustainable Future*; Springer: Singapore, 2017; pp. 43–84. [CrossRef]
2. Hoppe, T.; Butenko, A.; Heldeweg, M. Innovation in the European Energy Sector and Regulatory Responses to It: Guest Editorial Note. *Sustainability* **2018**, *10*, 416. [CrossRef]
3. Agyekum, E.B.; Nutakor, C.; Agwa, A.M.; Kamel, S. A Critical Review of Renewable Hydrogen Production Methods: Factors Affecting Their Scale-Up and Its Role in Future Energy Generation. *Membranes* **2022**, *12*, 173. [CrossRef] [PubMed]
4. Teng, Y.; Xu, Y.; Wang, X.; Christie, P. Function of Biohydrogen Metabolism and Related Microbial Communities in Environmental Bioremediation. *Front. Microbiol.* **2019**, *10*, 106. [CrossRef]
5. Fabris, M.; Abbriano, R.; Pernice, M.; Sutherland, D.L.; Commault, A.S.; Hall, C.C.; Labeeuw, L.; McCauley, J.I.; Kuzhiumparambil, U.; Ray, P.; et al. Emerging Technologies in Algal Biotechnology: Toward the Establishment of a Sustainable, Algae-Based Bioeconomy. *Front. Plant Sci.* **2020**, *11*, 279. [CrossRef] [PubMed]

6. Sheldon, R.A. Biocatalysis and biomass conversion: Enabling a circular economy. *Philos. Trans. R. Soc. A* **2020**, *378*, 20190274. [CrossRef] [PubMed]
7. Óhaiseadha, C.; Quinn, G.; Connolly, R.; Connolly, M.; Soon, W. Energy and Climate Policy—An Evaluation of Global Climate Change Expenditure 2011–2018. *Energies* **2020**, *13*, 4839. [CrossRef]
8. Alzate, C.A.C.; Serna-Loaiza, S.; Ortiz-Sanchez, M. Sustainable Biorefineries: What was Learned from the Design, Analysis and Implementation. *J. Sustain. Dev. Energy Water Environ. Syst.* **2020**, *8*, 88–117. [CrossRef]
9. Waldrop, M.M. News Feature: Microbes for better sewage treatment. *Proc. Natl. Acad. Sci. USA* **2021**, *118*, e2112863118. [CrossRef]
10. Zetterholm, J.; Bryngemark, E.; Ahlström, J.; Söderholm, P.; Harvey, S.; Wetterlund, E. Economic Evaluation of Large-Scale Biorefinery Deployment: A Framework Integrating Dynamic Biomass Market and Techno-Economic Models. *Sustainability* **2020**, *12*, 7126. [CrossRef]
11. Janssen, P.; Lambreva, M.D.; Plumeré, N.; Bartolucci, C.; Antonacci, A.; Buonasera, K.; Frese, R.; Scognamiglio, V.; Rea, G. Photosynthesis at the forefront of a sustainable life. *Front. Chem.* **2014**, *2*, 36. [CrossRef]
12. Li, G.; Hu, R.; Wang, N.; Yang, T.; Xu, F.; Li, J.; Wu, J.; Huang, Z.; Pan, M.; Lyu, T. Cultivation of microalgae in adjusted wastewater to enhance biofuel production and reduce environmental impact: Pyrolysis performances and life cycle assessment. *J. Clean. Prod.* **2022**, *355*, 131768. [CrossRef]
13. Li, G.; Zhang, J.; Li, H.; Hu, R.; Yao, X.; Liu, Y.; Zhou, Y.; Lyu, T. Towards high-quality biodiesel production from microalgae using original and anaerobically-digested livestock wastewater. *Chemosphere* **2021**, *273*, 128578. [CrossRef]
14. Rather, R.A.; Wani, A.W.; Mumtaz, S.; Padder, S.A.; Khan, A.H.; Almohana, A.I.; Almojil, S.F.; Alam, S.S.; Baba, T.R. Bioenergy: A foundation to environmental sustainability in a changing global climate scenario. *J. King Saud. Univ. Sci.* **2022**, *34*, 101734. [CrossRef]
15. Directive (EU) 2018/2001 of the European Parliament and of the Council of 11 December 2018 on the Promotion of the Use of Energy from Renewable Sources (Text with EEA Relevance). Available online: https://eur-lex.europa.eu/eli/dir/2018/2001/oj (accessed on 22 August 2022).
16. Vučurović, D.; Bajić, B.; Vučurović, V.; Jevtić-Mučibabić, R.; Dodić, S. Bioethanol Production from Spent Sugar Beet Pulp—Process Modeling and Cost Analysis. *Fermentation* **2022**, *8*, 114. [CrossRef]
17. Nordahl, S.L.; Devkota, J.P.; Amirebrahimi, J.; Smith, S.J.; Breunig, H.M.; Preble, C.V.; Satchwell, A.J.; Jin, L.; Brown, N.J.; Kirchstetter, T.W.; et al. Life-Cycle Greenhouse Gas Emissions and Human Health Trade-Offs of Organic Waste Management Strategies. *Environ. Sci. Technol.* **2020**, *54*, 9200–9209. [CrossRef]
18. Tapio, I.; Snelling, T.J.; Strozzi, F.; Wallace, R.J. The ruminal microbiome associated with methane emissions from ruminant livestock. *J. Anim. Sci. Biotechnol.* **2017**, *8*, 7. [CrossRef]
19. Conrad, R. The global methane cycle: Recent advances in understanding the microbial processes involved. *Environ. Microbiol. Rep.* **2009**, *1*, 285–292. [CrossRef]
20. Buan, N.R. Methanogens: Pushing the boundaries of biology. *Emerg. Top. Life Sci.* **2018**, *2*, 629–646. [CrossRef]
21. Hook, S.E.; Wright, A.-D.G.; McBride, B.W. Methanogens: Methane Producers of the Rumen and Mitigation Strategies. *Archaea* **2010**, *2010*, 945785. [CrossRef]
22. Srivastava, P.; Marjo, C.; Gerami, A.; Jones, Z.; Rahman, S. Surface Analysis of Coal Indicating Neutral Red Enhances the Precursor Steps of Methanogenesis. *Front. Microbiol.* **2020**, *11*. [CrossRef]
23. Ney, B.; Ahmed, F.H.; Carere, C.R.; Biswas, A.; Warden, A.C.; E Morales, S.; Pandey, G.; Watt, S.J.; Oakeshott, J.G.; Taylor, M.C.; et al. The methanogenic redox cofactor F420 is widely synthesized by aerobic soil bacteria. *ISME J.* **2017**, *11*, 125–137. [CrossRef]
24. Grinter, R.; Greening, C. Cofactor F420: An expanded view of its distribution, biosynthesis and roles in bacteria and archaea. *FEMS Microbiol. Rev.* **2021**, *45*, fuab021. [CrossRef] [PubMed]
25. Cavicchioli, R.; Ripple, W.J.; Timmis, K.N.; Azam, F.; Bakken, L.R.; Baylis, M.; Behrenfeld, M.J.; Boetius, A.; Boyd, P.W.; Classen, A.T.; et al. Scientists' warning to humanity: Microorganisms and climate change. *Nat. Rev. Microbiol.* **2019**, *17*, 569–586. [CrossRef] [PubMed]
26. Wang, Y.; Hu, P.; Yang, J.; Zhu, Y.-A.; Chen, D. C–H bond activation in light alkanes: A theoretical perspective. *Chem. Soc. Rev.* **2021**, *50*, 4299–4358. [CrossRef] [PubMed]
27. Meloni, E.; Martino, M.; Palma, V. A Short Review on Ni Based Catalysts and Related Engineering Issues for Methane Steam Reforming. *Catalysts* **2020**, *10*, 352. [CrossRef]
28. Wang, S.; Dames, E.E.; Davidson, D.F.; Hanson, R.K. Reaction Rate Constant of CH_2O + H = HCO + H_2 Revisited: A Combined Study of Direct Shock Tube Measurement and Transition State Theory Calculation. *J. Phys. Chem. A* **2014**, *118*, 10201–10209. [CrossRef] [PubMed]
29. Song, H.; Meng, X.; Wang, S.; Zhou, W.; Wang, X.; Kako, T.; Ye, J. Direct and Selective Photocatalytic Oxidation of CH_4 to Oxygenates with O_2 on Cocatalysts/ZnO at Room Temperature in Water. *J. Am. Chem. Soc.* **2019**, *141*, 20507–20515. [CrossRef]
30. Nisbet-Jones, P.B.R.; Fernandez, J.M.; Fisher, R.E.; France, J.L.; Lowry, D.; Waltham, D.A.; Maisch, C.A.W.; Nisbet, E.G. Is the destruction or removal of atmospheric methane a worthwhile option? *Philos. Trans. R. Soc. London. Ser. A Math. Phys. Eng. Sci.* **2022**, *380*, 20210108. [CrossRef]
31. Islam, J.; Shrestha, N.; Bathi, J.R.; Sani, R.K.; Gadhamshetty, V. Surface Modification Approaches for Methane Oxidation in Bioelectrochemical Systems. In *Bioelectrochemical Systems*; Springer: Singapore, 2020; pp. 343–374. [CrossRef]

32. Thauer, R.K.; Kaster, A.-K.; Seedorf, H.; Buckel, W.; Hedderich, R. Methanogenic archaea: Ecologically relevant differences in energy conservation. *Nat. Rev. Microbiol.* **2008**, *6*, 579–591. [CrossRef]
33. Dlugokencky, E.J.; Nisbet, E.G.; Fisher, R.; Lowry, D. Global atmospheric methane: Budget, changes and dangers. *Philos. Trans. R. Soc. London. Ser. A Math. Phys. Eng. Sci.* **2011**, *369*, 2058–2072. [CrossRef]
34. De Souza, R.F.B.; Florio, D.Z.; Antolini, E.; Neto, A.O. Partial Methane Oxidation in Fuel Cell-Type Reactors for Co-Generation of Energy and Chemicals: A Short Review. *Catalysts* **2022**, *12*, 217. [CrossRef]
35. Tse, T.J.; Wiens, D.J.; Reaney, M.J.T. Production of Bioethanol—A Review of Factors Affecting Ethanol Yield. *Fermentation* **2021**, *7*, 268. [CrossRef]
36. Renewable Fuels Association. *2022 Ethanol Industry Outlook*; Renewable Fuels Association: Washington, DC, USA, 2020.
37. Lawford, H.G.; Rousseau, J.D. Corn steep liquor as a cost-effective nutrition adjunct in high-performance zymomonas ethanol fermentations. *Appl. Biochem. Biotechnol.* **1997**, *63–65*, 287–304. [CrossRef]
38. Linde, M.; Galbe, M.; Zacchi, G. Simultaneous saccharification and fermentation of steam-pretreated barley straw at low enzyme loadings and low yeast concentration. *Enzym. Microb. Technol.* **2007**, *40*, 1100–1107. [CrossRef]
39. Choi, I.S.; Wi, S.G.; Kim, S.-B.; Bae, H.-J. Conversion of coffee residue waste into bioethanol with using popping pretreatment. *Bioresour. Technol.* **2012**, *125*, 132–137. [CrossRef]
40. Ballesteros, M.; Oliva, J.; Negro, M.; Manzanares, P. Ethanol from lignocellulosic materials by a simultaneous saccharification and fermentation process (SFS) with *Kluyveromyces marxianus* CECT 10875. *Process Biochem.* **2004**, *39*, 1843–1848. [CrossRef]
41. Pessani, N.K.; Atiyeh, H.K.; Wilkins, M.; Bellmer, D.D.; Banat, I. Simultaneous saccharification and fermentation of Kanlow switchgrass by thermotolerant *Kluyveromyces marxianus* IMB3: The effect of enzyme loading, temperature and higher solid loadings. *Bioresour. Technol.* **2011**, *102*, 10618–10624. [CrossRef]
42. Tanimura, A.; Nakamura, T.; Watanabe, I.; Ogawa, J.; Shima, J. Isolation of a novel strain of Candida shehatae for ethanol production at elevated temperature. *SpringerPlus* **2012**, *1*, 27. [CrossRef]
43. Xiros, C.; Christakopoulos, P. Enhanced ethanol production from brewer's spent grain by a *Fusarium oxysporum* consolidated system. *Biotechnol. Biofuels* **2009**, *2*, 4. [CrossRef]
44. Van Aalst, A.C.; de Valk, S.C.; van Gulik, W.M.; Jansen, M.L.; Pronk, J.T.; Mans, R. Pathway engineering strategies for improved product yield in yeast-based industrial ethanol production. *Synth. Syst. Biotechnol.* **2022**, *7*, 554–566. [CrossRef]
45. Lin, P.P.; Jaeger, A.J.; Wu, T.-Y.; Xu, S.C.; Lee, A.S.; Gao, F.; Chen, P.-W.; Liao, J.C. Construction and evolution of an *Escherichia coli* strain relying on nonoxidative glycolysis for sugar catabolism. *Proc. Natl. Acad. Sci. USA* **2018**, *115*, 3538–3546. [CrossRef] [PubMed]
46. Liu, X.; Unaegbunam, E.; Stuart, D.T. Co-Production of Isobutanol and Ethanol from Prairie Grain Starch Using Engineered *Saccharomyces cerevisiae*. *Fermentation* **2021**, *7*, 150. [CrossRef]
47. Kolesinska, B.; Fraczyk, J.; Binczarski, M.; Modelska, M.; Berlowska, J.; Dziugan, P.; Antolak, H.; Kaminski, Z.J.; Witonska, I.A.; Kregiel, D. Butanol Synthesis Routes for Biofuel Production: Trends and Perspectives. *Materials* **2019**, *12*, 350. [CrossRef] [PubMed]
48. Fuentes, J.L.; Garbayo, I.; Cuaresma, M.; Montero, Z.; González-Del-Valle, M.; Vílchez, C. Impact of Microalgae-Bacteria Interactions on the Production of Algal Biomass and Associated Compounds. *Mar. Drugs* **2016**, *14*, 100. [CrossRef]
49. Lucia, U.; Grisolia, G.; Energia, D.; Ferraris, G. Biofuels from Micro-Organisms: Thermodynamic Considerations on the Role of Electrochemical Potential on Micro-Organisms Growth. *Appl. Sci.* **2021**, *11*, 2591. [CrossRef]
50. Poudel, J.; Choi, J.H.; Oh, S.C. Process Design Characteristics of Syngas (CO/H_2) Separation Using Composite Membrane. *Sustainability* **2019**, *11*, 703. [CrossRef]
51. Ciliberti, C.; Biundo, A.; Albergo, R.; Agrimi, G.; Braccio, G.; De Bari, I.; Pisano, I. Syngas Derived from Lignocellulosic Biomass Gasification as an Alternative Resource for Innovative Bioprocesses. *Processes* **2020**, *8*, 1567. [CrossRef]
52. Znad, H.; Awual, R.; Martini, S. The Utilization of Algae and Seaweed Biomass for Bioremediation of Heavy Metal-Contaminated Wastewater. *Molecules* **2022**, *27*, 1275. [CrossRef]
53. Abeln, F.; Chuck, C.J. The history, state of the art and future prospects for oleaginous yeast research. *Microb. Cell Fact.* **2021**, *20*, 1–31. [CrossRef]
54. Jach, M.E.; Serefko, A.; Ziaja, M.; Kieliszek, M. Yeast Protein as an Easily Accessible Food Source. *Metabolites* **2022**, *12*, 63. [CrossRef]
55. Jiang, C.; Peces, M.; Andersen, M.H.; Kucheryavskiy, S.; Nierychlo, M.; Yashiro, E.; Andersen, K.S.; Kirkegaard, R.H.; Hao, L.; Høgh, J.; et al. Characterizing the growing microorganisms at species level in 46 anaerobic digesters at Danish wastewater treatment plants: A six-year survey on microbial community structure and key drivers. *Water Res.* **2021**, *193*, 116871. [CrossRef]
56. Adegboye, M.F.; Ojuederie, O.B.; Talia, P.M.; Babalola, O.O. Bioprospecting of microbial strains for biofuel production: Metabolic engineering, applications, and challenges. *Biotechnol. Biofuels* **2021**, *14*, 5. [CrossRef]
57. Eram, M.S.; Ma, K. Decarboxylation of Pyruvate to Acetaldehyde for Ethanol Production by Hyperthermophiles. *Biomolecules* **2013**, *3*, 578–596. [CrossRef]
58. Gambacorta, F.V.; Dietrich, J.J.; Yan, Q.; Pfleger, B.F. Rewiring yeast metabolism to synthesize products beyond ethanol. *Curr. Opin. Chem. Biol.* **2020**, *59*, 182–192. [CrossRef]
59. Koppolu, V.; Vasigala, V.K. Role of *Escherichia coli* in Biofuel Production. *Microbiol. Insights* **2016**, *9*, 29–35. [CrossRef]

60. Yang, S.; Fei, Q.; Zhang, Y.; Contreras, L.M.; Utturkar, S.M.; Brown, S.D.; Himmel, M.E.; Zhang, M. *Zymomonas mobilis* as a model system for production of biofuels and biochemicals. *Microb. Biotechnol.* **2016**, *9*, 699–717. [CrossRef]
61. Duwe, A.; Tippkötter, N.; Ulber, R. Lignocellulose-Biorefinery: Ethanol-Focused. *Adv. Biochem. Eng. Biotechnol.* **2019**, *166*, 177–215. [CrossRef]
62. Lennartsson, P.R.; Erlandsson, P.; Taherzadeh, M.J. Integration of the first and second generation bioethanol processes and the importance of by-products. *Bioresour. Technol.* **2014**, *165*, 3–8. [CrossRef]
63. Ruchala, J.; Kurylenko, O.O.; Dmytruk, K.V.; Sibirny, A.A. Construction of advanced producers of first- and second-generation ethanol in *Saccharomyces cerevisiae* and selected species of non-conventional yeasts (*Scheffersomyces stipitis*, *Ogataea polymorpha*). *J. Ind. Microbiol. Biotechnol.* **2020**, *47*, 109–132. [CrossRef]
64. Knothe, G.; Razon, L.F. Biodiesel fuels. *Prog. Energy Combust. Sci.* **2017**, *58*, 36–59. [CrossRef]
65. Raimo, N.; Malandrino, O.; Esposito, B.; Khatami, F.; Goharian, E. Beyond Profitable Shifts to Green Energies, towards Energy Sustainability. *Sustainability* **2022**, *14*, 4506. [CrossRef]
66. Mizik, T.; Gyarmati, G. Economic and Sustainability of Biodiesel Production—A Systematic Literature Review. *Clean Technol.* **2021**, *3*, 19–36. [CrossRef]
67. Pasha, M.K.; Dai, L.; Liu, D.; Guo, M.; Du, W. An overview to process design, simulation and sustainability evaluation of biodiesel production. *Biotechnol. Biofuels* **2021**, *14*, 129. [CrossRef] [PubMed]
68. Khan, M.I.; Shin, J.H.; Kim, J.D.; Khan, M.I.; Shin, J.H.; Kim, J.D. The promising future of microalgae: Current status, challenges, and optimization of a sustainable and renewable industry for biofuels, feed, and other products. *Microb. Cell Fact.* **2018**, *17*, 36. [CrossRef]
69. Leach, F.; Kalghatgi, G.; Stone, R.; Miles, P. The scope for improving the efficiency and environmental impact of internal combustion engines. *Transp. Eng.* **2020**, *1*, 100005. [CrossRef]
70. Papakonstantinou, E.; Mitsis, T.; Dragoumani, K.; Efthimiadou, A.; Bacopoulou, F.; Chrousos, G.P.; Eliopoulos, E.; Vlachakis, D. Materials of biological origin and biofuels: Small environmental footprint and epigenetic impact (Review). *Int. J. Epigenet.* **2021**, *1*, 1–8. [CrossRef]
71. Popp, J.; Harangi-Rákos, M.; Gabnai, Z.; Balogh, P.; Antal, G.; Bai, A. Biofuels and Their Co-Products as Livestock Feed: Global Economic and Environmental Implications. *Molecules* **2016**, *21*, 285. [CrossRef]
72. Mondal, M.; Goswami, S.; Ghosh, A.; Oinam, G.; Tiwari, O.N.; Das, P.; Gayen, K.; Mandal, M.K.; Halder, G.N. Production of biodiesel from microalgae through biological carbon capture: A review. *3 Biotech* **2017**, *7*, 1–21. [CrossRef]
73. Hindersah, R.; Kamaluddin, N.N.; Samanta, S.; Banerjee, S.; Sarkar, S. Role and perspective of Azotobacter in crops production. *SAINS Tanah J. Soil Sci. Agroclimatol.* **2020**, *17*, 170–179. [CrossRef]
74. Liu, A.; Contador, C.A.; Fan, K.; Lam, H.-M. Interaction and Regulation of Carbon, Nitrogen, and Phosphorus Metabolisms in Root Nodules of Legumes. *Front. Plant Sci.* **2018**, *9*, 1860. [CrossRef]
75. Encinar, J.M.; Pardal, A.; Sánchez, N.; Nogales, S. Biodiesel by Transesterification of Rapeseed Oil Using Ultrasound: A Kinetic Study of Base-Catalysed Reactions. *Energies* **2018**, *11*, 2229. [CrossRef]
76. Niehus, X.; Casas-Godoy, L.; Valadez, F.J.R.; Sandoval, G. Evaluation of *Yarrowia lipolytica* Oil for Biodiesel Production: Land Use Oil Yield, Carbon, and Energy Balance. *J. Lipids* **2018**, *2018*, 6393749. [CrossRef]
77. Caporusso, A.; Capece, A.; De Bari, I. Oleaginous Yeasts as Cell Factories for the Sustainable Production of Microbial Lipids by the Valorization of Agri-Food Wastes. *Fermentation* **2021**, *7*, 50. [CrossRef]
78. Lobo, C.B.; Tomás, M.S.J.; Viruel, E.; Ferrero, M.A.; Lucca, M.E. Development of low-cost formulations of plant growth-promoting bacteria to be used as inoculants in beneficial agricultural technologies. *Microbiol. Res.* **2019**, *219*, 12–25. [CrossRef]
79. Wang, H.; Peng, X.; Zhang, H.; Yang, S.; Li, H. Microorganisms-promoted biodiesel production from biomass: A review. *Energy Convers. Manag. X* **2021**, *12*, 100137. [CrossRef]
80. Zabermawi, N.M.; Alsulaimany, F.A.; El-Saadony, M.T.; El-Tarabily, K.A. New eco-friendly trends to produce biofuel and bioenergy from microorganisms: An updated review. *Saudi J. Biol. Sci.* **2022**. [CrossRef]
81. Patel, A.; Karageorgou, D.; Rova, E.; Katapodis, P.; Rova, U.; Christakopoulos, P.; Matsakas, L. An Overview of Potential Oleaginous Microorganisms and Their Role in Biodiesel and Omega-3 Fatty Acid-Based Industries. *Microorganisms* **2020**, *8*, 434. [CrossRef]
82. Moradian, J.M.; Fang, Z.; Yong, Y.-C. Recent advances on biomass-fueled microbial fuel cell. *Bioresour. Bioprocess.* **2021**, *8*, 1–13. [CrossRef]
83. Pandit, S.; Savla, N.; Sonawane, J.M.; Sani, A.M.; Gupta, P.K.; Mathuriya, A.S.; Rai, A.K.; Jadhav, D.A.; Jung, S.P.; Prasad, R. Agricultural Waste and Wastewater as Feedstock for Bioelectricity Generation Using Microbial Fuel Cells: Recent Advances. *Fermentation* **2021**, *7*, 169. [CrossRef]
84. Taylor-Cornejo, E. Empowering Undergraduates to Fight Climate Change with Soil Microbes. *DNA Cell Biol.* **2022**, *41*, 58–63. [CrossRef]
85. De_Richter, R.; Ming, T.; Davies, P.; Liu, W.; Caillol, S. Removal of non-CO_2 greenhouse gases by large-scale atmospheric solar photocatalysis. *Prog. Energy Combust. Sci.* **2017**, *60*, 68–96. [CrossRef]
86. Lamb, W.F.; Wiedmann, T.; Pongratz, J.; Andrew, R.; Crippa, M.; Olivier, J.G.J.; Wiedenhofer, D.; Mattioli, G.; Al Khourdajie, A.; House, J.; et al. A review of trends and drivers of greenhouse gas emissions by sector from 1990 to 2018. *Environ. Res. Lett.* **2021**, *16*, 073005. [CrossRef]

87. Jafarihaghighi, F.; Ardjmand, M.; Hassani, M.S.; Mirzajanzadeh, M.; Bahrami, H. Effect of Fatty Acid Profiles and Molecular Structures of Nine New Source of Biodiesel on Combustion and Emission. *ACS Omega* **2020**, *5*, 16053–16063. [CrossRef] [PubMed]
88. Lin, H.; Wang, Q.; Shen, Q.; Zhan, J.; Zhao, Y. Genetic engineering of microorganisms for biodiesel production. *Bioengineered* **2013**, *4*, 292–304. [CrossRef]
89. Gouveia, L.; Oliveira, A.C. Microalgae as a raw material for biofuels production. *J. Ind. Microbiol. Biotechnol.* **2009**, *36*, 269–274. [CrossRef]
90. Makareviciene, V.; Sendzikiene, E. Application of Microalgae Biomass for Biodiesel Fuel Production. *Energies* **2022**, *15*, 4178. [CrossRef]
91. Huang, Z.; Zhang, J.; Pan, M.; Hao, Y.; Hu, R.; Xiao, W.; Li, G.; Lyu, T. Valorisation of microalgae residues after lipid extraction: Pyrolysis characteristics for biofuel production. *Biochem. Eng. J.* **2022**, *179*, 108330. [CrossRef]
92. Li, G.; Bai, X.; Huo, S.; Huang, Z. Fast pyrolysis of LERDADEs for renewable biofuels. *IET Renew. Power Gener.* **2020**, *14*, 959–967. [CrossRef]
93. Rezić, T.; Filipović, J.; Šantek, B. Microalgae–A potential source of lipids for biodiesel production. *Hrvat. Časopis Prehrambenu Tehnol. Biotehnol. Nutr.* **2014**, *9*, 26–36.
94. Chisti, Y. Biodiesel from microalgae. *Biotechnol. Adv.* **2007**, *25*, 294–306. [CrossRef]
95. Meng, X.; Yang, J.; Xu, X.; Zhang, L.; Nie, Q.; Xian, M. Biodiesel production from oleaginous microorganisms. *Renew. Energy* **2009**, *34*, 1–5. [CrossRef]
96. Nigam, H.; Malik, A.; Singh, V. A novel nanoemulsion-based microalgal growth medium for enhanced biomass production. *Biotechnol. Biofuels* **2021**, *14*, 111. [CrossRef]
97. Dolganyuk, V.; Belova, D.; Babich, O.; Prosekov, A.; Ivanova, S.; Katserov, D.; Patyukov, N.; Sukhikh, S. Microalgae: A Promising Source of Valuable Bioproducts. *Biomolecules* **2020**, *10*, 1153. [CrossRef]
98. Subramaniam, R.; Dufreche, S.; Zappi, M.; Bajpai, R. Microbial lipids from renewable resources: Production and characterization. *J. Ind. Microbiol. Biotechnol.* **2010**, *37*, 1271–1287. [CrossRef]
99. Uma, V.S.; Usmani, Z.; Sharma, M.; Diwan, D.; Sharma, M.; Guo, M.; Tuohy, M.G.; Makatsoris, C.; Zhao, X.; Thakur, V.K.; et al. Valorisation of algal biomass to value-added metabolites: Emerging trends and opportunities. *Phytochem. Rev.* **2022**, 1–26. [CrossRef]
100. Ebhodaghe, S.O.; Imanah, O.E.; Ndibe, H. Biofuels from microalgae biomass: A review of conversion processes and procedures. *Arab. J. Chem.* **2022**, *15*, 103591. [CrossRef]
101. Jin, M.; Slininger, P.J.; Dien, B.S.; Waghmode, S.; Moser, B.R.; Orjuela, A.; da Costa Sousa, L.; Balan, V. Microbial lipid-based lignocellulosic biorefinery: Feasibility and challenges. *Trends Biotechnol.* **2015**, *33*, 43–54. [CrossRef]
102. Kayfeci, M.; Keçebaş, A.; Bayat, M. Hydrogen production. In *Solar Hydrogen Production Processes, Systems and Technologies*; Academic Press: Cambridge, MA, USA, 2019; pp. 45–83. [CrossRef]
103. Osman, A.I.; Mehta, N.; Elgarahy, A.M.; Hefny, M.; Al-Hinai, A.; Al-Muhtaseb, A.H.; Rooney, D.W. Hydrogen production, storage, utilisation and environmental impacts: A review. *Environ. Chem. Lett.* **2021**, *20*, 153–188. [CrossRef]
104. Sharma, S.; Agarwal, S.; Jain, A. Significance of Hydrogen as Economic and Environmentally Friendly Fuel. *Energies* **2021**, *14*, 7389. [CrossRef]
105. Nucleosynthesis and the origin of the elements. *Philos. Trans. R. Soc. London. Ser. A Math. Phys. Sci.* **1988**, *325*, 391–403. [CrossRef]
106. Zaikowski, L.; Wilkens, R.T.; Fisher, K. Science and the Concept of Evolution: From the Big Bang to the Origin and Evolution of Life. *Evol. Educ. Outreach* **2007**, *1*, 65–73. [CrossRef]
107. Haider, O. Nuclear Fusion: Holy Grail of Energy. In *Nuclear Fusion—One Noble Goal and a Variety of Scientific and Technological Challenges*; Girka, I., Ed.; IntechOpen: London, UK, 2019; ISBN 978-1-78985-788-7.
108. Croswell, K. News Feature: Tracing gold's cosmic origin story. *Proc. Natl. Acad. Sci. USA* **2021**, *118*, e2026110118. [CrossRef]
109. Chen, H.; Simoska, O.; Lim, K.; Grattieri, M.; Yuan, M.; Dong, F.; Lee, Y.S.; Beaver, K.; Weliwatte, S.; Gaffney, E.M.; et al. Fundamentals, Applications, and Future Directions of Bioelectrocatalysis. *Chem. Rev.* **2020**, *120*, 12903–12993. [CrossRef]
110. Clemente, J. Energy as a Foundation of Modern Life. *J. Energy Dev.* **2009**, *35*, 33–48.
111. Lambert, J.G.; Hall, C.A.S.; Balogh, S.; Gupta, A.; Arnold, M. Energy, EROI and quality of life. *Energy Policy* **2014**, *64*, 153–167. [CrossRef]
112. Jacobson, M.Z.; Delucchi, M.A.; Bauer, Z.A.F.; Goodman, S.C.; Chapman, W.E.; Cameron, M.A.; Bozonnat, C.; Chobadi, L.; Clonts, H.A.; Enevoldsen, P.; et al. 100% Clean and Renewable Wind, Water, and Sunlight All-Sector Energy Roadmaps for 139 Countries of the World. *Joule* **2017**, *1*, 108–121. [CrossRef]
113. Fox, W.K. Energy source. *Access Sci.* **2021**. [CrossRef]
114. Hora, H.; Eliezer, S.; Kirchhoff, G.; Nissim, N.; Wang, J.; Lalousis, P.; Xu, Y.; Miley, G.; Martinez-Val, J.; McKenzie, W. Road map to clean energy using laser beam ignition of boron-hydrogen fusion. *Laser Part. Beams* **2017**, *35*, 730–740. [CrossRef]
115. Holmlid, L.; Kotarba, A.; Stelmachowski, P. Production of ultra-dense hydrogen H(0): A novel nuclear fuel. *Int. J. Hydrogen Energy* **2021**, *46*, 18466–18480. [CrossRef]
116. Ibrahim, A.A. Hydrogen Production from Light Hydrocarbons. In *Advances in Hydrogen Generation Technologies*; Eyvaz, M., Ed.; IntechOpen: London, UK, 2018; ISBN 978-1-78923-535-7.
117. Mann, M.E. Do Global Warming and climate change represent a serious threat to our welfare and environment? *Soc. Philos. Policy* **2009**, *26*, 193–230. [CrossRef]

118. Su, X.; Zhao, W.; Xia, D. The diversity of hydrogen-producing bacteria and methanogens within an in situ coal seam. *Biotechnol. Biofuels* **2018**, *11*, 245. [CrossRef] [PubMed]
119. Sánchez-Bastardo, N.; Schlögl, R.; Ruland, H. Methane Pyrolysis for Zero-Emission Hydrogen Production: A Potential Bridge Technology from Fossil Fuels to a Renewable and Sustainable Hydrogen Economy. *Ind. Eng. Chem. Res.* **2021**, *60*, 11855–11881. [CrossRef]
120. Cuellar-Bermudez, S.P.; Garcia-Perez, J.S.; Rittmann, B.E.; Parra-Saldivar, R. Photosynthetic bioenergy utilizing CO_2: An approach on flue gases utilization for third generation biofuels. *J. Clean. Prod.* **2015**, *98*, 53–65. [CrossRef]
121. Davey, M.E.; O'Toole, G.A. Microbial Biofilms: From Ecology to Molecular Genetics. *Microbiol. Mol. Biol. Rev.* **2000**, *64*, 847–867. [CrossRef]
122. Smith, N.W.; Shorten, P.R.; Altermann, E.; Roy, N.C.; McNabb, W.C. The Classification and Evolution of Bacterial Cross-Feeding. *Front. Ecol. Evol.* **2019**, *7*, 153. [CrossRef]
123. Greening, C.; Geier, R.; Wang, C.; Woods, L.C.; Morales, S.E.; McDonald, M.J.; Rushton-Green, R.; Morgan, X.C.; Koike, S.; Leahy, S.C.; et al. Diverse hydrogen production and consumption pathways influence methane production in ruminants. *ISME J.* **2019**, *13*, 2617–2632. [CrossRef]
124. Anukam, A.; Mohammadi, A.; Naqvi, M.; Granström, K. A Review of the Chemistry of Anaerobic Digestion: Methods of Accelerating and Optimizing Process Efficiency. *Processes* **2019**, *7*, 504. [CrossRef]
125. Alberts, B.; Johnson, A. How Cells Obtain Energy from Food–Molecular Biology of the Cell–NCBI Bookshelf. In *Molecular Biology of the Cell*; Garland Science: New York, NY, USA, 2002.
126. Sharma, R.; Garg, P.; Kumar, P.; Bhatia, S.K.; Kulshrestha, S. Microbial Fermentation and Its Role in Quality Improvement of Fermented Foods. *Fermentation* **2020**, *6*, 106. [CrossRef]
127. Wainaina, S.; Lukitawesa; Awasthi, M.K.; Taherzadeh, M. Bioengineering of anaerobic digestion for volatile fatty acids, hydrogen or methane production: A critical review. *Bioengineered* **2019**, *10*, 437–458. [CrossRef]
128. Detman, A.; Bucha, M.; Treu, L.; Chojnacka, A.; Pleśniak, Ł.; Salamon, A.; Łupikasza, E.; Gromadka, R.; Gawor, J.; Gromadka, A.; et al. Evaluation of acidogenesis products' effect on biogas production performed with metagenomics and isotopic approaches. *Biotechnol. Biofuels* **2021**, *14*, 1–25. [CrossRef]
129. Karekar, S.; Stefanini, R.; Ahring, B. Homo-Acetogens: Their Metabolism and Competitive Relationship with Hydrogenotrophic Methanogens. *Microorganisms* **2022**, *10*, 397. [CrossRef]
130. González, J.; Sánchez, M.E.; Gómez, X. Enhancing Anaerobic Digestion: The Effect of Carbon Conductive Materials. *C J. Carbon Res.* **2018**, *4*, 59. [CrossRef]
131. Enzmann, F.; Mayer, F.; Rother, M.; Holtmann, D. Methanogens: Biochemical background and biotechnological applications. *AMB Express* **2018**, *8*, 1–22. [CrossRef]
132. Rodríguez-Reyes, J.J.; García-Depraect, O.; Castro-Muñoz, R.; León-Becerril, E. Dark Fermentation Process Response to the Use of Undiluted Tequila Vinasse without Nutrient Supplementation. *Sustainability* **2021**, *13*, 11034. [CrossRef]
133. Nosek, D.; Cydzik-Kwiatkowska, A. Microbial Structure and Energy Generation in Microbial Fuel Cells Powered with Waste Anaerobic Digestate. *Energies* **2020**, *13*, 4712. [CrossRef]
134. Bhagchandanii, D.D.; Babu, R.P.; Sonawane, J.M.; Khanna, N.; Pandit, S.; Jadhav, D.A.; Khilari, S.; Prasad, R. A Comprehensive Understanding of Electro-Fermentation. *Fermentation* **2020**, *6*, 92. [CrossRef]
135. Osman, A.I.; Deka, T.J.; Baruah, D.C.; Rooney, D.W. Critical challenges in biohydrogen production processes from the organic feedstocks. *Biomass Convers. Biorefin.* **2020**, 1–19. [CrossRef]
136. Barghash, H.; Okedu, K.E.; Al Balushi, A. Bio-Hydrogen Production Using Landfill Leachate Considering Different Photo-Fermentation Processes. *Front. Bioeng. Biotechnol.* **2021**, *9*, 644065. [CrossRef]
137. Moussa, R.N.; Moussa, N.; Dionisi, D. Hydrogen Production from Biomass and Organic Waste Using Dark Fermentation: An Analysis of Literature Data on the Effect of Operating Parameters on Process Performance. *Processes* **2022**, *10*, 156. [CrossRef]
138. Yates, M.; Strycharz-Glaven, S.; Golden, J.; Roy, J.; Tsoi, S.; Erickson, J.; El-Naggar, M.; Barton, S.C.; Tender, L. Characterizing Electron Transport through Living Biofilms. *J. Vis. Exp.* **2018**, *2018*, e54671. [CrossRef]
139. Nawaz, A.; Hafeez, A.; Abbas, S.Z.; Haq, I.U.; Mukhtar, H.; Rafatullah, M. A state of the art review on electron transfer mechanisms, characteristics, applications and recent advancements in microbial fuel cells technology. *Green Chem. Lett. Rev.* **2020**, *13*, 101–117. [CrossRef]
140. Yan, Z.; Hitt, J.L.; Turner, J.A.; Mallouk, T.E. Renewable electricity storage using electrolysis. *Proc. Natl. Acad. Sci. USA* **2020**, *117*, 12558–12563. [CrossRef] [PubMed]
141. Hernández, C.; Osma, J.F. Microbial Electrochemical Systems: Deriving Future Trends from Historical Perspectives and Characterization Strategies. *Front. Environ. Sci.* **2020**, *8*, 44. [CrossRef]
142. Schachinger, F.; Chang, H.; Scheiblbrandner, S.; Ludwig, R. Amperometric Biosensors Based on Direct Electron Transfer Enzymes. *Molecules* **2021**, *26*, 4525. [CrossRef] [PubMed]
143. Dębowski, M.; Dudek, M.; Zieliński, M.; Nowicka, A.; Kazimierowicz, J. Microalgal Hydrogen Production in Relation to Other Biomass-Based Technologies—A Review. *Energies* **2021**, *14*, 6025. [CrossRef]
144. Wang, K.; Khoo, K.S.; Chew, K.W.; Selvarajoo, A.; Chen, W.-H.; Chang, J.-S.; Show, P.L. Microalgae: The Future Supply House of Biohydrogen and Biogas. *Front. Energy Res.* **2021**, *9*, 158. [CrossRef]

145. Hallenbeck, P.C.; Lazaro, C.Z.; Sagir, E. CHAPTER 1. Photosynthesis and Hydrogen from Photosynthetic Microorganisms. *Compr. Ser. Photochem. Photobiol. Sci.* **2018**, *16*, 1–30. [CrossRef]
146. Bayro-Kaiser, V.; Nelson, N. Microalgal hydrogen production: Prospects of an essential technology for a clean and sustainable energy economy. *Photosynth. Res.* **2017**, *133*, 49–62. [CrossRef]
147. Burlacot, A.; Peltier, G. Photosynthetic Electron Transfer Pathways During Hydrogen Photoproduction in Green Algae: Mechanisms and Limitations. *Compr. Ser. Photochem. Photobiol. Sci.* **2018**, *16*, 189–212. [CrossRef]
148. Dudek, M.; Dębowski, M.; Nowicka, A.; Kazimierowicz, J.; Zieliński, M. The Effect of Autotrophic Cultivation of *Platymonas subcordiformis* in Waters from the Natural Aquatic Reservoir on Hydrogen Yield. *Resources* **2022**, *11*, 31. [CrossRef]
149. Bhardwaj, N.; Kumar, B.; Agrawal, K.; Verma, P. Current perspective on production and applications of microbial cellulases: A review. *Bioresour. Bioprocess.* **2021**, *8*, 1–34. [CrossRef]
150. Bhatia, S.K.; Jagtap, S.S.; Bedekar, A.A.; Bhatia, R.K.; Rajendran, K.; Pugazhendhi, A.; Rao, C.V.; Atabani, A.; Kumar, G.; Yang, Y.-H. Renewable biohydrogen production from lignocellulosic biomass using fermentation and integration of systems with other energy generation technologies. *Sci. Total Environ.* **2021**, *765*, 144429. [CrossRef]
151. Cabrol, L.; Marone, A.; Tapia-Venegas, E.; Steyer, J.-P.; Ruiz-Filippi, G.; Trably, E. Microbial ecology of fermentative hydrogen producing bioprocesses: Useful insights for driving the ecosystem function. *FEMS Microbiol. Rev.* **2017**, *41*, 158–181. [CrossRef]
152. Chu, C.-Y.; Zheng, J.-L.; Chen, T.-H.; Bhuyar, P. High Performance of Biohydrogen Production in Packed-Filter Bioreactor via Optimizing Packed-Filter Position. *Int. J. Environ. Res. Public Health* **2021**, *18*, 7462. [CrossRef]
153. Dawood, F.; Anda, M.; Shafiullah, G.M. Hydrogen Production for Energy: An Overview. *Int. J. Hydrogen Energy* **2020**, *45*, 3847–3869. [CrossRef]
154. El-Shafie, M.; Kambara, S.; Hayakawa, Y. Hydrogen Production Technologies Overview. *J. Power Energy Eng.* **2019**, *7*, 107–154. [CrossRef]
155. Mah, A.X.Y.; Ho, W.S.; Bong, C.P.C.; Hassim, M.H.; Liew, P.Y.; Asli, U.A.; Kamaruddin, M.J.; Chemmangattuvalappil, N.G. Review of hydrogen economy in Malaysia and its way forward. *Int. J. Hydrogen Energy* **2019**, *44*, 5661–5675. [CrossRef]
156. Xie, Q.; Lu, Y.; Tang, L.; Zeng, G.; Yang, Z.; Fan, C.; Wang, J.; Atashgahi, S. The mechanism and application of bidirectional extracellular electron transport in the field of energy and environment. *Crit. Rev. Environ. Sci. Technol.* **2020**, *51*, 1924–1969. [CrossRef]
157. Li, X.; Abu-Reesh, I.M.; He, Z. Development of Bioelectrochemical Systems to Promote Sustainable Agriculture. *Agriculture* **2015**, *5*, 367–388. [CrossRef]
158. Elhenawy, S.; Khraisheh, M.; AlMomani, F.; Al-Ghouti, M.; Hassan, M.K. From Waste to Watts: Updates on Key Applications of Microbial Fuel Cells in Wastewater Treatment and Energy Production. *Sustainability* **2022**, *14*, 955. [CrossRef]
159. Santoro, C.; Garcia, M.J.S.; Walter, X.A.; You, J.; Theodosiou, P.; Gajda, I.; Obata, O.; Winfield, J.; Greenman, J.; Ieropoulos, I. Urine in Bioelectrochemical Systems: An Overall Review. *ChemElectroChem* **2020**, *7*, 1312–1331. [CrossRef]
160. Pareek, A.; Dom, R.; Gupta, J.; Chandran, J.; Adepu, V.; Borse, P.H. Insights into renewable hydrogen energy: Recent advances and prospects. *Mater. Sci. Energy Technol.* **2020**, *3*, 319–327. [CrossRef]
161. Jose, V. Photobiological hydrogen as a renewable fuel. *Holist. Approach Environ.* **2021**, *11*, 67–71. [CrossRef]
162. Chandrasekhar, K.; Lee, Y.-J.; Lee, D.-W. Biohydrogen Production: Strategies to Improve Process Efficiency through Microbial Routes. *Int. J. Mol. Sci.* **2015**, *16*, 8266–8293. [CrossRef]
163. Sakurai, H.; Masukawa, H.; Kitashima, M.; Inoue, K. Photobiological hydrogen production: Bioenergetics and challenges for its practical application. *J. Photochem. Photobiol. C Photochem. Rev.* **2013**, *17*, 1–25. [CrossRef]
164. Vardar-Schara, G.; Maeda, T.; Wood, T. Metabolically engineered bacteria for producing hydrogen via fermentation. *Microb. Biotechnol.* **2008**, *1*, 107–125. [CrossRef]
165. Manyi-Loh, C.E.; Mamphweli, S.N.; Meyer, E.L.; Okoh, A.I.; Makaka, G.; Simon, M. Microbial Anaerobic Digestion (Bio-Digesters) as an Approach to the Decontamination of Animal Wastes in Pollution Control and the Generation of Renewable Energy. *Int. J. Environ. Res. Public Health* **2013**, *10*, 4390–4417. [CrossRef]
166. Bozdag, G.O.; Libby, E.; Pineau, R.; Reinhard, C.T.; Ratcliff, W.C. Oxygen suppression of macroscopic multicellularity. *Nat. Commun.* **2021**, *12*, 2838. [CrossRef] [PubMed]
167. Sotelo-Navarro, P.X.; Poggi-Varaldo, H.M. Hydrogen from Dark Fermentation of the Organic Fraction of Waste Diapers: Optimization Based on Response Surface Experiments. *Front. Energy Res.* **2021**, *9*, 18. [CrossRef]
168. Marshall, I.P.; Berggren, D.R.; Azizian, M.F.; Burow, L.C.; Semprini, L.; Spormann, A.M. The Hydrogenase Chip: A tiling oligonucleotide DNA microarray technique for characterizing hydrogen-producing and -consuming microbes in microbial communities. *ISME J.* **2011**, *6*, 814–826. [CrossRef] [PubMed]
169. Khanna, N.; Lindblad, P. Cyanobacterial Hydrogenases and Hydrogen Metabolism Revisited: Recent Progress and Future Prospects. *Int. J. Mol. Sci.* **2015**, *16*, 10537–10561. [CrossRef]
170. Veaudor, T.; Blanc-Garin, V.; Chenebault, C.; Diaz-Santos, E.; Sassi, J.-F.; Cassier-Chauvat, C.; Chauvat, F. Recent Advances in the Photoautotrophic Metabolism of Cyanobacteria: Biotechnological Implications. *Life* **2020**, *10*, 71. [CrossRef]
171. Lima-Melo, Y.; Kılıç, M.; Aro, E.-M.; Gollan, P.J. Photosystem I Inhibition, Protection and Signalling: Knowns and Unknowns. *Front. Plant Sci.* **2021**, *12*, 791124. [CrossRef]

172. Allakhverdiev, S.I.; Kreslavski, V.D.; Thavasi, V.; Zharmukhamedov, S.K.; Klimov, V.V.; Nagata, T.; Nishihara, H.; Ramakrishna, S. Hydrogen photoproduction by use of photosynthetic organisms and biomimetic systems. *Photochem. Photobiol. Sci.* **2009**, *8*, 148–156. [CrossRef] [PubMed]
173. Zimorski, V.; Mentel, M.; Tielens, A.G.M.; Martin, W.F. Energy metabolism in anaerobic eukaryotes and Earth's late oxygenation. *Free Radic. Biol. Med.* **2019**, *140*, 279–294. [CrossRef]
174. Key, S.; Ma, J.K.-C.; Drake, P.M. Genetically modified plants and human health. *J. R. Soc. Med.* **2008**, *101*, 290–298. [CrossRef]
175. Gregory, T.R. Understanding Natural Selection: Essential Concepts and Common Misconceptions. *Evol. Educ. Outreach* **2009**, *2*, 156–175. [CrossRef]
176. Zhang, J.; Liu, G.; Carvajal, A.I.; Wilson, R.H.; Cai, Z.; Li, Y. Discovery of a readily heterologously expressed Rubisco from the deep sea with potential for CO_2 capture. *Bioresour. Bioprocess.* **2021**, *8*, 1–16. [CrossRef]
177. Bay, S.K.; Waite, D.W.; Dong, X.; Gillor, O.; Chown, S.L.; Hugenholtz, P.; Greening, C. Chemosynthetic and photosynthetic bacteria contribute differentially to primary production across a steep desert aridity gradient. *ISME J.* **2021**, *15*, 3339–3356. [CrossRef] [PubMed]
178. Adam, N.; Perner, M. Microbially Mediated Hydrogen Cycling in Deep-Sea Hydrothermal Vents. *Front. Microbiol.* **2018**, *9*, 2873. [CrossRef]
179. Jordaan, K.; Lappan, R.; Dong, X.; Aitkenhead, I.J.; Bay, S.K.; Chiri, E.; Wieler, N.; Meredith, L.K.; Cowan, D.A.; Chown, S.L.; et al. Hydrogen-Oxidizing Bacteria Are Abundant in Desert Soils and Strongly Stimulated by Hydration. *mSystems* **2020**, *5*, e01131-20. [CrossRef]
180. Stabnikova, O.; Stabnikov, V.; Marinin, A.; Klavins, M.; Klavins, L.; Vaseashta, A. Microbial Life on the Surface of Microplastics in Natural Waters. *Appl. Sci.* **2021**, *11*, 11692. [CrossRef]
181. Tremblay, P.-L.; Lovley, D.R. Role of the NiFe Hydrogenase Hya in Oxidative Stress Defense in Geobacter sulfurreducens. *J. Bacteriol.* **2012**, *194*, 2248–2253. [CrossRef]
182. Kruse, S.; Goris, T.; Wolf, M.; Wei, X.; Diekert, G. The NiFe Hydrogenases of the Tetrachloroethene-Respiring Epsilonproteobacterium *Sulfurospirillum multivorans*: Biochemical Studies and Transcription Analysis. *Front. Microbiol.* **2017**, *8*, 444. [CrossRef]
183. Ding, H.; Liu, H.; Chu, W.; Wu, C.; Xie, Y. Structural Transformation of Heterogeneous Materials for Electrocatalytic Oxygen Evolution Reaction. *Chem. Rev.* **2021**, *121*, 13174–13212. [CrossRef] [PubMed]
184. Kumar, R.; Kumar, P. Future Microbial Applications for Bioenergy Production: A Perspective. *Front. Microbiol.* **2017**, *8*, 450. [CrossRef]
185. Sahni, S.; Singh, M.K.; Narang, A. Sustainable Solution for Future Energy Challenges Through Microbes. *Energy* **2021**, 231–249. [CrossRef]
186. Park, S.-G.; Rajesh, P.; Sim, Y.-U.; Jadhav, D.A.; Noori, T.; Kim, D.-H.; Al-Qaradawi, S.Y.; Yang, E.; Jang, J.-K.; Chae, K.-J. Addressing scale-up challenges and enhancement in performance of hydrogen-producing microbial electrolysis cell through electrode modifications. *Energy Rep.* **2022**, *8*, 2726–2746. [CrossRef]
187. Germscheidt, R.L.; Moreira, D.E.B.; Yoshimura, R.G.; Gasbarro, N.P.; Datti, E.; dos Santos, P.L.; Bonacin, J.A. Hydrogen Environmental Benefits Depend on the Way of Production: An Overview of the Main Processes Production and Challenges by 2050. *Adv. Energy Sustain. Res.* **2021**, *2*, 2100093. [CrossRef]
188. Pulyaeva, V.N.; A Kharitonova, N.; Kharitonova, E.N. Advantages and Disadvantages of the Production and Using of Liquid Biofuels. *IOP Conf. Ser. Mater. Sci. Eng.* **2020**, *976*, 012031. [CrossRef]

energies

MDPI

Review

An Analysis of Research Trends in the Sustainability of Production Planning

Mohamed Saeed Khaled [1], Ibrahim Abdelfadeel Shaban [1,2,*], Ahmed Karam [3,4,*], Mohamed Hussain [1], Ismail Zahran [1] and Mohamed Hussein [2]

1 Mechanical Engineering Department, Faculty of Engineering, Helwan University, Helwan, Cairo 11795, Egypt; mohamedsaeed@h-eng.helwan.edu.eg (M.S.k.); Mohamed.hussien@h-eng.helwan.edu.eg (M.H.); ismail.zahran@h-eng.helwan.edu.eg (I.Z.)

2 Department of Building and Real Estate, The Hong Kong Polytechnic University, Hong Kong 999666, China; mohamed.hussein@connect.polyu.hk

3 Mechanical Engineering Department, Faculty of Engineering at Shoubra, Benha University, Banha, Cairo 11629, Egypt

4 Department of the Built Environment, Aalborg University, 9220 Aalborg, Denmark

* Correspondence: ibrahim.shaban@polyu.edu.hk (I.A.S.); akam@build.aau.dk (A.K.)

Abstract: Sustainability has become of great interest in many fields, especially in production systems due to the continual increase in the scarcity of raw materials and environmental awareness. Recent literature has given significant attention to considering the three sustainability pillars (i.e., environmental, economic, and social sustainability) in solving production planning problems. Therefore, the present study conducts a review of the literature on sustainable production planning to analyze the relationships among different production planning problems (e.g., scheduling, lot sizing, aggregate planning, etc.) and the three sustainability pillars. In addition, we analyze the identified studies based on the indicators that define each pillar. The results show that the literature most frequently addresses production scheduling problems while it lacks studies on aggregate production planning problems that consider the sustainability pillars. In addition, there is a growing trend towards obtaining integrated solutions of different planning problems, e.g., combining production planning problems with maintenance planning or energy planning. Additionally, around 45% of the identified studies considered the integration of the economic and the environmental pillars in different production planning problems. In addition, energy consumption and greenhouse gas emissions are the most frequent sustainability indicators considered in the literature, while less attention has been given to social indicators. Another issue is the low number of studies that have considered all three sustainability pillars simultaneously. The finidings highlight the need for more future research towards holistic sustainable production planning approaches.

Keywords: sustainability; production planning; sustainability indicators; sustainability objectives; review; sustainable production planning

Citation: Khaled, M.S.; Shaban, I.A.; Karam, A.; Hussain, M.; Zahran, I.; Hussein, M. An Analysis of Research Trends in the Sustainability of Production Planning. *Energies* **2022**, *15*, 483. https://doi.org/10.3390/en15020483

Academic Editor: Dimitrios A. Georgakellos

Received: 29 November 2021
Accepted: 6 January 2022
Published: 11 January 2022

1. Introduction

Production planning is the process of making a set of decisions or a plan to ensure the correct and efficient flow of production processes according to specific objectives. These objectives mostly focus on achieving the desired product quality with the least possible production cost within the planned production schedules [1]. Production planning is a complex task that includes many decision-making problems related to various production stages such as aggregate production planning, lot sizing, and scheduling [2,3]. For example, aggregate production planning aims to match plant capacity with demand while considering lower costs [4,5]. Due to increasing the world population and production capacities, the resources of our planet are excessively consumed. Methods to safeguard these resources from vanishing are necessarily required [6,7]. Furthermore, increase of global temperatures and changes in weather patterns have increased the severity of the issue [8]. In 2015, the United Nations proposed 17 sustainable development goals. Since then, sustainability has become of great interest in many fields, especially in production systems, because the scarcity of raw materials and environmental regulations are continuously increasing. This has made production planning one of the most important research topics to support the three sustainability pillars (3Ps) (i.e., environmental, economic, and social sustainability) [9] as most of the sustainability issues can be found through all production planning stages such as aggregate production planning, scheduling, etc. [10]. In addition, the eruption of COVID-19 has caused several socioeconomic disruptions in the manufacturing and industrial firm sectors [11]. These sectors have adopted several strategies and polices to reduce the undesirable impacts. One of these strategies is to modify their production plans to apply the social distancing [12]. However, these strategies have led to many drawbacks because of the complexity of the production planning process. Moreover, production planning is connected to the product life cycle through process planning, product design, and recycling, and it is also connected with social aspects of employees and customers. Hence, many scholars have studied the production planning problems while considering the 3Ps of sustainability [9,10].

Considering at least one of the sustainability 3Ps in the traditional production planning extends its scope towards sustainable production planning [13], as shown in Figure 1. The 3Ps of sustainability could be achieved by minimizing energy consumption, greenhouse gas emissions, and increasing health and safety or training of employees [13]. For example, a low carbon process design strategy is considered a sustainable production planning objective [14]. Wichmann et al. [15] studied lot-sizing and scheduling operations to improve energy consumption by minimizing the machining time. In addition, it is of great importance that a company makes joint or integrated decisions combining various aspects of a planning process such as pricing, retailer selection, labor time, etc., while considering a more sustainable environment [16]. Some studies have reported difficulties in combining management and planning requirements with the 3Ps of sustainability [17], especially social sustainability. In this regard, it is important to revise the theoretical background of sustainability, sustainable development, production planning, and sustainable production planning.

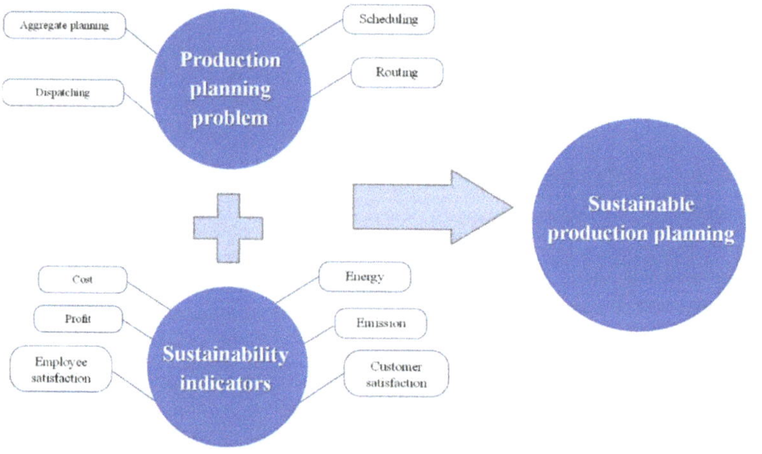

Figure 1. Converting traditional production planning to sustainable production planning.

Gaps in the Existing Reviews and Contributions

In recent years, some scholars have conducted review studies relevant to the sustainability aspects in manufacturing and production processes. Table 1 summarizes the previous reviews and presents the number of papers reviewed, the covered period, and the objective of each review. We found that only three review studies [18] considered production planning as a tool for implementing the 3Ps of sustainability. Giret, et al. [18] covered the period from 2008 to 2014 and considered only one production planning problem, i.e., scheduling. They found that the link between tactical and operational levels wasvery neglected and further studies areneeded, especially for planning activities. As for the objectives of these problems, they reported that the input-oriented energy parameters were the focus of most studies, while less attention had been given to social sustainability. Biel and Glock [19] mainly focused on energy-oriented production planning. They also reported an increase in the research of energy-oriented production planning approaches. Moreover, they concluded that most articles mainly focused on job allocation and sequencing more than any other planning problem. Bóna and Korkulu [20] addressed only one production problem, i.e., lot-sizing and its impact on social sustainability. Thus, the contribution of this work stems from the existing literature gaps. The existing literature lacks a holistic review study that analyzes the relationships among different production planning problems (e.g., scheduling, lot sizing, aggregate planning, etc.) and 3Ps of sustainability. Unlike previous review studies, the present review study focuses on all production planning problems and considers all pillars of sustainability, i.e., economic, environmental, and social sustainability.

Table 1. Classification of previous review studies.

Reference	Focus	Objective	Covered Period	No of Articles
[18]	Sustainable manufacturing operations scheduling	- Addressing sustainable manufacturing from a scheduling perspective - Classifying sustainable operation scheduling according to the orientation of the approach, the method of scheduling, and multi-objective considered	2008–2014	45
[19]	Energy-efficient production planning	- Addressing sustainable production planning but only from the economical perspective, specifical energy - Addressing decision support models that integrate energy considerations - Classifying reviewed articles mainly according to the type of production planning problem	Until 2015	89
[13]	Decision support system for sustainable manufacturing	- Addressing different decision-making methods and different sustainable indicators used in sustainable manufacturing from a product and production life cycle perspective - The resulted papers were categorized by methods, sustainable indicators, and life cycle phase.	2007–2017	23
[21]	Tools available for implementing sustainable development goals	- A scoping methodology was used to address tools available for sustainable development goals. - The review sought three main properties of each tool nature or type of the tool, purpose of the tool, and background to its development. - The resulting studies were categorized based on three main categories: mapping tools, reporting tools, and aligning tools.	2000–2018	50

Table 1. *Cont.*

Reference	Focus	Objective	Covered Period	No of Articles
[22]	Sustainable consumption and production	- Addressing a comparison of sustainable production and consumption considering differences and challenges between developed and developing countries	1998–2018	90
[20]	Social sustainability lot sizing	- Addressing ergonomics as a sustainable social objective in lot-sizing problems	Until 2019	36

The remainder of this paper is organized as follows: In Section 2, we present an overview of the different production planning problems and explain the sustainability pillars used to classify the relevant studies; in Section 3, we discuss the review methodology; in Section 4, we present the results; in Section 5, we suggest implications for future research; in Section 6, we present the conclusions.

2. Production Planning and Sustainability Pillars

2.1. Production Planning Problems

A thriving production process mainly depends on appropriate allocation of available resources [23]. Production planning is the ultimate tool for meeting increasing customer requirements, diversity of products, and a decrease in resources [24,25]. It can also enable utilizing available resources to obtain the desired quality at the least possible cost [26]. Hence, production planning plays a vital role in the production process by increasing its efficiency [27]. Production planning is considered to be a non-isolated function that depends on multiple parameters. Hence, information obtained from procurement and selling or the parameters assumed by manufacturing, engineering, finance, and material management, even marketing functions, are crucial to production planning [28] and are connected with all production stages. These parameters can be divided into multiple steps and various categories [1], as shown in Figure 2., in which we categorized production planning problems into several categories based on a categorization used by [19] and another introduced in [1] to ease the classification of the sampled articles. This does not mean that this is the only existing classification of the production planning problems, but we classified them based on the articles included in this review. For example, aggregate production planning is a more concentrated and compact version of production planning that is only concerned with a shorter period of time [29] and has a specific objective of matching plant capacity with demand while considering lower costs [4,5].

The other production planning problems can be listed as lot-sizing, scheduling, routing, loading, dispatching, and controlling [30]. Some of these steps have substeps that form problems and constraints of their own. Rasmi, et al. [31] stated that aggregate production planning was a primary step in defining other secondary parameters such as production rates, inventory levels, and workforce requirements. Biel and Glock [19] considered scheduling and capacity planning as essential tasks for performing the planning process. In addition, scheduling can have subprocesses such as operation scheduling, order scheduling, and shop scheduling [30].

Consequently, due to its connection with various parameters of a production process, production planning controls the flow of a production process [2] and ensures the smoothness of such a flow to reach the desired product [32]. Kiran [1] considered it to be the brain and the nervous system of the production program. Production planning has also been recognized as the process that ensures the availability of all materials, as well as helps ensure assembly at the right time, at the right place, and in the right quantities [33], in other words, a balance between the required orders (capacity) and the produced units [34]. Consequently, production planning is considered to be an intermediate step that connects the design of a product and its manufacturing to reach the product use and recycling, when described from a product life cycle point of view [35].

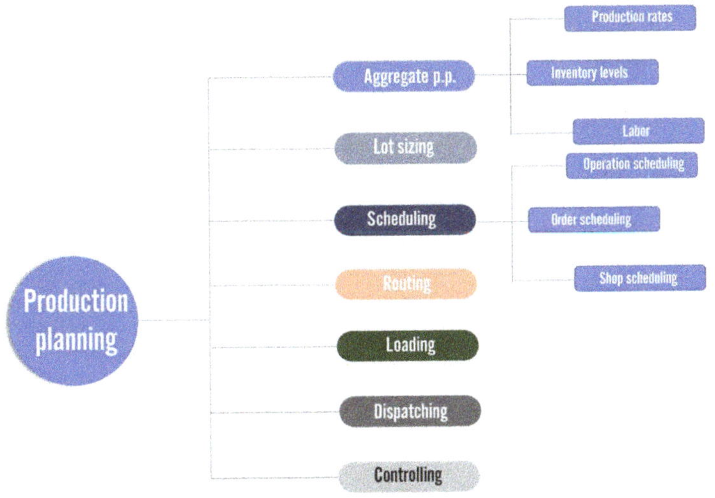

Figure 2. Production planning categorization.

2.2. Sustainability Pillars and Their Indicators

The growth in awareness of environmental consciousness has led sustainability to gain more attention during the last century [36]. Hence, the United Nations has found a specific commission for sustainability issues. This commission was formerly known as the "World Commission on Environment and Development", and then it was renamed as the "Brundtland Commission". The commission focused mainly on studying the capability of the environment to maintain stability through the 21st century. In this connection, the commission wrote a report called "Our Common Future" which established the basis of sustainable development and sustainability. In that report, sustainable development was defined as "the development that fulfils the needs of the present without compromising the ability of future generations to meet their own needs" [37,38]. In 2015, the 2030 agenda of the United Nations for Sustainable Development proposed 17 sustainable development goals, which are shown in Table 2 [39]. This table shows the extent to which sustainability goals have evolved through the years. These goals of sustainability were defined by three main pillars termed the triple bottom line (TBL), i.e., economic, environmental, and social [40], and they were considered to be the foundations to build up the generalized definition of sustainability [41,42]. Hence, this table can be considered to be a listed form of a written explanations for these three pillars. In addition, each pillar contains subterms called indicators that can define each pillar of sustainability [31,43]. Articles by [13,44] provided a similar categorization of indicators for three pillars of sustainability. The present work follows the categorization proposed in [11] and we adapted it for this review. As shown in Figure 3, the sustainability pillars are considered in production planning problems in two ways: Either a production planning problem that tackles a single sustainability pillar or a production planning problem that tackles at least two integrated pillars. For more details, each pillar contains some indictors which can be addressed in production planning problems. For example, the economic pillar addresses only two indicators, i.e., cost and profit, neglecting the investment subpillar which was not found in the sample studies of sustainable production planning problems. The environmental pillar includes three indicators, i.e., material, energy, and greenhouse gases [45]. The social well-being pillar addresses three responsibilities towards the customer, the employee, and the whole community [46]. Furthermore, at least two of these pillars are integrated.

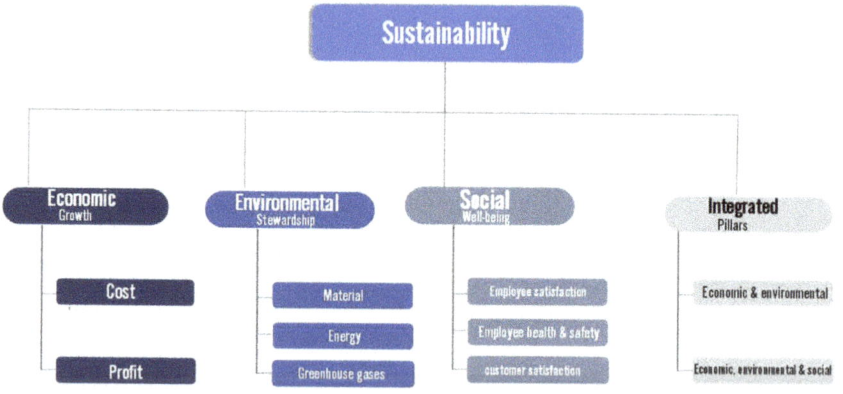

Figure 3. The categorization of sustainability indicators.

Based on the above understandings, sustainable production planning aims at decreasing the negative environmental impact while preserving energy for less consumption and a safer economic impact for stakeholders [47]. The indicator categorization shown in Figure 3 is used to classify and discuss the existing literature on sustainable production planning, which enables providing a clear and better understanding of trends and possible shortcomings in the existing literature.

Table 2. United Nations proposed 17 goals in the 2030 agenda for sustainable development [39].

Goal 1.	End poverty in all its forms everywhere
Goal 2.	End hunger, achieve food security and improve nutrition and promote sustainable agriculture
Goal 3.	Ensure healthy lives and promote well-being for all at all ages
Goal 4.	Ensure inclusive and equitable quality education and promote lifelong learning opportunities for all
Goal 5.	Achieve gender equality and empower all women and girls
Goal 6.	Ensure availability and sustainable management of water and sanitation for all
Goal 7.	Ensure access to affordable, reliable, sustainable, and modern energy for all
Goal 8.	Promote sustained, inclusive, and sustainable economic growth, full and productive employment, and decent work for all
Goal 8.	Build resilient infrastructure, promote inclusive and sustainable industrialization, and foster innovation
Goal 10.	Reduce inequality within and among countries
Goal 11.	Make cities and human settlements inclusive, safe, resilient, and sustainable
Goal 12.	Ensure sustainable consumption and production patterns
Goal 13.	Take urgent action to combat climate change and its impacts
Goal 14.	Conserve and sustainably use the oceans, seas, and marine resources for sustainable development
Goal 15.	Protect, restore and promote sustainable use of terrestrial ecosystems, sustainably manage forests, combat desertification, and halt and reverse land degradation and halt biodiversity loss
Goal 16.	Promote peaceful and inclusive societies for sustainable development, provide access to justice for all, and build effective, accountable, and inclusive institutions at all levels
Goal 17.	Strengthen the means of implementation and revitalize the Global Partnership for Sustainable Development

3. Research Methodology

In this research, a systemic literature review was conducted on sustainable production planning studies using a methodology adapted from [13] with the following steps:

Step One, define the research scope The main scope and objective of this review mainly focused on the application of production planning approaches to achieve sustainable goals. **Step two, select the search keywords** This step aimed at finding the most suitable keywords for the required review. Two sets of keywords were used. The first set included three keywords: production planning, production control, and planning, while the second set considered two keywords: sustainable and sustainability. These two sets resulted in six different combinations of search keywords. The authors used the Scopus database to perform the search, because it has one of the widest search library [13]. The search process resulted in identifying 560 articles.

Step three, define the inclusion and exclusion criteria This step aimed at identifying the most relevant articles among the identified 560 articles. Hence, the following inclusion and exclusion criteria were used:

- Only peer-reviewed articles published in English were considered.
- Only engineering, decision, and environmental sciences were considered.
- A time frame condition from 2011 to 2021 was added.
- The production planning problem needed to have at least one sustainable objective.
- Any framework related to production planning was considered, such as joint production planning and pricing or hybrid manufacturing remanufacturing systems addressing production planning.

Step four, screen the identified articles This step applied the inclusion and exclusion criteria and reduced the number of related articles to 36 articles. Then, a backward review for the resulted articles is conducted to find any missing articles. The final set of the identified articles included 45 articles and three review articles. Then, the 45 research articles were categorized into a two-dimensional classification based on production planning problems and sustainability pillars. In addition, the problems' solution methods were discussed.

4. Results and Discussions

In this section, the identified studies are discussed. Their classification is based on a two-dimensional classification, i.e., sustainability indicators (see Figure 3) and production planning problems, as illustrated in Figure 2.

Table 3 presents an overview of the identified articles. In addition, each article is assigned to its corresponding sustainability indicator and production planning problem used. Figure 4 reflects the analysis in Table 3 and shows the percentage of studies using the sustainable pillars. For example, Satyro et al. [48] considered the economic pillar in a holistic production planning approach, whereas Xiao et al. [49] integrated the economic and environmental pillars while solving a routing problem. For more details, the 45 articles were classified as follows: 9 articles addressed economic sustainability indicators, 12 articles addressed environmental indicators, and only 1 article addressed social indicators [50], while the remaining 23 articles used integrated indicators of two or more sustainability pillars. Three of these articles used the integration of social, economic, and environmental indicators [2,30,49], while 20 articles used economic and environmental indicators together.

In Table 3, the identified articles were also categorized according to the type of production planning problems. For example, the 12 articles addressing environmental indicators addressed different production planning problems as follows: Three articles considered scheduling problems [51–53], five articles considered hybrid methods of integrating more than one production planning problem [14,54–57], and two articles considered the lot-sizing problem [7,58].

In the following sections, we thoroughly discuss the identified articles shown in Table 3.

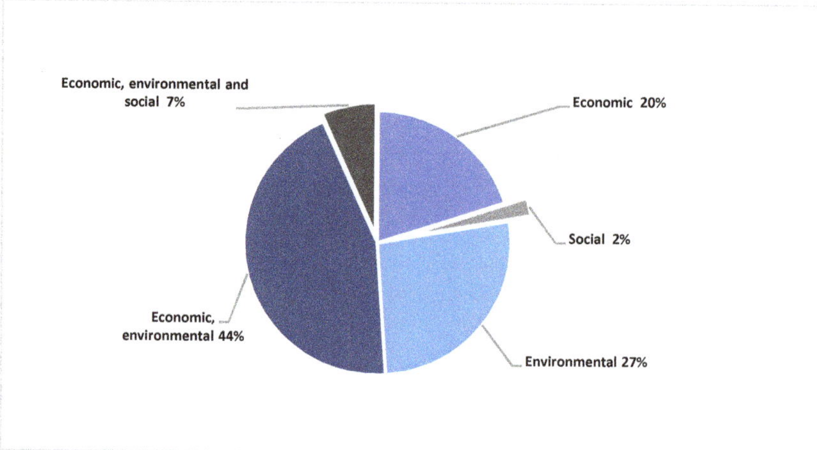

Figure 4. Percentage of studies that considered each sustainability pillar or an integration of various pillars.

Table 3. The classification of the identified articles.

Sustainability Pillars and Indicators	Economic			Environmental			Social		Production Planning Problem
Reference	Cost	Profit	Investment	Material	Energy	GHG	Employee Satisfaction	Customer Satisfaction	
[48]	✓		✓						Holistic approach
[59]	✓								* Hybrid
[60]		✓							Hybrid
[61]	✓								Hybrid
[62]	✓								Hybrid
[63]		✓							Scheduling
[64]	✓								Hybrid
[65]	✓								Routing
[66]	✓								** Other
[54]						✓			Hybrid
[14]						✓			Hybrid
[51]					✓				Scheduling
[58]						✓			Lot sizing
[55]				✓					Hybrid
[56]					✓	✓			Hybrid
[52]					✓				Scheduling
[57]					✓				Hybrid
[67]					✓				Other
[68]					✓				Other

Table 3. *Cont.*

Sustainability Pillars and Indicators	Economic			Environmental			Social		Production Planning Problem
Reference	**Cost**	**Profit**	**Investment**	**Material**	**Energy**	**GHG**	**Employee Satisfaction**	**Customer Satisfaction**	
[53]					✓				Scheduling
[7]				✓					Lot sizing
[50]							✓		Other
[2]	✓				✓		✓		Scheduling
[49]	✓				✓				Routing
[16]	✓	✓				✓			Hybrid
[69]	✓				✓				Hybrid
[70]	✓				✓				Dispatching
[71]	✓				✓				Scheduling
[47]	✓				✓	✓			Hybrid
[15]	✓				✓				Hybrid
[72]	✓				✓	✓			Hybrid
[31]	✓	✓			✓	✓	✓	✓	Aggregate production planning
[73]	✓					✓			Hybrid
[74]	✓				✓				Hybrid
[75]	✓				✓	✓			Dispatching
[76]		✓				✓			Hybrid
[77]		✓				✓	✓		Hybrid
[78]	✓				✓				Hybrid
[79]	✓				✓				Holistic approach
[80]		✓				✓			Holistic approach
[81]		✓			✓				Other
[82]	✓				✓				Scheduling
[83]	✓				✓				Scheduling
[84]	✓					✓			Scheduling
[85]	✓				✓	✓			Other

* Hybrid, more than one production planning problem used, and production planning problem combined with other planning processes; ** Other, other planning operations such as energy planning and shipment planning.

4.1. Economic Sustainability Pillar

In this section, we discuss the identified studies that considered any production planning problem with an economic objective, i.e., minimizing the cost or maximizing the profit. The economic pillar is the second most studied objective after the environmental objective, specifically the energy indicator. Energy is also the most studied single indicator [11] and is mostly driven by cost. As presented in Table 3, nine papers addressed the economic perspective using different production planning approaches.

The most recent study was by Satyro, et al. [48] who used a multi-correspondence analysis to find the production planning variables with the most affect over achieving economic sustainability. Their study used a systematic questionnaire to analyze the whole production planning process of six companies. They found that the effect of production planning could vary according to the size of the industry. With industries of more than 9000 employees, the implementation of production planning was not on an operational level as compared with that of smaller companies that varied from 400 to 5000 employees. However, their analysis was based on a small number of companies, and thus, the findings could not be generalized to any company, therefore, more confirmation and a wider set of companies was needed. Lage Junior and Godinho Filho [59] integrated two different production planning stages, i.e., scheduling and routing in a remanufacturing system. Their proposed model determined the optimum number of products to be disassembled to reduce the total cost expected from stochastic routing. However, their proposed model did not consider the number of products to be disassembled, which would affect the material recovery rate. Farahani and Rahmani [60] proposed a hybrid system in which they used production planning process, distribution planning process, and facility location-allocation of a crude oil network, while using net present value as a sustainable objective. The model was formed as mixed-integer linear programming solved by IBM LOG CPLEX. In addition, Yildirim and Nezami [64] developed a hybrid model that used lot sizing and preventive maintenance to decrease machine degradation over time, electrical cost, and operational cost. The proposed model determined lot sizes while satisfying the demand to determine the suitable preventive maintenance plan based on the machine up-and-down time. They introduced a coherence between production planning presented by lot-sizing and preventive maintenance.

4.2. Social Sustainability Pillar

Social sustainability is the most neglected pillar among the 3Ps of sustainability [2,13]. Zarte, et al. [13] conducted a review on addressing sustainable objectives through different decision-making methods. They found that regardless of the decision-making method used, the social sustainability pillar was always the least addressed pillar among the 3Ps. In a similar context, but from another point of view, the human factor has been shown to be the most neglected factor in planning objectives [86], which also proves that the social pillar has been neglected. Relatively, as shown in Figure 4, social sustainability is the least addressed sustainability pillar, either as a single objective or integrated with economic and environmental objectives. Cattaruzza, et al. [50] introduced a packaging and shipping problem that used production planning, workforce, and demand peaks to achieve the ideal number of employees who could process a set of orders to enhance employee satisfaction and development.

4.3. Environmental Sustainability Pillar

The environmental pillar is the most frequent sustainability pillar considered in the literature. As shown in Figure 4, the energy indicator specifically has the most attention. Energy, as an environmental sustainability indicator, can be addressed in two different ways, either as a cost where the objective is to minimize the overall cost or as a resource consumption [15].

Zheng, et al. [54] introduced a lagrangian algorithm to solve a production planning problem with stochastic demands. The proposed lagrangian algorithm could obtain nearly the same optimal solution with less than a 1% difference as compared with the solution calculated by the CPLEX solver. They conducted their research on a real-life case study facing issues with inventory and customer demand. The model considered customer demand as a stochastic demand, because the product was a special order product since different customers could order special requirements in different periods that needed to be met.

A joint production planning with pricing model was introduced by Zhang, et al. [16], who used pricing, production planning, and retailer selection to develop a model. Their model could help firms in making optimal joint decisions. In this study, the Stackelberg game theory was used to formulate the model in which the manufacturer was a leader, and the retailers were followers. Additionally, the model considered an emission control constraint. A nested genetic algorithm and the Stackelberg game model were used to solve the problem. However, the model did not consider that the retailers might be more influential than the manufacturers with respect to refusing the proposed solutions and pricing. In addition, they ignored the influence of other competitors.

In [58], the authors discussed another joint production planning model that considered the supplier and manufacturer as two separate parties, each of whom had revenue preferences. The proposed model assumed a centralized system where the supplier was considered to be a subsidiary to the manufacturer, having a single profit function for the whole system. The model was very informative about the influence of reducing carbon emissions on profitability. Nevertheless, the problem did not consider stochastic demand rates and dealt with only one supplier, which was not the most applicable case.

Rubaiee and Yildirim [51] introduced a fully sustainable framework using a scheduling problem to reduce total completion time to reduce energy costs. The reduction of total completion time was achieved by simply changing the on-off modes of machines to produce more energy-efficient machine scheduling. The developed model was solved using different methods, i.e., the weighted sum method and two different ant colony-based algorithms.

Another study used scheduling problems to achieve environmental sustainability [87]. This study introduced a framework to enable the decision-maker to decide on the best schedule that was less time consuming and more energy efficient. The on-off mode of machines was also used by Liu, et al. [52]. However, instead of working on total completion time for less energy consumption, the objective was to decrease the machine non-processing time; they integrated the problem of scheduling and the on-off modes of machines which resulted in a multi-objective model. The objective was to switch off under-utilized resources. The study proposed a novel genetic algorithm based on a non-sorted genetic algorithm (NSGA II) to solve the resulting model. Nevertheless, the model was not tested in a broader set of job shop cases, therefore, it could not be generalized on every job shop instance.

4.4. Integration of Economic and Environmental Sustainability Pillars

The reduction of total energy consumption always results in a decrease in greenhouse gas emissions [2]. Thus, most economic-oriented problems have an additional environmental objective of reducing greenhouse gas emissions with the increase in the interest of simultaneously addressing economic and environmental pillars [76]. Hence, most of the identified articles that focused on energy consumption considered greenhouse gas emissions while solving the model [19,57]. In order to have a clearer view over each specific indicator such as energy, greenhouse gas, profit, etc., Figure 5 was constructed, which shows the percentage of production planning problems used to solve each indicator either when combined with other indicators or when addressed as a single indicator. Hence, we found that most of studies were energy-oriented studies.

Banasik, et al. [74] proposed an analytical study to prove that implementing uncertainty measures could reduce the difference between actual and expected planning solutions. Their study used a real case to compare the actual results of using a deterministic model versus a two-stage stochastic model. The comparison showed a decrease in the difference between expected and actual results, and also showed a decrease in environmental impact and an increase in profit from using a deterministic model. Another approach for achieving both the economic pillar and the environmental pillar was presented in [56]. The authors used intelligent data collection and processing to simulate future energy consumption situations and used it in production planning and decision making. Similarly, the work in [81] introduced an approach to determine energy consumption values using energy measurement methods and reference cycles. Afterwards, these consumption values were employed to calculate the energy demands for better and efficient planning. Medini and Boucher [80] aimed at balancing forecasted sales and volumes produced in a diverse manufacturing environment while considering environmental and economic sustainability indicators. Thus, they introduced the impact of product diversity on environmental and economic sustainability indicators. Total completion time is an interesting area of research. Liu, et al. [84] introduced a mathematical model that could simultaneously decrease total completion time and greenhouse gas emissions. However, their study assumed that arrival times were deterministic parameters which practically is not the case.

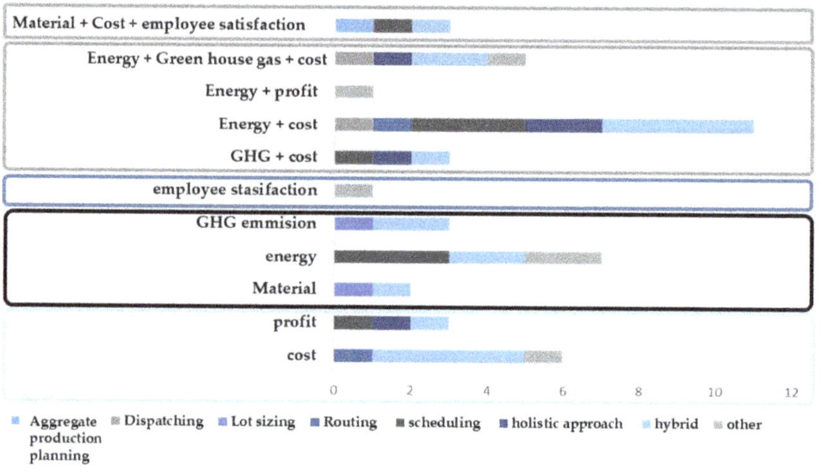

Figure 5. The number of production planning problems' studies and their consideration of various sustainability indicators.

4.5. Integration of Economic, Environmental, and Social Sustainability Pillars

In addition to environmental and economic pillars, Dal Borgo and Meneghetti [73] addressed the social sustainability pillar by considering the learning forgetting phenomena of the working personnel. They could form learning forgetting curves to be used as a framework to develop a production and shipment plan. The results showed that the consideration of the learning forgetting phenomenon could decrease the excessive overtime and stress that workers faced addressing the social sustainability pillar. In addition, this could achieve a full load transport by determining the panels that could be stacked together and then produced consecutively. This led to a decrease in greenhouse gas emissions indirectly and a decrease in shipping costs directly.

Zarte, et al. [2] used a fuzzy optimization to consider all the three main pillars of sustainability in production planning. The authors proposed a fuzzy interference model that combined multiple qualitative and quantitative input variables. This model could assess production sustainability, contrary to a traditional mathematical approach that required input and output measurements to validate the model. However, their proposed model neglected some production planning tasks such as inventory management maintenance, quality control, and product refurbishment.

4.6. International Cases in Production Planning for Sustainability

In this subsection, we consider the international cases which applied production planning for sustainability, as shown in Table 4. The table consists of four columns; the first column presents the study, the second column includes the country case, and the third and the fourth columns describe the targeted sustainability pillar in a production problem. For example, China was involved in most of the literature with seven articles. Four articles aimed at integrating the economic and environmental sustainability pillars [49,70,85,86], while three articles were aimed at the environmental sustainability pillar [12,47,57]. The case of Germany was referred to in four articles. Two articles considered integration of economic, environmental, and social sustainability pillars, one article used scheduling problems [2] and the other article used a hybrid method of integrating between APP and routing problems [77]. Regarding the two remaining articles that referred to Germany, one article used the economic objective [56], and the other article integrated the economic and environmental sustainability pillars while using energy planning [81]. In addition, different production planning problems integrated with different sustainability pillars were implemented in several studies and were implemented in many other countries, such as a hybrid production planning model with the environmental pillar Korea [55], and other production planning problems with integrated sustainability pillars in France [50], Italy [70], Turkey [31], and others.

Table 4. The classification of the identified articles based on each article.

Reference	Country	Sustainability Pillar	Production Planning Problem
[48]	Brazil	Economic	Holistic approach
[66]	Germany	Economic	Other
[61]	Canada	Economic	Hybrid
[54]	China	Environmental	Hybrid
[14]	China	Environmental	Hybrid
[58]	China	Environmental	Lot sizing
[55]	Korea	Environmental	Hybrid
[7]	Germany	Environmental	Lot sizing
[50]	France	Social	Other
[2]	Germany	Integration (economic, environmental and social)	Scheduling
[49]	China	Integration (economic and environmental)	Routing
[69]	China	Integration (economic and environmental)	Hybrid
[70]	Italy	Integration (economic and environmental)	Dispatching
[71]	Ireland	Integration (economic and environmental)	Scheduling
[31]	Turkey	Integration (economic, environmental and social)	Aggregate P.P.
[76]	U.S.A.	Integration (economic and environmental)	Hybrid
[77]	Germany	Integration (economic, environmental and social)	Hybrid
[78]	U.A.E.	Integration (economic and environmental)	Hybrid
[80]	France	Integration (economic and environmental)	Holistic approach
[81]	Germany	Integration (economic and environmental)	Other
[84]	China	Integration (economic and environmental)	Scheduling
[85]	China	Integration (economic and environmental)	Other

5. Implications for Future Research

In this section, we introduce an analysis of the identified sample articles and the fields that require further research and need more attention. Figure 6 summarizes the analysis of the identified articles, which was genuinely created based on the data extracted from the studied articles in each sustainability pillar and its corresponding production planning problem. The chart in Figure 6 provides a summary of the different production planning problems and outlines the extent to which these studies are mixed with sustainability pillars. Hence, we can see that aggregate production planning was the least used production planning problem either as a single problem or in a hybrid system. In contrast, scheduling received the most attention from scholars. This implies a requirement for using aggregate planning in achieving sustainability goals. In addition, existing studies [31,88] on aggregate production planning have addressed multiple sustainability pillars because aggregate production planning enables determining the levels of both workforce and production. Hence, aggregate production planning can be useful in applying multiple sustainability pillars, especially the social sustainability pillar, as it is directly connected with the workforce.

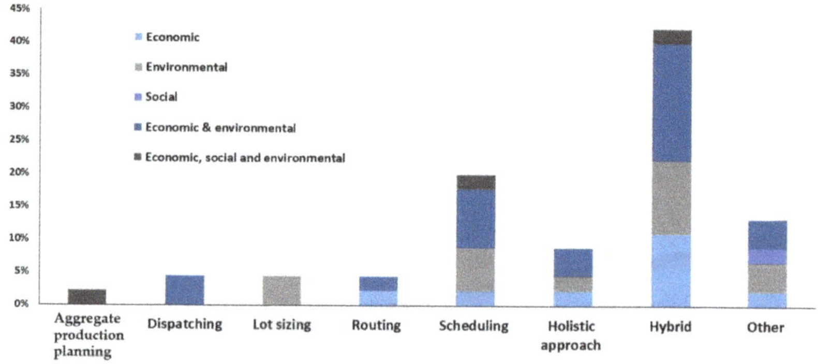

Figure 6. Production planning problems concerning sustainability pillars.

Regarding sustainability indicators, as shown in Figure 5, since the identified studies were relatively few, no studies were carried out on investment as a sustainability indicator for the economic pillar. Almost all of the studies that considered the economic pillar used the cost indicator, while neglecting both of the indicators of profit and investment. In addition, the studies in the existing literature lacked methods for measuring the impact of various sustainability pillars. For example, environmental life cycle assessment was used to measure the environmental impact, and social life cycle assessment was used for measuring social impact [85], while life cycle assessment was integrated between both [89]. Magrassi, et al. [90] developed an optimization model that integrated a proposed decision support system with life cycle assessment to measure environmental impact. Such research and models need to be integrated into production planning for a quantitative measure of sustainable impacts. Another noticeable issue was the reason behind choosing a sustainable indicator. Choosing sustainable indicators through production planning studies was mainly based on the addressed objective function of the problem, which was mostly decreasing energy costs. Because decreasing cost is mostly accompanied by fewer emissions, the economic pillar, thus, was mostly accompanied by the environmental pillar. This identifies another gap in the research, since this connection between energy consumption and harmful emissions is a point of debate.

Zarte, et al. [2] reported that it was useless to add emissions as an indicator to an energy reduction model as long as consuming more energy produced more emissions. In comparison, Biel and Glock [19] found that the relation between CO_2 emissions and energy consumption needed to be more realistic and could not always be considered to be linear. The stochastic modeling techniques in the identified articles found in this review are very scarce [55,74,75], but have more promising and actual values than deterministic models. Hence, this debate needs more attention and research. In a similar context, the reason for choosing a specific production planning problem for a specific sustainability pillar was not defined in most of the articles, but, as mentioned earlier, choosing a sustainable pillar was based on the problem and the required objective. Therefore, more research is needed on finding and assigning the suitable production planning problem with the suitable sustainability pillar.

Social sustainability was the least addressed sustainability pillar either as a single pillar or integrated with other pillars. Economic and environmental sustainability received much more attention. This issue needs more attention because some production planning problems are suitable for addressing the social pillar, such as aggregate production planning, which is concerned with the workforce. Thus, sustainable production planning should give more attention to considering social indicators such as customer satisfaction and employee health and safety.

Another interesting area for future research is integrating sophisticated computational tools into physical production systems which can be called cyber-physical systems [91].

Simply put, systems with embedded computers that enable a real-time connection between workstations and decision support systems, which can be used in various applications such as monitoring systems for intelligent consumption monitoring and smart electrical grids. An interesting study by Rossit, et al. [91] aimed at improving scheduling problems used these systems but did not consider a sustainable objective. Future research might also benefit from the findings of other research fileds with respect to the social sustainability pillers. For example, considering the flexibility in the use of human resources as proposed in the recent literature on the project managemen [92].

On the practical level, the findings provide an easy-to-understand guideline for practitioners to better understand the different pillars of sustainability and their inherent challenges, and how they can be realized in different production planning problems. In addition, the findings can show practitioners how to align their production systems with the 17 goals proposed by the United Nations, in order to gain a competitive advantage over their competitors. Furthermore, practitioners are recommended to develop integrated solutions of different production planning problems in order to achieve a production system in which sustainability pillars are accounted for at every stage.

Lastly, this research introduced a full review considering all sustainable production planning problems in addition to considering all sustainability pillars. The classification showed that the most used solving method among the identified sample articles was genetic algorithms. Nonetheless, this review lacked an inclusive study on each optimization method and its relation with a sustainable objective and the optimization method used. Sustainable production planning is already a complex problem with many parameters. As a result, there is a need for future research to address various optimization methods considering sustainable production planning and the suitability of sustainable objectives.

Although there were some review articles in the literature on sustainable production planning [18–20], each one of them was dedicated to addressing specific production planning problems and their integration with certain indicators of the 3Ps. For instance, Giret, et al. [18] reviewed studies on the production scheduling problem and their consideration of economic and environmental sustainability indicators. Biel and Glock [19] investigated studies related to energy consumption (as an environmental sustainability indicator) in different production planning problems. Recently, Bóna and Korkulu [20] discussed the consideration of ergonomic issues, as a social sustainability indicator, in previous studies on the lot-sizing production planning problem. Unlike previous review studies, this research contributes to the theoretical knowledge by providing a more holistic and comprehensive review of sustainable production planning. This study explores the consideration of various indicators of the sustainability 3Ps when solving different production planning problems that has not been observed well by extant literature. Moreover, this study highlights some theoretical research gaps that the current literature has not yet addressed properly and has ignored some of their critical aspects, as discussed in this section. This research provides researchers, research and development (R&D) centers, and policymakers with a holistic reference on sustainable production planning. First, researchers and R&D centers could benefit from the identified research gaps and the updated overview of the sustainability issues in the production planning field. In addition, the information provided in this research could guide sustainability policymakers to the critical areas that require more efficient policy formulation to further promote sustainable production.

6. Conclusions

This review addresses the studies that considered the three sustainability pillars in production planning. This review considers all production planning stages, sustainability, and indicators of 3Ps in a time frame from 2011 to 2021. The review shows that most of the studies implemented more than one sustainability pillar simultaneously; however, the addressed dual sustainability pillars are always considered to be connected. Another issue is the low number of studies that considered all three sustainability pillars, which shows the need for more attention towards holistic sustainable production planning. In addition, the literature analysis indicates an increasing inclination towards integrating multiple production planning problems with the objective of providing comprehensive production planning solutions. The consideration of the social pillar is still limited either as a single pillar or when integrated with other sustainability pillars. The review shows that few studies considered the 3Ps of sustainability in cyber-physical systems. These applications could assist in multiple sustainable production planning problems. Hence, more attention is required to study the contributions, success factors, and barriers to using cyber-physical systems for more sustainable production planning.

The study results should be considered in light of some limitations. Firstly, some articles related to the study topic might be missed. Secondly, the classification of the included articles based on the production planning problems and the three sustainability pillars depended on the authors' subjective judgements. To mitigate the impact of subjective opinions, recent text mining techniques could be used in the future. However, such techniques cannot provide an in-depth analysis and classification of the included documents.

Author Contributions: Conceptualization, M.S.k. and I.A.S.; review methodology and identified artilces, M.S.k. and I.A.S.; manuscript structure and organization, M.S.k., I.A.S. and A.K.; results and discussion, M.S.k., I.A.S., A.K., M.H. (Mohamed Hussain), I.Z. and M.H. (Mohamed Hussein); writing—original draft preparation, M.S.k., I.A.S., M.H. (Mohamed Hussain) and I.Z.; writing—review and editing, M.S.k., I.A.S., A.K., M.H. (Mohamed Hussain), I.Z. and M.H. (Mohamed Hussein); funding acquisition, A.K. All authors have read and agreed to the published version of the manuscript.

Funding: This research received no external funding.

Institutional Review Board Statement: Not applicable.

Informed Consent Statement: Not applicable.

Data Availability Statement: The search queries used in Scopus during the study are available from the corresponding authors on reasonable requests.

Conflicts of Interest: The authors declare no conflict of interest.

References

1. Kiran, D.R. Elements of production planning and control. In *Production Planning and Control*; Elsevier: Amsterdam, The Netherlands, 2019; pp. 1–20.
2. Zarte, M.; Pechmann, A.; Nunes, I. Fuzzy Inference Model for Decision Support in Sustainable Production Planning Processes—A Case Study. *Sustainability* **2021**, *13*, 1355. [CrossRef]
3. Takahashi, K.; Onosato, M.; Tanaka, F. A comprehensive approach for managing feasible solutions in production planning by an interacting network of Zero-Suppressed Binary Decision Diagrams. *J. Comput. Des. Eng.* **2015**, *2*, 105–112. [CrossRef]
4. Kiran, D.R. Aggregate planning. In *Production Planning and Control*; Elsevier: Amsterdam, The Netherlands, 2019; pp. 303–316.
5. Ganesh, K.; Punniyamoorthy, M. Optimization of continuous-time production planning using hybrid genetic algorithms-simulated annealing. *Int. J. Adv. Manuf. Technol.* **2004**, *26*, 148–154. [CrossRef]
6. Fung, Y.-N.; Choi, T.-M.; Liu, R. Sustainable planning strategies in supply chain systems: Proposal and applications with a real case study in fashion. *Prod. Plan. Control.* **2019**, *31*, 883–902. [CrossRef]
7. Lehtoranta, S.; Nissinen, A.; Mattila, T.; Melanen, M. Industrial symbiosis and the policy instruments of sustainable consumption and production. *J. Clean. Prod.* **2011**, *19*, 1865. [CrossRef]
8. Tokede, O.O.; Roetzel, A.; Ruge, G. A holistic life cycle sustainability evaluation of a building project. *Sustain. Cities Soc.* **2021**, *73*, 103107. [CrossRef]
9. Zarte, M.; Pechmann, A.; Nunes, I.L. Indicator framework for sustainable production planning and controlling. *Int. J. Sustain. Eng.* **2018**, *12*, 149–158. [CrossRef]
10. Akbar, M.; Irohara, T. Scheduling for sustainable manufacturing: A review. *J. Clean. Prod.* **2018**, *205*, 866–883. [CrossRef]
11. Zhang, H. The Impact of COVID-19 on Global Poduction. Available online: https://voxeu.org/article/impact-covid-19-global-production (accessed on 27 December 2021).
12. Butt, A.S. Strategies to mitigate the impact of COVID-19 on supply chain disruptions: A multiple case analysis of buyers and distributors. *Int. J. Logist. Manag.* **2021**. [CrossRef]
13. Zarte, M.; Pechmann, A.; Nunes, I.L. Decision support systems for sustainable manufacturing surrounding the product and production life cycle–A literature review. *J. Clean. Prod.* **2019**, *219*, 336–349. [CrossRef]

14. Zheng, J.; Zhou, X.; Yu, Y.; Wu, J.; Ling, W.; Ma, H. Low carbon, high efficiency and sustainable production of traditional manufacturing methods through process design strategy: Improvement process for sand casting defects. *J. Clean. Prod.* **2019**, *253*, 119917. [CrossRef]
15. Wichmann, M.G.; Johannes, C.; Spengler, T.S. An extension of the general lot-sizing and scheduling problem (GLSP) with time-dependent energy prices. *J. Bus. Econ.* **2019**, *89*, 481–514. [CrossRef]
16. Zhang, L.L.; Gang, D.U.; Jun, W.U.; Yujie, M.A. Joint production planning, pricing and retailer selection with emission control based on Stackelberg game and nested genetic algorithm. *Expert Syst. Appl.* **2020**, *161*, 113733. [CrossRef]
17. Jabbour, A.B.L.D.S.; Song, M.; Filho, M.G. Sustainability implications for operations management: Building the bridge through exemplar case studies. *Prod. Plan. Control.* **2019**, *31*, 841–844. [CrossRef]
18. Giret, A.; Trentesaux, D.; Prabhu, V. Sustainability in manufacturing operations scheduling: A state of the art review. *J. Manuf. Syst.* **2015**, *37*, 126–140. [CrossRef]
19. Biel, K.; Glock, C. Systematic literature review of decision support models for energy-efficient production planning. *Comput. Ind. Eng.* **2016**, *101*, 243–259. [CrossRef]
20. Korkulu, S.; Bóna, K. Ergonomics as a Social Component of Sustainable Lot-sizing: A Review. *Period. Polytech. Soc. Manag. Sci.* **2019**, *27*, 1–8. [CrossRef]
21. Grainger-Brown, J.; Malekpour, S. Implementing the Sustainable Development Goals: A Review of Strategic Tools and Frameworks Available to Organisations. *Sustainability* **2019**, *11*, 1381. [CrossRef]
22. Wang, C.; Ghadimi, P.; Lim, M.K.; Tseng, M.-L. A literature review of sustainable consumption and production: A comparative analysis in developed and developing economies. *J. Clean. Prod.* **2018**, *206*, 741–754. [CrossRef]
23. Zadeh, A.H.; Afshari, H.; Khorshid-Doust, R.R. Integration of process planning and production planning and control in cellular manufacturing. *Prod. Plan. Control* **2013**, *25*, 840–857. [CrossRef]
24. Stevenson, M.; Hendry, L.C.; Kingsman, B.G. A review of production planning and control: The applicability of key concepts to the make-to-order industry. *Int. J. Prod. Res.* **2005**, *43*, 869–898. [CrossRef]
25. Hees, A.; Schutte, C.S.L.; Reinhart, G. A production planning system to continuously integrate the characteristics of reconfigurable manufacturing systems. *Prod. Eng.* **2017**, *11*, 511–521. [CrossRef]
26. Oluyisola, O.E.; Sgarbossa, F.; Strandhagen, J.O. Smart Production Planning and Control: Concept, Use-Cases and Sustainability Implications. *Sustainability* **2020**, *12*, 3791. [CrossRef]
27. Wang, C.N.; Nhieu, N.L.; Tran, T.T.T. Stochastic chebyshev goal programming mixed integer linear model for sustainable global production planning. *Mathematics* **2021**, *9*, 483. [CrossRef]
28. Stecca, G. Production Planning. In *CIRP Encyclopedia of Production Engineering*; Springer: Berlin/Heidelberg, Germany, 2016; pp. 1–4.
29. Sakallı, Ü.S.; Baykoç, Ö.F.; Birgören, B. A possibilistic aggregate production planning model for brass casting industry. *Prod. Plan. Control.* **2010**, *21*, 319–338. [CrossRef]
30. Kiran, D.R. Routing, scheduling, and loading. In *Production Planning and Control*; Elsevier: Amsterdam, The Netherlands, 2019; pp. 317–330.
31. Rasmi, S.A.B.; Kazan, C.; Türkay, M. A multi-criteria decision analysis to include environmental, social, and cultural issues in the sustainable aggregate production plans. *Comput. Ind. Eng.* **2019**, *132*, 348–360. [CrossRef]
32. Wang, C.; Liu, X.-B. Integrated production planning and control: A multi-objective optimization model. *J. Ind. Eng. Manag.* **2013**, *6*, 815–830. [CrossRef]
33. Bhosale, K.C.; Pawar, P.J. Production planning and scheduling problem of continuous parallel lines with demand uncertainty and different production capacities. *J. Comput. Des. Eng.* **2020**, *7*, 761–774. [CrossRef]
34. Kim, S.-C.; Kim, Y.-W. Workforce Information Database System to Support Production Planning in Construction Projects. *J. Civ. Eng. Manag.* **2012**, *18*, 867–878. [CrossRef]
35. Inoue, M.; Lindow, K.; Stark, R.; Tanaka, K.; Nahm, Y.-E.; Ishikawa, H. Decision-making support for sustainable product creation. *Adv. Eng. Inform.* **2012**, *26*, 782–792. [CrossRef]
36. Lima, F.V.; Li, S.; Mirlekar, G.V.; Sridhar, L.N.; Ruiz-Mercado, G. Modeling and Advanced Control for Sustainable Process Systems. In *Sustainability in the Design, Synthesis and Analysis of Chemical Engineering Processes*; Elsevier: Amsterdam, The Netherlands, 2016; pp. 115–139.
37. Brundtland, G.H. *Report of the World Commission on Environment and Development: "Our Common Future"*; United Nations Genral Assemply Oxford University Press: Oxford, UK, 1987.
38. Ray, A.; De, A.; Mondal, S.; Wang, J. Selection of best buyback strategy for original equipment manufacturer and independent remanufacturer–Game theoretic approach. *Int. J. Prod. Res.* **2020**, *59*, 5495–5524. [CrossRef]
39. United Nations. *Transforming Our World: The 2030 Agenda for Sustainable Development*; UN: New Yord, NY, USA, 2015.
40. Ebrahimi, A.; Khakpour, R.; Saghiri, S. Sustainable setup stream mapping (3SM): A systematic approach to lean sustainable manufacturing. *Prod. Plan. Control.* **2021**, *32*, 1–19. [CrossRef]
41. Suzanne, E.; Absi, N.; Borodin, V. Towards circular economy in production planning: Challenges and opportunities. *Eur. J. Oper. Res.* **2020**, *287*, 168–190. [CrossRef]
42. Choudhary, A.; De, A.; Ahmed, K.; Shankar, R. An integrated fuzzy intuitionistic sustainability assessment framework for manufacturing supply chain: A study of UK based firms. *Ann. Oper. Res.* **2021**, 1–44. [CrossRef]

43. Kono, J.; Ostermeyer, Y.; Wallbaum, H. Investigation of regional conditions and sustainability indicators for sustainable product development of building materials. *J. Clean. Prod.* **2018**, *196*, 1356–1364. [CrossRef]
44. Joung, C.B.; Carrell, J.; Sarkar, P.; Feng, S.C. Categorization of indicators for sustainable manufacturing. *Ecol. Indic.* **2013**, *24*, 148–157. [CrossRef]
45. Goswami, M.; De, A.; Habibi, M.K.K.; Daultani, Y. Examining freight performance of third-party logistics providers within the automotive industry in India: An environmental sustainability perspective. *Int. J. Prod. Res.* **2020**, *58*, 7565–7592. [CrossRef]
46. Goswami, M.; Daultani, Y.; De, A. Decision modeling and analysis in new product development considering supply chain uncertainties: A multi-functional expert based approach. *Expert Syst. Appl.* **2020**, *166*, 114016. [CrossRef]
47. Dellnitz, A.; Braschczok, D.; Ostmeyer, J.; Hilbert, M.; Kleine, A. Energy costs vs. carbon dioxide emissions in short-term production planning. *J. Bus. Econ.* **2020**, *90*, 1383–1407. [CrossRef]
48. Satyro, W.C.; Spinola, M.D.M.; de Almeida, C.M.B.; Giannetti, B.F.; Sacomano, J.B.; Contador, J.C. Sustainable industries: Production planning and control as an ally to implement strategy. *J. Clean. Prod.* **2020**, *281*, 124781. [CrossRef]
49. Xiao, Y.; Zhang, H.; Jiang, Z.; Gu, Q.; Yan, W. Multiobjective optimization of machining center process route: Tradeoffs between energy and cost. *J. Clean. Prod.* **2020**, *280*, 124171. [CrossRef]
50. Cattaruzza, D.; Brotcorne, L.; Semet, F.; Tounsi, B. A three-phase matheuristic for the Packaging and Shipping Problem. *Appl. Math. Model.* **2018**, *64*, 713–732. [CrossRef]
51. Rubaiee, S.S.; Yildirim, M.B. An energy-aware multiobjective ant colony algorithm to minimize total completion time and energy cost on a single-machine preemptive scheduling. *Comput. Ind. Eng.* **2018**, *127*, 240–252. [CrossRef]
52. Liu, Y.; Dong, H.; Lohse, N.; Petrovic, S. A multi-objective genetic algorithm for optimisation of energy consumption and shop floor production performance. *Int. J. Prod. Econ.* **2016**, *179*, 259–272. [CrossRef]
53. Nezami, F.G.; Ghazinezami, A.; Krishnan, K.K. Sustainable development in manufacturing systems. In *Measuring Sustainable Development and Green Investments in Contemporary Economies*; IGI Global: Hershey, PA, USA, 2017; pp. 201–224.
54. Zheng, M.; Li, W.; Liu, Y.; Liu, X. A Lagrangian heuristic algorithm for sustainable supply chain network considering CO2 emission. *J. Clean. Prod.* **2020**, *270*, 122409. [CrossRef]
55. Lee, H. Development of Sustainable Recycling Investment Framework Considering Uncertain Demand and Nonlinear Recycling Cost. *Sustainability* **2019**, *11*, 3891. [CrossRef]
56. Sucic, B.; Al-Mansour, F.; Pusnik, M.; Vuk, T. Context sensitive production planning and energy management approach in energy intensive industries. *Energy* **2016**, *108*, 63–73. [CrossRef]
57. Biel, K.; Glock, C.H. On the use of waste heat in a two-stage production system with controllable production rates. *Int. J. Prod. Econ.* **2016**, *181*, 174–190. [CrossRef]
58. Qiao, A.; Choi, S.; Wang, X. Lot size optimisation in two-stage manufacturer-supplier production under carbon management constraints. *J. Clean. Prod.* **2019**, *224*, 523–535. [CrossRef]
59. Junior, M.L.; Filho, M.G. Master disassembly scheduling in a remanufacturing system with stochastic routings. *Central Eur. J. Oper. Res.* **2015**, *25*, 123–138. [CrossRef]
60. Farahani, M.; Rahmani, D. Production and distribution planning in petroleum supply chains regarding the impacts of gas injection and swap. *Energy* **2017**, *141*, 991–1003. [CrossRef]
61. Aljuneidi, T.; Bulgak, A.A. Designing a Cellular Manufacturing System featuring remanufacturing, recycling, and disposal options: A mathematical modeling approach. *CIRP J. Manuf. Sci. Technol.* **2017**, *19*, 25–35. [CrossRef]
62. Aljuneidi, T.; Bulgak, A.A. A mathematical model for designing reconfigurable cellular hybrid manufacturing-remanufacturing systems. *Int. J. Adv. Manuf. Technol.* **2016**, *87*, 1585–1596. [CrossRef]
63. Ben-Awuah, E. Simultaneous Production Scheduling and Waste Management Optimization for an Oil Sands Application. *J. Environ. Informatics* **2015**. [CrossRef]
64. YYildirim, M.B.; Nezami, F.G. Integrated maintenance and production planning with energy consumption and minimal repair. *Int. J. Adv. Manuf. Technol.* **2014**, *74*, 1419–1430. [CrossRef]
65. Tsai, P.-F. A Label Correcting Algorithm for Partial Disassembly Sequences in the Production Planning for End-of-Life Products. *Math. Probl. Eng.* **2012**, *2012*, 1–13. [CrossRef]
66. Lanza, G.; Peters, S.; Herrmann, H.-G. Dynamic optimization of manufacturing systems in automotive industries. *CIRP J. Manuf. Sci. Technol.* **2012**, *5*, 235–240. [CrossRef]
67. Anderluh, A.; Nolz, P.C.; Hemmelmayr, V.C.; Crainic, T.G. Multi-objective optimization of a two-echelon vehicle routing problem with vehicle synchronization and 'grey zone' customers arising in urban logistics. *Eur. J. Oper. Res.* **2019**, *289*, 940–958. [CrossRef]
68. Fysikopoulos, A.; Pastras, G.; Alexopoulos, T.; Chryssolouris, G. On a generalized approach to manufacturing energy efficiency. *Int. J. Adv. Manuf. Technol.* **2014**, *73*, 1437–1452. [CrossRef]
69. Wang, Y.; Liu, X.; Zhang, H.; Liu, Y.; Cui, P.; Zhu, Z.; Ma, Y.; Gao, J. Comprehensive 3E analysis and multi-objective optimization of a novel process for CO_2 capture and separation process from syngas. *J. Clean. Prod.* **2020**, *274*, 122871. [CrossRef]
70. Materi, S.; D'Angola, A.; Renna, P. A dynamic decision model for energy-efficient scheduling of manufacturing system with renewable energy supply. *J. Clean. Prod.* **2020**, *270*, 122028. [CrossRef]
71. Eccher, C.; Geraghty, J. Incorporating sustainable criteria in a dynamic multi-objective recommendation planning tool for a continuous manufacturing process: A dairy case study. *J. Manuf. Syst.* **2020**, *55*, 159–170. [CrossRef]

72. Venkata Deepthi, T.; Ramakotaiah, K.; Krishnaveni, K. Research on performance of multi-skilled workers for sustainable production planning in seru production systems. *Int. J. Innov. Technol. Explor. Eng.* **2019**, *8*, 1016–1028. [CrossRef]
73. Borgo, E.D.; Meneghetti, A. Production and shipment planning for Project Based Enterprises: Exploiting learning-forgetting phenomena for sustainable assembly of Curtain Walls. *Comput. Ind. Eng.* **2019**, *131*, 488–501. [CrossRef]
74. Banasik, A.; Kanellopoulos, A.; Bloemhof-Ruwaard, J.M.; Claassen, G.D.H. Accounting for uncertainty in eco-efficient agri-food supply chains: A case study for mushroom production planning. *J. Clean. Prod.* **2019**, *216*, 249–256. [CrossRef]
75. Iqbal, A.; Al-Ghamdi, K.A. Energy-efficient cellular manufacturing system: Eco-friendly revamping of machine shop configuration. *Energy* **2018**, *163*, 863–872. [CrossRef]
76. Halati, A.; He, Y. Intersection of economic and environmental goals of sustainable development initiatives. *J. Clean. Prod.* **2018**, *189*, 813–829. [CrossRef]
77. Hahn, G.J.; Brandenburg, M. A sustainable aggregate production planning model for the chemical process industry. *Comput. Oper. Res.* **2018**, *94*, 154–168. [CrossRef]
78. Awad, M.I.; Hassan, N.M. Joint decisions of machining process parameters setting and lot-size determination with environmental and quality cost consideration. *J. Manuf. Syst.* **2018**, *46*, 79–92. [CrossRef]
79. Biel, K.; Glock, C.H. Prerequisites of efficient decentralized waste heat recovery and energy storage in production planning. *J. Bus. Econ.* **2016**, *87*, 41–72. [CrossRef]
80. Medini, K.; Boucher, X. An approach for steering products and services offering variety towards economic and environmental sustainability. *CIRP J. Manuf. Sci. Technol.* **2016**, *15*, 65–73. [CrossRef]
81. Bornschlegl, M.; Bregulla, M.; Franke, J. Methods-Energy Measurement–An approach for sustainable energy planning of manufacturing technologies. *J. Clean. Prod.* **2016**, *135*, 644–656. [CrossRef]
82. Liu, Y.; Dong, H.; Lohse, N.; Petrovic, S. Reducing environmental impact of production during a Rolling Blackout policy—A multi-objective schedule optimisation approach. *J. Clean. Prod.* **2015**, *102*, 418–427. [CrossRef]
83. Liu, Y.; Dong, H.; Lohse, N.; Petrovic, S.; Gindy, N. An investigation into minimising total energy consumption and total weighted tardiness in job shops. *J. Clean. Prod.* **2014**, *65*, 87–96. [CrossRef]
84. Liu, C.; Yang, J.; Lian, J.; Li, W.; Evans, S.; Yin, Y. Sustainable performance oriented operational decision-making of single machine systems with deterministic product arrival time. *J. Clean. Prod.* **2014**, *85*, 318–330. [CrossRef]
85. Lin, Q.; Wu, Q.; Huang, G.; Zhai, M. An interval parameter optimization model for sustainable power systems planning under uncertainty. *Int. J. Electr. Power Energy Syst.* **2013**, *54*, 631–641. [CrossRef]
86. Grosse, E.H.; Calzavara, M.; Glock, C.; Sgarbossa, F. Incorporating human factors into decision support models for production and logistics: Current state of research. *IFAC-PapersOnLine* **2017**, *50*, 6900–6905. [CrossRef]
87. Yildirim, M.B.; Mouzon, G. Single-Machine Sustainable Production Planning to Minimize Total Energy Consumption and Total Completion Time Using a Multiple Objective Genetic Algorithm. *IEEE Trans. Eng. Manag.* **2011**, *59*, 585–597. [CrossRef]
88. Cheraghalikhani, A.; Khoshalhan, F.; Mokhtari, H. Aggregate production planning: A literature review and future research directions. *Int. J. Ind. Eng. Comput.* **2019**, 309–330. [CrossRef]
89. Valente, C.; Møller, H.; Johnsen, F.M.; Saxegård, S.; Brunsdon, E.R.; Alvseike, O.A. Life cycle sustainability assessment of a novel slaughter concept. *J. Clean. Prod.* **2020**, *272*, 122651. [CrossRef]
90. MMagrassi, F.; Del Borghi, A.; Gallo, M.; Strazza, C.; Robba, M. Optimal Planning of Sustainable Buildings: Integration of Life Cycle Assessment and Optimization in a Decision Support System (DSS). *Energies* **2016**, *9*, 490. [CrossRef]
91. Rossit, D.A.; Tohmé, F.; Frutos, M. Production planning and scheduling in Cyber-Physical Production Systems: A review. *Int. J. Comput. Integr. Manuf.* **2019**, *32*, 385–395. [CrossRef]
92. Karam, A.; Attia, E.; Duquenne, P. A MILP model for an integrated project scheduling and multi-skilled workforce allocation with flexible working hours. *IFAC-PapersOnLine* **2017**, *50*, 13964–13969. [CrossRef]

MDPI
St. Alban-Anlage 66
4052 Basel
Switzerland
Tel. +41 61 683 77 34
Fax +41 61 302 89 18
www.mdpi.com

Energies Editorial Office
E-mail: energies@mdpi.com
www.mdpi.com/journal/energies

www.ingramcontent.com/pod-product-compliance
Lightning Source LLC
Chambersburg PA
CBHW040757250926
51850CB00007BA/1013
9783036575445